Corrosion Engineering

Corrosion Engineering

Edited by
Theodore Schneider

WILLFORD PRESS

www.willfordpress.com

Published by Willford Press,
118-35 Queens Blvd., Suite 400,
Forest Hills, NY 11375, USA

ISBN: 978-1-64728-342-1

Cataloging-in-Publication Data

Corrosion engineering / edited by Theodore Schneider.
    p. cm.
Includes bibliographical references and index.
ISBN 978-1-64728-342-1
1. Corrosion and anti-corrosives. 2. Materials--Biodegradation. 3. Materials--Deterioration. I. Schneider, Theodore.
TA462 .C67 2022
604.7--dc23

For information on all Willford Press publications
visit our website at www.willfordpress.com

Contents

Preface..VII

Chapter 1   **Electrochemical and Microstructural Analysis of FeS Films from Acidic Chemical Bath at Varying Temperatures, pH, and Immersion Time**............................1
Ladan Khaksar, Gary Whelan and John Shirokoff

Chapter 2   **Simulation of the Ill-Posed Problem of Reinforced Concrete Corrosion Detection using Boundary Element Method**..10
Syarizal Fonna, Israr M. Ibrahim, M. Ridha, Syifaul Huzni and A. K. Ariffin

Chapter 3   **Effect of Stress Corrosion on Relaxation of Large Diameter BGFRP Bars**........................15
Guowei Li, Sidi Kabba Bakarr, Jingqiu Wang, Xue Liu and Chengyu Hong

Chapter 4   **Hot Corrosion of SrTiO$_3$ Perovskite in Na$_2$SO$_4$+ 50wt.%V$_2$O$_5$ and Na$_2$SO$_4$+ 10 wt.% NaCl Environments at 900°C**..23
M. Krishna Prasad, K. Srinivasa Rao, Madhusudhan Reddy and Gosipathala Sreedhar

Chapter 5   **Experimental Assessment of Rebar Corrosion in Concrete Slab using Ground Penetrating Radar (GPR)**..30
Ahmad Zaki, Megat Azmi Megat Johari, Wan Muhd Aminuddin Wan Hussin and Yessi Jusman

Chapter 6   **A Comparative Study of Hydrogen-Induced Cracking Resistances of API 5L B and X52MS Carbon Steels**..40
Rodrigo Monzon Figueredo, Mariana Cristina de Oliveira, Leandro Jesus de Paula, Heloisa Andréa Acciari and Eduardo Norberto Codaro

Chapter 7   **Experimental Investigation into Corrosion Effect on Mechanical Properties of High Strength Steel Bars under Dynamic Loadings**..47
Hui Chen, Jinjin Zhang, Jin Yang and Feilong Ye

Chapter 8   **Regression Analysis of Bond Parameters between Corroded Rebar and Concrete based on Reported Test Data**..59
H. J. Zhou, Y. F. Zhou, Y. N. Xu, Z. Y. Lin, F. Xing and L. X. Li

Chapter 9   **Tribocorrosion of Passive Materials: A Review on Test Procedures and Standards**..77
A. López-Ortega, J. L. Arana and R. Bayón

Chapter 10   **Monitoring the Interaction of Two Heterocyclic Compounds on Carbon Steel by Electrochemical Polarization, Noise, and Quantum Chemical Studies**..101
Vinod P. Raphael, Shaju K. Shanmughan and Joby Thomas Kakkassery

Chapter 11   **Study on the Electrochemical Performance of Sacrificial Anode Interfered by Alternating Current Voltage**..111
Qingmiao Ding, Xiao Chu, Tao Shen and Xiaoxiao Yu

Chapter 12   **The Discrete Wavelet Transform and its Application for Noise Removal in Localized Corrosion Measurements**...118
Rogelio Ramos, Benjamin Valdez-Salas, Roumen Zlatev, Michael Schorr Wiener and Jose María Bastidas Rull

Chapter 13   **Corrosion of Reinforced Concrete Structures Submerged by the 2004 Tsunami in West Aceh**...125
Herdi Susanto, Syifaul Huzni and Syarizal Fonna

Chapter 14   **Effect of Grain Size on the Stress Corrosion Cracking of Ultrafine Grained Cu-10 wt% Zn Alloy in Ammonia**...134
Takuma Asabe, Muhammad Rifai, Motohiro Yuasa and Hiroyuki Miyamoto

Chapter 15   **Developing Field Test Procedures for Chloride Stress Corrosion Cracking in the Arabian Gulf**...142
Hanan Farhat

Chapter 16   **Seismic Behavior of Corroded RC Bridges: Review and Research Gaps**...150
Kaveh Andisheh, Allan Scott and Alessandro Palermo

Chapter 17   **The Inhibition Effect of Sodium Glutarate towards Carbon Steel Corrosion in Neutral Aqueous Solutions**...172
G. Chan-Rosado and M. A. Pech-Canul

Chapter 18   **Adaptive Corrosion Protection System using Continuous Corrosion Measurement, Parameter Extraction, and Corrective Loop**...184
Jasbir N. Patel, Andre Chang, Haleh Shahbazbegian and Bozena Kaminska

Chapter 19   **Experimental and Runge–Kutta Method Simulation to Investigate Corrosion Kinetics of Mild Steel in Sulfuric Acid Solutions**...195
Ismaeel M. Alwaan

Chapter 20   **Pitting Corrosion of the Resistance Welding Joints of Stainless Steel Ventilation Grille Operated in Swimming Pool Environment**...201
Mirosław Szala and Daniel Łukasik

Chapter 21   **Electrolyte Composition for Distinguishing Corrosion Mechanisms in Steel Alloy Screening**...208
Ingmar Bösing, Jorg Thöming and Michael Baune

Permissions

List of Contributors

Index

Preface

The natural process which leads to the conversion of a refined metal to a form that is more chemically-stable like sulfide, hydroxide and oxide is known as corrosion. It includes the chemical and electrochemical reaction of materials with their environment. The field which is concerned with the control and prevention of corrosion is referred to as corrosion engineering. The two main types of corrosion that are central to corrosion engineering are external and internal corrosion. Underwater external corrosion, microbial corrosion, high temperature corrosion, underwater soil side corrosion, atmospheric corrosion and pitting corrosion are a few types of external corrosion. A few examples of internal corrosion are water pipe corrosion, oil pipe corrosion, water tank reservoir corrosion and gas pipe corrosion. This book outlines the processes and applications of corrosion engineering in detail. While understanding the long-term perspectives of the topics, the book makes an effort in highlighting their impact as a modern tool for the growth of the discipline. Students, researchers, experts and all associated with this field will benefit alike from this book.

This book has been the outcome of endless efforts put in by authors and researchers on various issues and topics within the field. The book is a comprehensive collection of significant researches that are addressed in a variety of chapters. It will surely enhance the knowledge of the field among readers across the globe.

It gives us an immense pleasure to thank our researchers and authors for their efforts to submit their piece of writing before the deadlines. Finally in the end, I would like to thank my family and colleagues who have been a great source of inspiration and support.

Editor

Electrochemical and Microstructural Analysis of FeS Films from Acidic Chemical Bath at Varying Temperatures, pH, and Immersion Time

Ladan Khaksar,[1] Gary Whelan,[1] and John Shirokoff[2]

[1]Department of Mechanical Engineering, Faculty of Engineering and Applied Science, Memorial University of Newfoundland,
 St. John's, NL, Canada A1B 3X5
[2]Department of Process Engineering, Faculty of Engineering and Applied Science, Memorial University of Newfoundland,
 St. John's, NL, Canada A1B 3X5

Correspondence should be addressed to John Shirokoff; shirokof@mun.ca

Academic Editor: Yu Zuo

The corrosion resistance and corrosion products of 4130 alloy steel have been investigated by depositing thin films of iron sulfide synthesized from an acidic chemical bath. Tests were conducted at varying temperatures (25°C–75°C), pH levels (2–4), and immersion time (24–72 hours). The corrosion behavior was monitored by linear polarization resistance (LPR) method. X-ray Diffraction (XRD), Energy Dispersive X-ray (EDX) spectroscopy, and Scanning Electron Microscopy (SEM) have been applied to characterize the corrosion products. The results show that, along with the formation of an iron sulfide protective film on the alloy surface, increasing temperature, increasing immersion time, and decreasing pH all directly increase the corrosion rate of steel in the tested experimental conditions. It was also concluded that increasing temperature causes an initial increase of the corrosion rate followed by a large decrease due to transformation of the iron sulfide crystalline structure.

1. Introduction

The corrosion of steel in aqueous environments containing hydrogen sulfide (H_2S) is of great interest to the oil and gas industry [1–5]. Unlike carbon dioxide corrosion, H_2S corrosion always involves the formation of corrosion products that are predominantly iron sulfide (FeS) compounds with various phases. These corrosion product films should be characterized to illustrate the corrosion mechanism. It has been reported that the formation of the FeS generally controls the H_2S corrosion [6]. However, there is still debate on how the initial corrosion product layers form.

It is well known that surface scale formation is one of the most important factors that influences the corrosion rate [7]. The scale slows down the corrosion process by presenting a diffusion barrier for the species involved in the corrosion process and by covering and preventing the underlying steel from further dissolution. The scale growth depends primarily on the kinetics of scale formation [8].

H_2S corrosion on the metal surface is also strongly dependent on the type of corrosion product films formed on the surface of the metal during the corrosion process. The precipitation rate or the formation of these films depends on various environmental factors and the concentration of species. The stability, protectiveness, and adherence of these films determine the nature and the rate of corrosion [9, 10]. It is important to note that, in contrast to one single type of iron carbonate formed in CO_2 corrosion, many types of FeS may form during H_2S corrosion such as amorphous ferrous sulfide, mackinawite, cubic ferrous sulfide, smythite, greigite, pyrrhotite, troilite, pyrite, and marcasite [11–18].

In aqueous solutions of H_2S, two mechanisms were proposed for the formation of FeS films, namely, dissolution of iron followed by precipitation of FeS and sulfide ion adsorption followed by direct film formation [19].

The first proposed theory is a possible mechanism for FeS formation in that the FeS layer is formed by precipitation only when its concentration reaches the solubility limit, analogous to how precipitation equilibrium governs the mechanism of iron carbonate formation. However, if this is to be true, the kinetics of FeS formation must be much faster than that of iron carbonate. In cases where FeS is highly undersaturated in the bulk, it can still be formed on the steel surface. This is suspected to be due to the high surface pH caused by consumption of hydronium ions by corrosion as well as locally high ferrous ion concentration, resulting in supersaturation of FeS on the steel surface. Therefore, FeS forms relatively fast on the steel surface, irrespective of the bulk conditions [20–22]. Another possible theory has been proposed by Shoesmith et al., which describes the idea that the first layer of mackinawite is generated by a direct, solid-state reaction between the steel surface and H_2S [2, 19]. Mackinawite then grows with time. The corrosion product layer growth rate depends upon the corrosion rate as well as the water chemistry with regard to pH, temperature, and so forth. It has been found that when the thickness of FeS reaches a critical value, this corrosion product layer cracks due to the development of internal stresses [6, 23]. More corrosive species such as H_2S or hydrogen ions diffuse through the now porous FeS layer and attack the steel surface. More FeS is then formed by either solid-state reaction between steel and H_2S akin to what happened initially or precipitation of FeS due to local FeS supersaturation. This direct, solid-state reaction theory is supported by other researches [24, 25].

How FeS initially forms is pertinent, because it can help to better predict the H_2S corrosion. However, until now research efforts have not achieved agreement on this subject. The situation is complicated by the variety of types of FeS that can be formed. Depending on the conditions relating to the corrosion environments, mackinawite, pyrrhotite, greigite, smythite, marcasite, and pyrite are the six naturally occurring FeS minerals [5, 19].

Most of the previous studies in this area are conducted at high temperatures and usually in gaseous H_2S environment. In the present study, all the experiments are performed at lower temperature in an aqueous solution because the real temperature of some oil and gas production and pipelines is below 100°C. In this study, FeS films have been synthesized on the metal alloy surface without the presence of H_2S in the solution. Rather, FeS was formed by chemical bath deposition of iron and sulfur ions at acidic pH levels under varying environmental conditions.

2. Experimental Procedure

2.1. Material and Sample Preparation.
According to NACE MR0175/ISO 15156, the most common steel alloy for tubulars and tubular components in sour service is UNS G41XX0, formerly AISI 41XX [26]. 4130 steel is among the most common alloys used in industry. This steel typically consists of 0.80–1.1 Cr, 0.15–0.25 Mo, 0.28–0.33 C, 0.40–0.60 Mn, 0.035 P, 0.040 S, 0.15–0.35 Si, and balanced Fe. The working electrode was machined from the parent material into cylinders having dimensions of approximately 9 mm length and diameter. Prior to the experiments, all specimens were polished with Coated Abrasive Manufacturers Institute

(CAMI) grit designations 320, 600, and 1000 corresponding to average particle diameters 36.0, 16.0, and 10.3 microns and finally 6-micron grit silicon carbide paper and then cleansed with deionized water until a homogenous surface was observed. Following this, the specimens were quickly dried using cold air to avoid oxidation.

2.2. Electrolyte Solution Preparation and Synthesis of FeS Films.
Due to the inherent safety concerns associated with H_2S gas, an alternative method of FeS film deposition was employed [27]. The alternative method provided an acidic electrolyte solution which has the potential to form thin FeS layer on the steel surface like what happens in the sour oil pipeline.

This acidic chemical bath contains 6.25 g iron (II) chloride (0.15 M), 12.60 g urea (1 M), and 31.55 g thioacetamide (2 M). Deionized water was used as the solvent in every experiment. Each reagent was mixed with 210 mL of deionized water, stirred with a magnetic stir rod for 30 minutes, and mixed together under stirring for additional two hours to achieve a clear solution.

The mechanism of FeS formation in this acidic bath is the slow release of iron and sulfur ions within solution followed by the deposition of these ions on the alloy surface. The iron and sulfur ions are provided from iron (II) chloride and thioacetamide, respectively. The formation of FeS films from this acidic bath is dependent on whether the deposition rate of the ionic product of iron and sulfur is higher than solubility of FeS. Adding urea to the solution adjusted the balance between hydrolysis and deposition. The proposed reactions for this mechanism are described as follow [27]:

$$FeCl_2 \longrightarrow Fe^{2+} + 2Cl^- \qquad (1)$$

$$CH_3CSNH_2 + H_2O \longleftrightarrow S^{2-} + CH_3CONH_2 + 2H^+ \qquad (2)$$

$$CO\,(NH_2)_2 + H_2O \longleftrightarrow 2NH_3 + CO_2 \qquad (3)$$

$$NH_3 + H_2O \longleftrightarrow NH_4^+ + OH^- \qquad (4)$$

$$Fe^{2+} + S^{2-} \longleftrightarrow FeS \qquad (5)$$

Finally, the overall reaction would be written as

$$
\begin{aligned}
Fe^{2+} + CH_3CSNH_2 + CO\,(NH_2)_2 + 2H_2O \\
\longrightarrow FeS + CH_3CONH_2 + 2NH_4^+ + CO_2
\end{aligned}
\qquad (6)
$$

2.3. Corrosion Tests.
Experiments were conducted in a multiport glass cell with a three-electrode setup at atmospheric pressure based on the ASTM G5-94 standard for potentiostatic anodic polarization measurements [28]. A graphite rod was used as the counter electrode (CE) and saturated silver/silver chloride (Ag/AgCl) was used as the reference electrode (RE). In order to investigate the electrochemical characteristic of the corrosion films formed on the steel alloy, the specimens subjected to corrosion were used as working electrodes (WE).

An Ivium Compactstat Potentiostat monitoring system was used to perform electrochemical corrosion measurements. Linear Polarization Resistance (LPR) technique was

TABLE 1: Experimental conditions.

Condition number	Temperature (°C)	pH	Immersion time (hour)
1	50	4	24
2	50	4	48
3	50	4	72
4	25	4	24
5	50	4	24
6	75	4	24
7	50	2	24
8	50	3	24
9	50	4	24

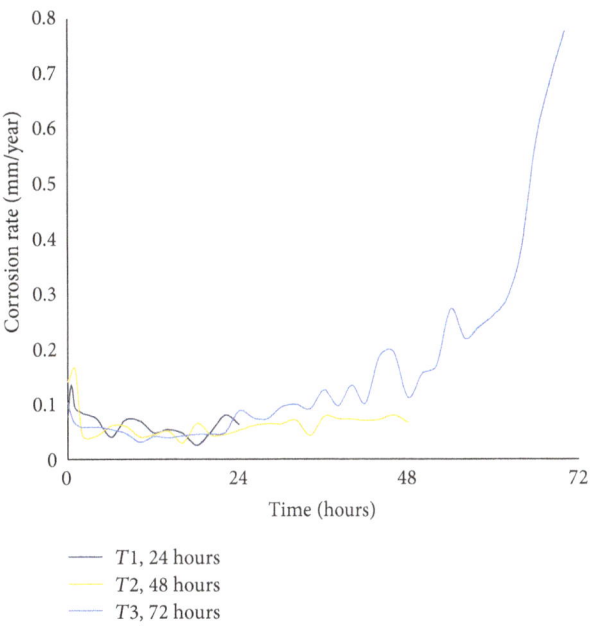

— $T1$, 24 hours
— $T2$, 48 hours
— $T3$, 72 hours

FIGURE 1: Corrosion rate with time at pH 4 and 50°C.

used to investigate the corrosion rate. The applied potential range for the LPR measurements was from −0.02 V to 0.02 V with a scanning rate of 0.125 mV/s. All the measurements were conducted by setting the potentiostat to take measurements at 0.5, 1, 2, 4, . . . to 24, 48, or 72 hours depending on the test. Prior to start of each test, the sample was immersed in the solution for 55 minutes in accordance with ASTM G5-82 [28]. The pH was adjusted by adding deoxygenated hydrochloric acid.

Table 1 describes the experimental conditions. Three series of experiments were conducted to investigate the effect of temperature, immersion time, and pH on the corrosion behavior of FeS films.

2.4. Surface Morphology Observation and Corrosion Product Analysis. Upon completion of corrosion testing, morphological characterization of the surface was conducted using FEI Quanta 400 Scanning Electronic Microscope (SEM) with Bruker Energy Dispersive X-ray (EDX) spectroscopy. The SEM was operating at 15 kV, with a working distance of 15 mm and beam current of 13 nA. The crystal structure and chemical composition of the corrosion products were characterized by X-ray Diffraction (XRD) using a Rigaku Ultima IV X-ray diffractometer operating at 40 kV and 44 mA and SEM-EDX to confirm the chemical elements.

3. Results and Discussion

3.1. Effect of Immersion Time on the Corrosion Mechanism and Products. Figure 1 shows the effect of 24, 48, and 72 hours of immersion time on the corrosion rate of the specimens at 50°C and pH 4. During a corrosion process, the rate of the reaction is determined by the corrosion mechanism. Growth of a corrosion film limits the rate of further corrosion by acting as a diffusion barrier for the species involved in the process. Gradually the corrosion rate decreases and the underlying steel is protected from further dissolution [8, 19]. Figure 1 indicates that in this experiment the results of LPR measurements did not agree well with the idea of a decrement of corrosion rate by increase of exposed time to the solution. It shows that corrosion rate is increasing gradually

by increasing the immersion time which could be explained as follows:

(1) The corrosion rate is significantly greater than the rate of film formation on the surface.

(2) The corrosion product has weak adherence to the alloy surface causing it to detach and expose the unprotected alloy to the corrosive solution and increase the possibility of localize corrosion on the surface.

The diffraction spectra in Figure 2 were search-matched to the XRD computer database (i.e., contains powder diffraction files (PDF) from the joint committee on powder diffraction standards (JCPDS) and international center for diffraction data (ICDD)). Figures 2(a)–2(d) identified 006–0696 iron Fe (alpha-Fe body centered cubic (bcc) crystal type), and Figure 2(e) identified both 006–0696 iron Fe (alpha-Fe bcc) and 015–0037 mackinawite FeS (tetragonal FeS crystal type). These PDF numbers and names appear in the top right corner of each diffraction spectra and corresponding line positions are superimposed onto the spectral peaks in each figure.

Figure 2 shows the results of crystal structure characterization of the steel alloy surfaces with powder- (P-) XRD. From Figures 2(a)–2(c) it is apparent that XRD results primarily indicated elemental Fe consistent with the uncorroded sample in Figure 2(d); this is likely a result of inadequate film thickness for detection by a P-XRD spectrometer. The thin nature of the corrosion film on the surface of the steel alloy is consistent with literature discussing the deposition of FeS using the indicated chemical bath alternative to H_2S exposure [24].

As shown in Figure 2(e), there was a small amount of mackinawite detected by the P-XRD spectrometer on the surface of the sample exposed to 75°C for 24 hours at 4 pH.

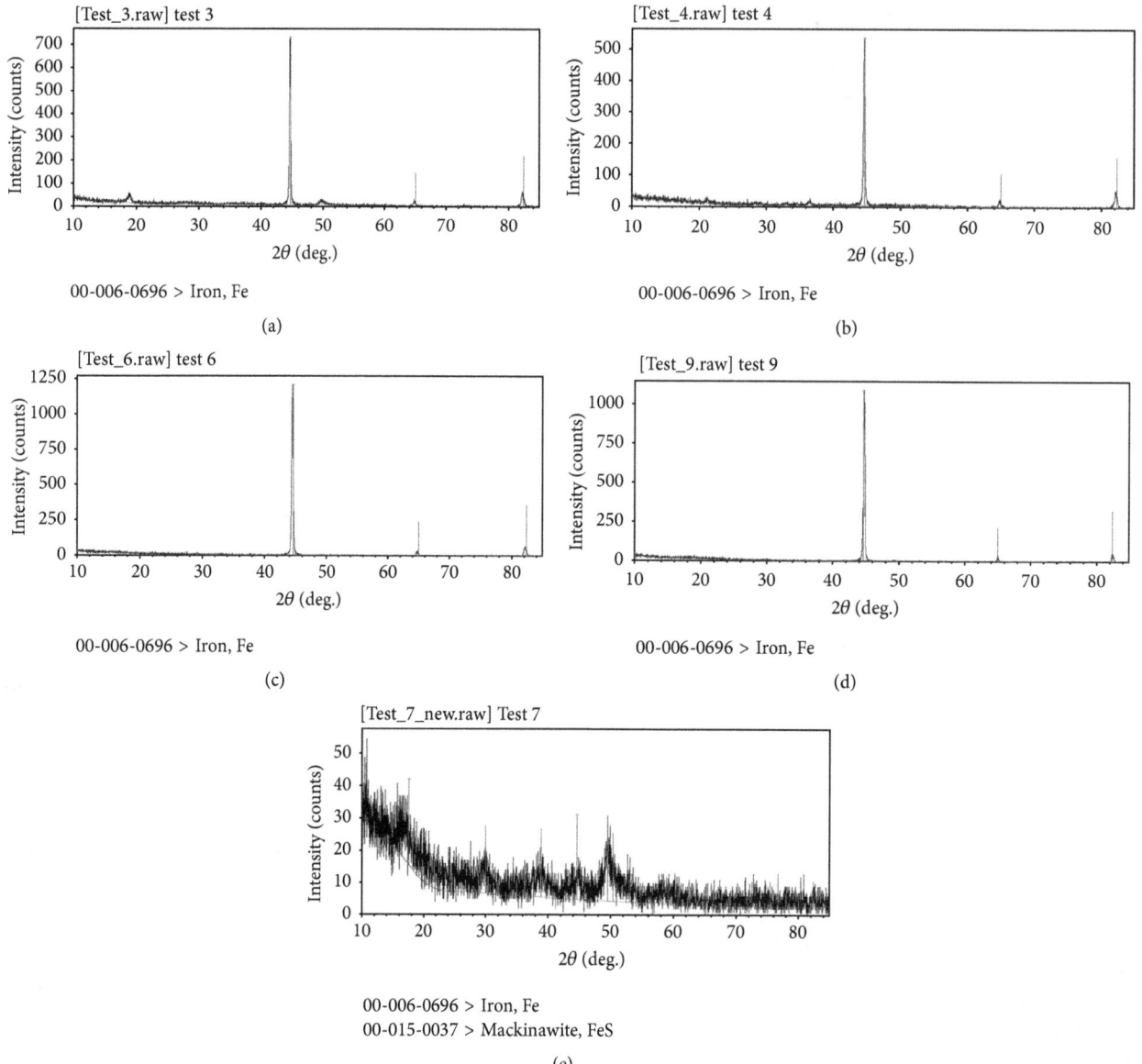

FIGURE 2: P-XRD analysis on the 4130 alloy surface at (a) 50°C, 4 pH, and 48 hours, (b) 25°C, 4 pH, and 24 hours, (c) 50°C, 2 pH, and 24 hours, (d) initial condition (uncorroded sample), and (e) 75°C, 4 pH, and 24 hours.

This result suggests that the film thickness is increased at high temperatures. In lieu of thin film XRD analysis, P-XRD may be able to detect thicker corrosion layers formed at relatively high temperatures.

Figure 3 shows the SEM images of corrosion product films formed under varying immersion time. After 24-hour immersion time, a uniform layer of corrosion product, consisting of small tetragonal mackinawite, covered the surface [29]. As shown in Figure 3(a), this thick corrosion layer is loose and full of blister and cracks, causing the corrosion rate to accelerate by increasing the diffusion of electrochemical reaction species such as Fe^{2+} through the alloy surface. As has been mentioned in other researches, this initial mackinawite layer is easily cracked and peeled off due to stress as a result of the volume effect [30]. This failure of the initial corrosion

layer will gradually increase the corrosion rate and expose more unprotected area to the solution.

Figure 3(b) shows the EDX analysis results of corrosion product films after 24-hour immersion. These results indicated that most of the corrosion products are iron-rich compounds such as mackinawite, which generally has lower corrosion resistance compared to sulfur-rich compounds such as troilite. The corrosion resistance of FeS follows a sequence of mackinawite < troilite and < pyrrhotite < pyrite [24].

After 48-hour immersion, the corrosion scale cracks become more severe and hexagonal crystals form beside the cracks as shown in Figure 3(c). The EDX results of these hexagonal crystals indicate high sulfur content in their chemical composition as shown in Figure 3(d).

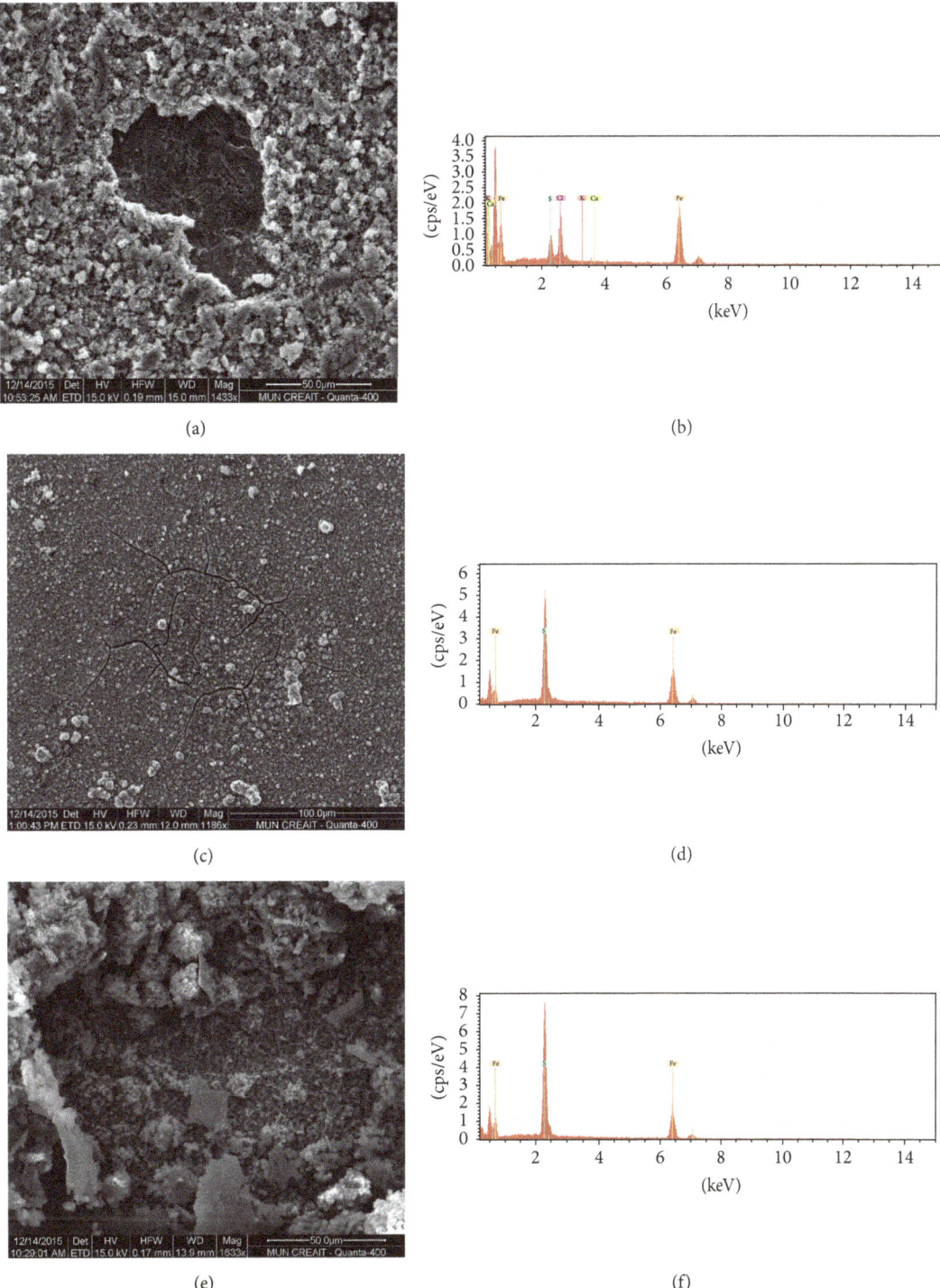

FIGURE 3: SEM analysis of corrosion products of 4130 alloy after (a) 24-, (c) 48-, and (e) 72-hour immersion at pH 4 and 50°C and EDX analysis of corrosion products of 4130 alloy after (b) 24-, (d) 48-, and (f) 72-hour immersion at pH 4 and 50°C.

Figure 3(e) shows that, after 72 hours of immersion, the initial corrosion product film has cracked and peeled off the surface of the specimen and a newly formed corrosion scale has been integrated. Larger, hexagonal shaped corrosion products formed on top of the new scale as mainly troilite crystals formed near the end of the 72 hours.

Generally, it could be said that by increase of immersion time more corrosion resistant products such as troilite replaced the initially formed mackinawite on the alloy surface. This is supported by the EDX results that indicate the major corrosion product varied from iron-rich mackinawite to sulfur-rich troilite, in Figures 3(b), 3(d), and 3(f). Despite

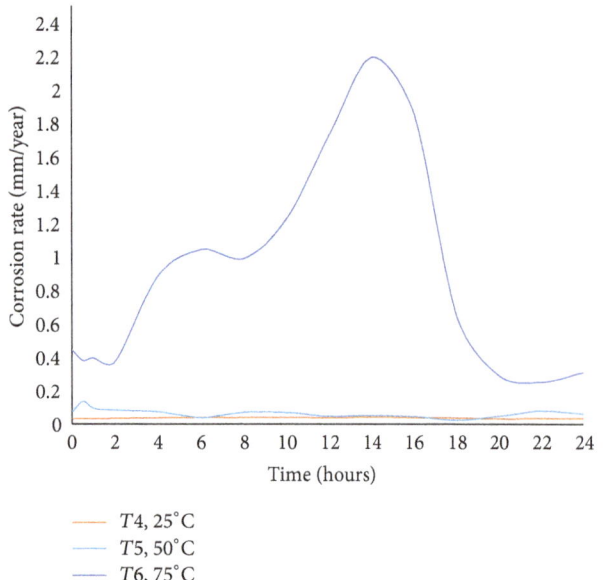

FIGURE 4: Corrosion rate with temperature at pH 4 and 24 hours.

FIGURE 5: SEM image of corrosion products on surface of 4130 alloy after 24-hour immersion at pH 4 and 75°C.

the nucleation of stable troilite crystals on the metal surface, the results of LPR measurements showed that between 48 and 72 hours the corrosion rate dramatically increased from 0.0662 to 0.779. This increase could be explained by localized fracture of the corrosion film due to weak adhesion of the scale on the surface. This provides a path for sulfide to penetrate and attack the substrate of metal surface.

3.2. Effect of Temperature on the Corrosion Mechanism and Products. Figure 4 shows the effect of increasing temperature on the corrosion rate of specimens over the course of 24 hours at pH 4.

It can be observed that, during first 12 hours of increasing temperature from 25°C to 75°C, the corrosion rate dramatically increased which can be explained by the following reasons:

(1) Increasing the temperature could accelerate the diffusion of species involved in electrochemical reactions.

(2) Temperature could affect the concentration of corrosion species by preferentially evaporating one or more species out of the solution, which could affect the corrosion reaction.

It has been confirmed by previous research that temperature generally accelerates most of the chemical, electrochemical, and transporting processes occurring during the corrosion process and also both cathodic reactions and anodic currents which were measured increased with increasing temperature [31].

During the final 12 hours of testing at 75°C, the corrosion rate significantly decreases from 2.2 to 0.25 mm/year, which could be related to transformation of mackinawite crystalline structure to a more resistant troilite crystalline structure. The SEM images in Figure 5 show significant fracturing of the

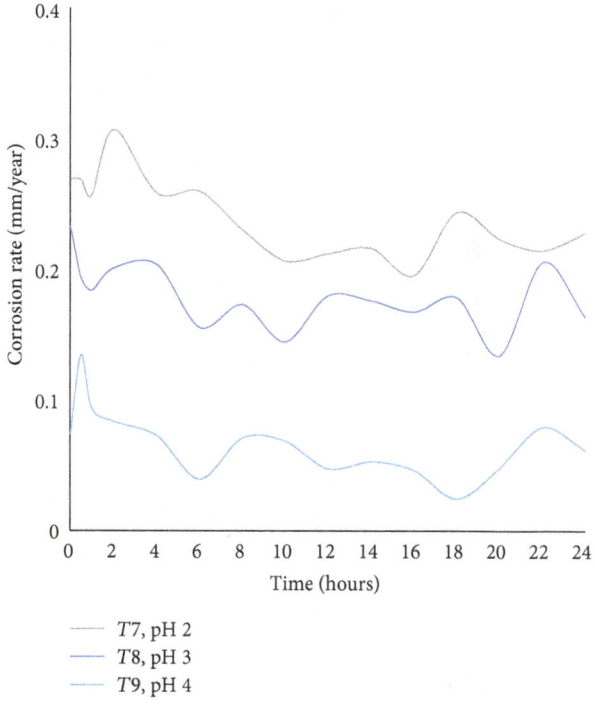

FIGURE 6: Corrosion rate with pH at 50°C and 24 hours.

surface film at 75°C which explains the initial higher corrosion rate due to the diffusion of species into nonprotective mackinawite followed by the decreased corrosion rate due to formation of the protective troilite crystalline structure on the alloy surface.

3.3. Effect of pH on the Corrosion Mechanism and Products. Figure 6 shows the effect of pH on the corrosion rate of specimens immersed for 24 hours at 50°C.

The results show that decreasing pH from 4 to 2 slightly increases the corrosion rate. The protective nature and composition of the corrosion product depend greatly on the pH of the solution. At lower values of pH (<3), iron is dissolved

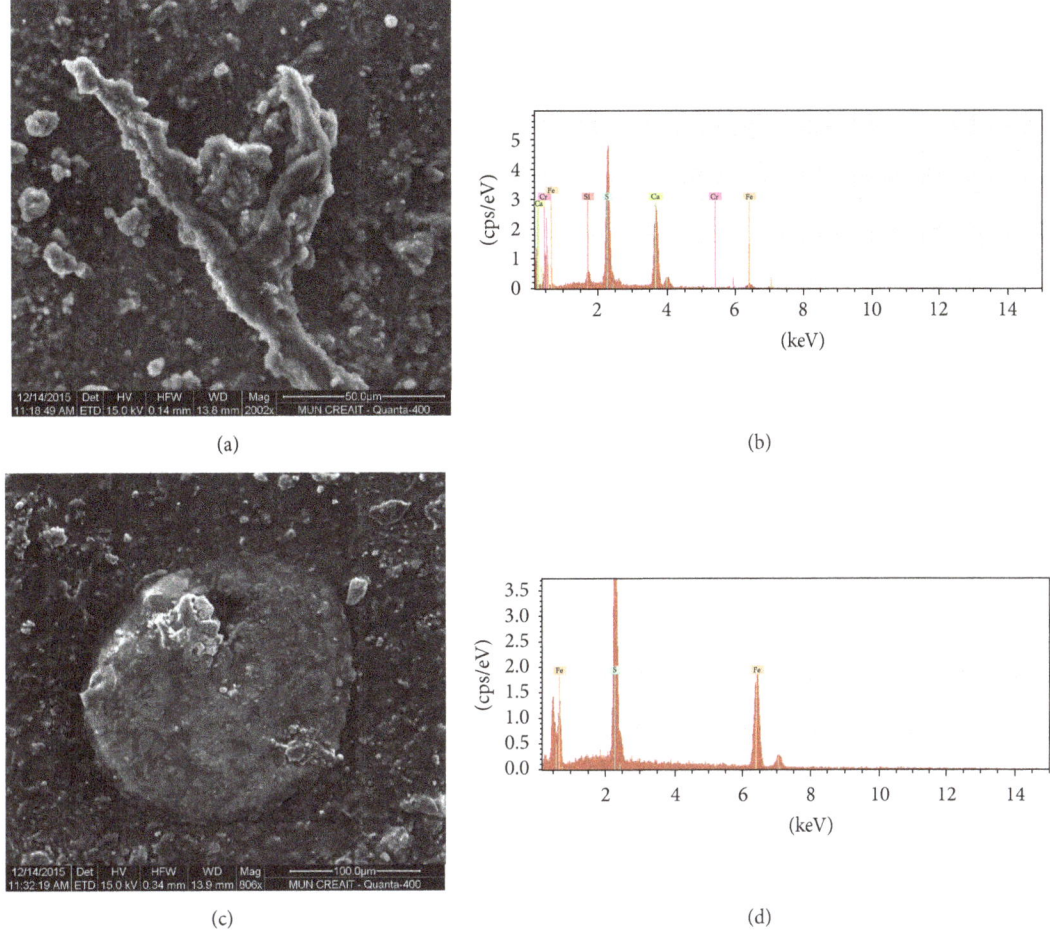

(a)

(b)

(c)

(d)

FIGURE 7: ((a) and (c)) SEM images of corrosion products on surface of 4130 alloy after 24-hour immersion at 50°C and pH 2 and((b) and (d)) EDX analysis of corrosion products on surface of 4130 alloy after 24-hour immersion at 50°C and pH 2.

and FeS is mostly inhibited from precipitating on the metal surface due to very high solubility of FeS phases [32]. Figure 7(a) shows the SEM image of corrosion products on the surface of a specimen after 24 hours at pH 2 and 50°C. As can be observed, the corrosion products are loose and detached from the surface. This could result in the products being easily removed by shear stress. The EDX results as shown in Figure 7(b) indicate a high presence of sulfur compounds and a low presence of iron compounds on the surface.

The SEM results in Figure 7(c) of the specimen immersed in the solution at pH 2 also show the presence of a pit on the surface. This is another reason for the higher corrosion rates seen at low pH. The corrosion pit shown in Figure 7(c) has a brittle cap covering the substrate. EDX analysis indicates that this cap is primarily sulfide as shown in Figure 7(d).

At pH 3, the top surface layer displayed a flaky structure as seen in Figure 8. Parts of the layer had spalled off and revealed the presence of much smaller crystallites under the outer layer. It is likely that this layer is the result of the immediate precipitation of Fe^{2+} released by corrosion [32]. At pH values from 3 to 4, an inhibitive effect of the corrosion mechanism is seen due to the formation of a marcasite FeS protective

FIGURE 8: SEM image of corrosion products on surface of 4130 alloy after 24-hour immersion at pH 3 and 50°C.

film on the electrode surface. At pH 3, small crystals were observed on areas where the outer layer had spalled off as shown in Figure 8. At pH 4, the surface was mostly covered with a much denser layer as shown in Figure 9.

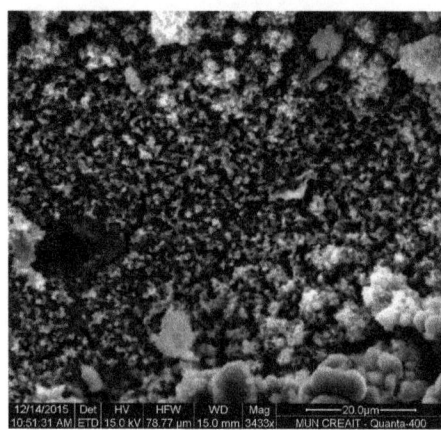

FIGURE 9: SEM image of corrosion products on surface of 4130 alloy after 24-hour immersion at pH 4 and 50°C.

4. Conclusions

The results of this research indicated that acidic chemical bath deposition could be successfully applied to investigate the formation and growth of FeS thin films under varying experimental conditions. Due to the inherent safety concerns associated with sour corrosion experiments in laboratories, this acidic chemical bath deposition method could be applied as a substitute for H_2S in certain experiments to characterize formation and transformation of FeS corrosion products.

Other primary findings of this research are as follows:

(i) Increase of immersion time gradually increases the corrosion rate of 4130 chromium alloy steel in this experiment, resulting from localize fracture of corrosion layer despite transformation of FeS crystalline structures from iron-rich mackinawite to sulfur-rich troilite compounds during the corrosion process.

(ii) Increase of pH directly decreases the corrosion rate of 4130 alloy steel in this experiment resulting from the formation of a more resistant FeS film at higher values of pH.

(iii) Increase of temperature from 25°C to 75°C causes an increase in the corrosion rate of 4130 alloy steel, likewise resulting from the transformation of FeS crystalline structure during the corrosion process.

Competing Interests

The authors declare that they have no competing interests.

Acknowledgments

The research in this paper is supported by the Suncor Reservoir Souring Initiative at Memorial University of Newfoundland.

References

[1] D. W. Shoesmith, P. Taylor, M. G. Bailey, and D. G. Owen, "The formation of ferrous monosulfide polymorphs during the corrosion of iron by aqueous hydrogen sulfide at 21°C," *Journal of the Electrochemical Society*, vol. 127, no. 5, pp. 1007–1015, 1980.

[2] D. W. Shoesmith, "Formation, transformation and dissolution of phases formed on surfaces," in *Proceedings of the Electrochemical Society Meeting*, Ottawa, Canada, November 1981.

[3] S. N. Smith and E. J. Wright, "Prediction of minimum H_2S levels required for slightly sour corrosion," in *Proceedings of the Conference on Corrosion*, Paper No. 11, NACE International, 1994.

[4] S. N. Smith and E. J. Wright, "Prediction of corrosion in slightly sour environments," in *Proceedings of the Conference on Corrosion*, Paper no. 02241, NACE International, 2002.

[5] J. S. Smith and J. D. A. Miller, "Nature of sulphides and their corrosive effect on ferrous metals: a review," *British Corrosion Journal*, vol. 10, no. 3, pp. 136–143, 1975.

[6] H. Fang, *Investigation of localized corrosion of carbon steel in H_2S environments [Ph.D. thesis]*, Ohio University, Athens, Ohio, USA, 2012.

[7] W. Sun and S. Nesic, "A mechanistic model of H_2S corrosion of mild steel," Paper 07655, NACE International, CORROSION/2007, St. Pete Beach, Fla, USA, 2007.

[8] W. Sun, S. Nesic, and S. Papavinasan, "Kinetics of iron sulfide and mixed iron sulfide/carbonate scale precipitation in CO_2/H_2S Corrosion," CORROSION/2007, NACE International, paper no. 06644, 2007.

[9] M. Koteeswaran, *CO_2 and H_2S corrosion in oil pipelines [M.S. thesis]*, University of Stavanger, Stavanger, Norway, June 2010.

[10] W. Sun, *Kinetics of iron carbonate and iron sulfide scale formation in CO_2/H_2S corrosion [Ph.D. thesis]*, Ohio University, Athens, Ohio, USA, 2006.

[11] P. Taylor, "The stereochemistry of iron sulfides—a structural rationale for the crystallization of some metastable phases from aqueous solution," *American Mineralogist*, vol. 65, pp. 1026–1030, 1980.

[12] M. Bonis, M. Girgis, K. Goerz, and R. MacDonald, "Weight loss corrosion with H_2S: using past operations for designing future facilities," CORROSION/2006, NACE International, paper no. 06122, 2006.

[13] D. Rickard and G. W. Luther, "Chemistry of iron sulfides," *Chemical Reviews*, vol. 107, no. 2, pp. 514–562, 2007.

[14] A. R. Lennie and D. J. Vaughan, "Spectroscopic studies of iron sulfide formation and phase relations at low temperatures," *Mineral Spectroscopy*, vol. 5, pp. 117–131, 1996.

[15] L. Smith and B. Craig, "Practical corrosion control measures for elemental sulfur," in *Proceedings of the Conference on Corrosion*, Paper No. 05646, pp. 1–20, NACE International, 2005.

[16] B. N. Brown, *The influence of sulfides on localized corrosion of mild steel [Ph.D. thesis]*, Ohio University, 2013.

[17] T. Laitinen, "Localized corrosion of stainless steel in chloride, sulfate and thiosulfate containing environments," *Corrosion Science*, vol. 42, no. 3, pp. 421–441, 2000.

[18] J. Kvarekval, "Morphology of localised corrosion attacks in sour environments," CORROSION/2007, NACE International, paper no. 07659, 2007.

[19] K. J. Lee, *A mechanistic modeling of CO_2 corrosion of mild steel in the presence of H_2S [Ph.D. thesis]*, Ohio University, 2004.

[20] N. G. Harmandas and P. G. Koutsoukos, "The formation of iron(II) sulfides in aqueous solutions," *Journal of Crystal Growth*, vol. 167, no. 3-4, pp. 719–724, 1996.

Electrochemical and Microstructural Analysis of FeS Films from Acidic Chemical Bath at Varying Temperatures...

9

[21] J. Amri and J. Kvarekvål, "Simultation of solid state growth of iron sulfide in sour corrosion conditions," Paper 11076, NACE International, CORROSION/2011, St. Pete Beach, Fla, USA, 2011.

[22] N. S. P. Obuka, O. N. Celestine, G. R. O. Ikwu, Chukwumuanya, and E. Okechukwu, "Review of corrosion kinetics and thermodynamics of CO_2 and H_2S corrosion effects and associated prediction/evaluation on oil and gas pipeline system," *International Journal of Scientific & Technology Research*, vol. 1, no. 4, pp. 156–162, 2012.

[23] A. G. Wikjord, T. E. Rummery, F. E. Doern, and D. G. Owen, "Corrosion and deposition during the exposure of carbon steel to hydrogen sulphide-water solutions," *Corrosion Science*, vol. 20, no. 5, pp. 651–671, 1980.

[24] S. N. Smith, "A proposed mechanism for corrosion in slightly sour oil and gas production," in *Proceedings of the 12th International Corrosion Congress*, vol. 4, pp. 2695–2706, NACE International, Houston, Tex, USA, September 1993.

[25] D. Rickard, "Kinetics of FeS precipitation: part 1. Competing reaction mechanisms," *Geochimica et Cosmochimica Acta*, vol. 59, no. 21, pp. 4367–4379, 1995.

[26] Petroleum and Natural Gas Industries, "Materials for use in H_2S containing environments in oil and gas production," Tech. Rep. NACE MR0175/ISO 15156-1, 2001.

[27] M. Saeed Akhtar, A. Alenad, and M. Azad Malik, "Synthesis of mackinawite FeS thin films from acidic chemical baths," *Materials Science in Semiconductor Processing*, vol. 32, pp. 1–5, 2015.

[28] ASTM-G5-82, "Standard reference method for making potentiostatic and potentiodynamic anodic polarisation measurements," Annul B. ASTM Standards vol. 03.02, 1982, Reapproved as ASTM-65-87 and as ASTM-65-94, pp. 511–521, 1982.

[29] P. Bai, S. Zheng, H. Zhao, Y. Ding, J. Wu, and C. Chen, "Investigations of the diverse corrosion products on steel in a hydrogen sulfide environment," *Corrosion Science*, vol. 87, pp. 397–406, 2014.

[30] M. Liu, J. Wang, W. Ke, and E.-H. Han, "Corrosion behavior of X52 anti-H_2S pipeline steel exposed to high H_2S concentration solutions at 90∘C," *Journal of Materials Science and Technology*, vol. 30, no. 5, pp. 504–510, 2014.

[31] Y. Zheng, B. Brown, and S. Nesic, "Electrochemical study and modeling of H_2S corrosion of mild steel," in *Proceedings of the Corrosion*, Paper no. 2406, pp. 1–22, NACE International, 2013.

[32] B. Valery, *Effect of pre-exposure of sulfur and iron sulfide on H_2S Corrosion at different temperatures [M.S. thesis]*, University of Stavanger, 2011.

Simulation of the Ill-Posed Problem of Reinforced Concrete Corrosion Detection Using Boundary Element Method

Syarizal Fonna,[1] **Israr M. Ibrahim,**[2] **M. Ridha,**[1] **Syifaul Huzni,**[1] **and A. K. Ariffin**[3]

[1]*Department of Mechanical Engineering, Syiah Kuala University, Jalan Tgk Syech Abdul Rauf 7, Banda Aceh 23111, Indonesia*
[2]*Tsunami & Disaster Mitigation Research Center (TDMRC), Syiah Kuala University, Jalan Tgk Abdul Rahman, Gp. Pie, Meuraxa District, Banda Aceh 23111, Indonesia*
[3]*Department of Mechanical and Materials Engineering, Universiti Kebangsaan Malaysia, 43600 Bangi, Selangor, Malaysia*

Correspondence should be addressed to Syarizal Fonna; syarizal.fonna@unsyiah.ac.id

Academic Editor: Jerzy A. Szpunar

Many studies have suggested that the corrosion detection of reinforced concrete (RC) based on electrical potential on concrete surface was an ill-posed problem, and thus it may present an inaccurate interpretation of corrosion. However, it is difficult to prove the ill-posed problem of the RC corrosion detection by experiment. One promising technique is using a numerical method. The objective of this study is to simulate the ill-posed problem of RC corrosion detection based on electrical potential on a concrete surface using the Boundary Element Method (BEM). BEM simulates electrical potential within a concrete domain. In order to simulate the electrical potential, the domain is assumed to be governed by Laplace's equation. The boundary conditions for the corrosion area and the noncorrosion area of rebar were selected from its polarization curve. A rectangular reinforced concrete model with a single rebar was chosen to be simulated using BEM. The numerical simulation results using BEM showed that the same electrical potential distribution on the concrete surface could be generated from different combinations of parameters. Corresponding to such a phenomenon, this problem can be categorized as an ill-posed problem since it has many solutions. Therefore, BEM successfully simulates the ill-posed problem of reinforced concrete corrosion detection.

1. Introduction

Rebar corrosion is one of the main causes of reinforced concrete (RC) premature failures [1–3]. Reports of these premature failures can be found in various publications. The failures include the collapse of Silver Bridge in USA, 1967 [4], the collapse of highway overpass in Canada, 2006 [5], and the collapse of Atlantis Water Adventure, *Taman Impian Jaya Ancol* in Indonesia, 2011 [6]. Recent failure due to corrosion was reported in March 2015: the porch of a building collapsed in Albany, USA [7]. Thus, It is important to conduct periodic evaluation, monitoring and early detection for RC corrosion [8–10].

The half-cell potential technique is among the conventional methods that are used in the field to detect or evaluate the RC corrosion [11, 12]. This technique follows the procedure as described in ASTM C876 to evaluate corrosion of an RC structure. However, the method only provides the probability of corrosion [13, 14] and needs a considerable amount of measurement data to generate an accurate potential map [11, 15]. Therefore, it is important to understand the nature of the RC corrosion problem before the development of other methods and/or improvement of conventional techniques to detect RC corrosion. Many workers have proposed methods based on inverse analysis to detect RC corrosion [13, 15, 16] since the nature of RC corrosion implies an ill-posed problem. However, it is difficult to prove the ill-posed problem of RC corrosion via experiments. Thus, using a numerical method to prove the ill-posed problem of RC corrosion is very promising.

Many researchers have explored a numerical method termed the Boundary Element Method (BEM) to simulate the corrosion phenomenon. The corrosion was modeled by the Laplace equation in BEM [16–18]. Thus, BEM can potentially be utilized to simulate the ill-posed problem of RC corrosion. The purpose of this paper is to simulate the ill-posed problem of RC corrosion problem by using BEM.

2. Basic Idea to Simulate the Ill-Posed Problem of RC Corrosion

The ill-posed problem is a problem that has one of the following criteria; that is, the problem has no unique solution or many solutions, and small error would give high disturbance to the solution [19]. The motivation for utilizing BEM to simulate the ill-posed problem of RC corrosion came from the actual condition that interpretation of the half-cell potential technique is merely based on electrical potential data on the RC surface, as mentioned in ASTM C876. Previous researchers have pointed out that the electrical potential on the RC surface is influenced not only by rebar corrosion but also by other parameters [20].

Furthermore, it has been suggested that the variation of some parameters could give similar electrical potential profiles on the RC surface, which should indicate an ill-posed problem. By simulating similar electrical potentials resulting from different parameter combinations, the ill-posed problem of RC corrosion can be proven. This ill-posed problem might lead to misleading conclusions in the detection of RC corrosion by the half-cell potential technique.

Since BEM has the capability of obtaining electrical potential and current density within an evaluated domain, it is proposed in this paper that BEM is also capable to be used to simulate the ill-posed problem of RC corrosion. The basic idea for this purpose was to compare the electrical potential on an RC surface obtained by BEM, which came from various combinations of parameters.

3. RC Corrosion Modeling in BEM

The RC model with single reinforcing steel as given in Figure 1(a) was considered. There is corrosion located in the reinforcing steel. This RC model was simplified into a 2D model, as shown in Figure 1(b), which also displays the boundary conditions for the model.

In developing BEM for RC corrosion simulation, the electrical potential field (ϕ) within the whole RC domain (Ω) is mathematically governed by the Laplace equation as given in [17, 18, 21]

$$\nabla^2 \phi = 0 \quad \text{in } \Omega. \tag{1}$$

The relationship between electrical potential and current density (i) for the domain should follow [21, 22]

$$i = -\kappa \frac{\partial \phi}{\partial \mathbf{n}} \left(A/m^2 \right), \tag{2}$$

where κ is the concrete conductivity, $\mathbf{n}$ is the outward normal unit, and $\partial/\partial \mathbf{n}$ is the derivative in the normal direction.

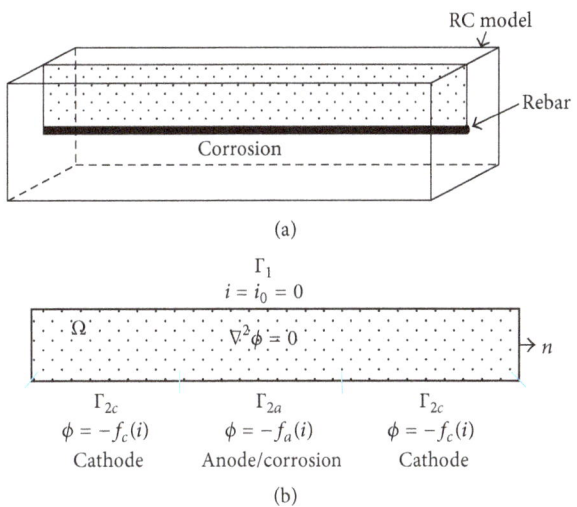

FIGURE 1: (a) RC model with single rebar; (b) 2D model of RC for BEM simulation.

The boundary conditions for the RC model are given in Figure 1(b) and are written as

$$i = i_0 = 0 \ \left(A/m^2\right) \quad \text{on } \Gamma_1$$

$$\phi = -f_a(i) \ (V) \quad \text{on } \Gamma_{2a} \tag{3}$$

$$\phi = -f_c(i) \ (V) \quad \text{on } \Gamma_{2c},$$

where i on the concrete surface (Γ_1) is constant and is considered equal to zero, due to the low conductivity of concrete. ϕ on any point of the rebar or reinforcing steel (Γ_2) is given by a function of i that is generated from the polarization curve, that is, $f_a(i)$ for the corroded part (anode) and $f_c(i)$ noncorroded part (cathode) areas. The polarization curve was measured experimentally.

BEM is formulated to solve the RC model; hence, the electrical potential and current density on concrete and rebar surfaces can be obtained. The procedure of BEM formulation for the RC corrosion case can be found in [21].

4. Numerical Simulation and Discussion

In order to simulate the ill-posed problem of RC corrosion detection using BEM, an RC model was considered, as given in Figure 2. The model consists of a single rebar and corrosion. The corrosion size and rebar length, respectively, were c cm and 50 cm. The concrete cover depth for the model was t cm, while concrete conductivity for the model was $\kappa \ \Omega^{-1} \cdot m^{-1}$. The boundary conditions for the model were the same as those already stated. The polarization curve for anode and cathode was obtained from [15, 23].

Ten combinations of parameters were selected for the simulation using BEM. Those parameters were corrosion size (c), concrete cover (t), concrete conductivity (κ), and corrosion intensity at the anode part of rebar that was generated from polarization curves of the rebar in concrete. The combinations of parameters are listed in Table 1. For all

TABLE 1: 10 combinations of parameters for evaluating the nature of the RC corrosion detection problem.

Parameter combinations	Corrosion size (c), cm	Concrete cover (t), cm	Concrete conductivity (κ), $\Omega^{-1} \cdot m^{-1}$	Corrosion intensity, V (versus SCE)
1	6	5	0.007	$\phi_{a1} = 0.6 - 10i$
2	10	5	0.007	$\phi_{a1} = 0.6 - 10i$
3	14	5	0.007	$\phi_{a1} = 0.6 - 10i$
4	6	5	1	$\phi_{a1} = 0.6 - 10i$
5	6	5	0.1	$\phi_{a1} = 0.6 - 10i$
6	6	5	0.01	$\phi_{a1} = 0.6 - 10i$
7	6	1	0.007	$\phi_{a1} = 0.6 - 10i$
8	6	10	0.007	$\phi_{a1} = 0.6 - 10i$
9	6	5	0.007	$\phi_{a2} = 0.5 - 10i$
10	6	5	0.007	$\phi_{a3} = 0.4 - 10i$

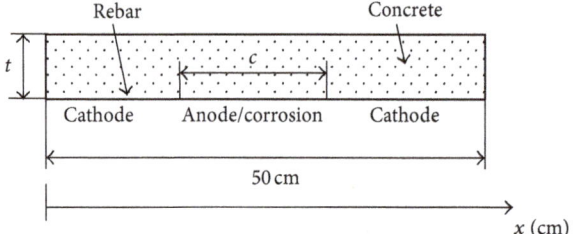

FIGURE 2: RC model for evaluating the nature of RC corrosion detection problem.

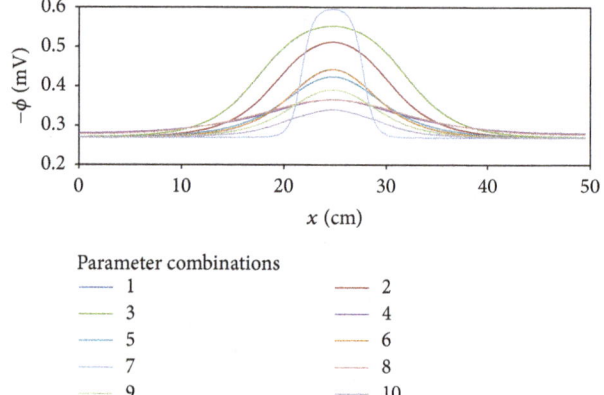

Parameter combinations
— 1   — 2
— 3   — 4
— 5   — 6
— 7   — 8
— 9   — 10

FIGURE 3: Electrical potential profiles on concrete surface for all combinations of parameters.

combinations, the cathode part of rebar was represented by its polarization curve in

$$\phi_c = 0.27 - 10i \text{ (V) versus SCE.} \qquad (4)$$

BEM was applied to simulate the electrical potential on the surface of RC for 10 combinations of parameters in Table 1. The simulation result is given in Figure 3, which shows the electrical potential profile on the RC surface for all parameter combinations. It shows that the electrical potential is generally higher above the corroded part than the cathode part.

Figure 4 shows how the corrosion size would affect the electrical potential profile on RC surface. Figure 4 also shows that the larger corrosion size would give a higher electrical potential on the RC surface. Also, the peak of the electrical potential profile becomes wider for larger corrosion size. Corresponding to half-cell potential technique, higher electrical potential on the RC surface means higher corrosion risk. It can be said that combination number 3 had a higher corrosion risk than combination numbers 1 and 2. This might mislead corrosion evaluation, because even though the corrosion size of combination number 3 was larger than others, the corrosion rate for both could still be similar, since they both had same corrosion intensity, as given in Table 1. Also, the boundary conditions for the cathode part were similar for all combinations.

The electrical potential profiles on the RC surface were influenced by concrete conductivity, as shown in Figure 5. It shows that the electrical potential profile will flatten by increasing conductivity of concrete. This characteristic is consistent with the investigation that was conducted by Pour-Ghaz et al. [20]. The phenomenon can also lead to misleading corrosion evaluation using the half-cell potential technique based on ASTM C876. For example, using combination number 6 will result in classifying the corrosion as severe corrosion risk level (< −380 mV versus SCE), while combination number 4 falls into the high corrosion risk level (−380 to −230 mV versus CSE) for the same corrosion. However, the actual corrosion for the combination was the same, that is, in terms of size and intensity.

Figure 6 shows the influence of concrete cover depth to electrical potential on the surface of concrete. The electrical potential above the corroded part would decrease by increasing the depth of concrete cover for the same corrosion. It was also similar to the work of Pour-Ghaz et al. [20]. Thus, similar to other parameters, the cover depth must be included in the analysis for detecting corrosion risk level based on ASTM

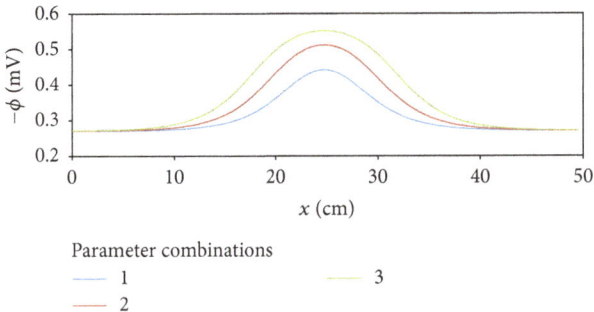

FIGURE 4: Electrical potential profiles on concrete surface for different corrosion sizes.

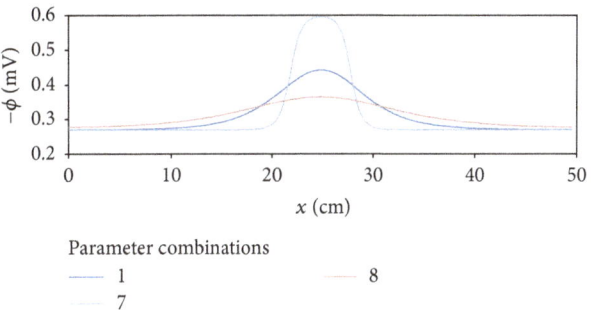

FIGURE 6: Electrical potential profile on concrete surface for different concrete cover depths.

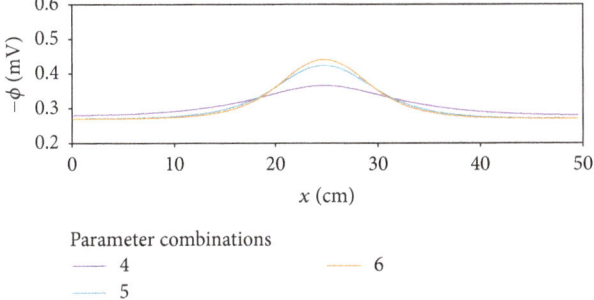

FIGURE 5: Electrical potential profile on concrete surface for different concrete conductivities.

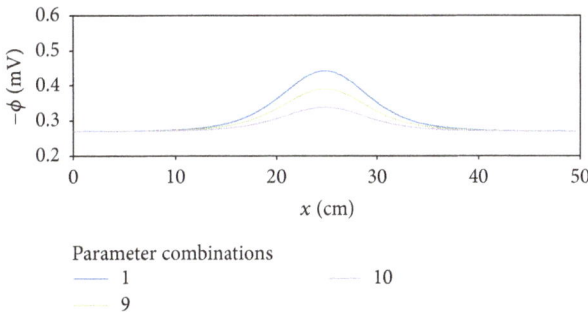

FIGURE 7: Electrical potential profile on concrete surface for different corrosion intensities.

C876, in order to eliminate false positives and negatives in interpretation of electrical potential data.

Moreover, the corrosion intensity would affect the electrical potential of the concrete surface above the corroded rebar part, as shown in Figure 7. It shows that the higher corrosion intensity would bring higher electrical potential values on the concrete surface above the corroded part for the same corrosion size. This could be true, because higher corrosion intensity might lead to higher corrosion rate, and thus the electrical potential should be higher too.

By comparing electrical potential profiles for all parameter combinations, it was found that some profiles are similar, as given in Figure 8. This figure shows that electrical potential on the concrete surface for combination number 1 was almost similar to combination number 6, and combination number 4 was almost identical to combination number 8. This suggests that there exist combinations of parameters that may give the same electrical potential profile on the concrete surface.

From the presented results, it can be concluded that there are many solutions for the rebar corrosion problem. It has been shown that several similar electrical potential profiles on the concrete surface can be generated from several combinations of parameters. Therefore, it would be difficult to evaluate actual rebar corrosion only by using electrical potential data on the concrete surface based on ASTM C876.

According to Kabanikhin [19], such a phenomenon was categorized into an ill-posed problem. The conventional method, such as direct method, is insufficient to solve the problem. One promising method to solve the ill-posed problem is inverse analysis [24]. Several researchers have explored

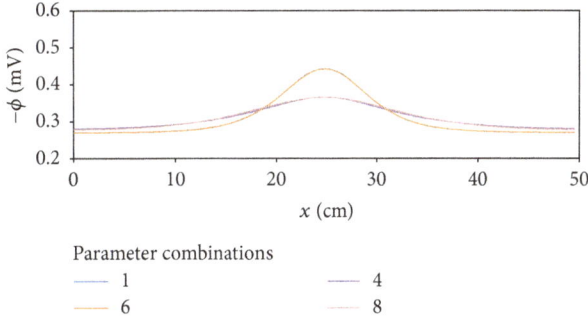

FIGURE 8: Electrical potential profile on concrete surface, showing that combination 1 is similar to 6 and that 4 is similar to 8.

the application of inverse analysis to solve the rebar corrosion detection in concrete [13, 16, 21, 22], and the method has the potential to be applied in solving the RC corrosion detection problem.

5. Conclusions

The simulation of the ill-posed problem of rebar corrosion detection using the Boundary Element Method (BEM) has been presented in this paper. BEM was used to simulate electrical potential within concrete domain, especially on the concrete surface. The numerical simulation results with 10 parameter combinations showed that the electrical potential on the RC surface was not solely influenced by corrosion, but also by other parameters, such as concrete conductivity and cover depth. Furthermore, some combination of parameters might give the same electrical potential profile on the concrete surface. This phenomenon is categorized as an ill-posed problem, since there exist many solutions to the problem. Therefore, the detection of rebar corrosion using only electrical potential data on the concrete surface, as suggested in ASTM C876, might mislead corrosion evaluation due to the ill-posedness of the problem.

Competing Interests

The authors declare that they have no competing interests.

References

[1] H. A. Elfergani, R. Pullin, and K. M. Holford, "Damage assessment of corrosion in prestressed concrete by acoustic emission," *Construction and Building Materials*, vol. 40, pp. 925–933, 2013.

[2] G. Qiao, T. Liu, Y. Hong, and J. Ou, "Optimization design of a corrosion monitoring sensor by FEM for RC structures," *IEEE Sensors Journal*, vol. 11, no. 9, pp. 2111–2112, 2011.

[3] B. Elsener, "Corrosion rate of steel in concrete—measurements beyond the tafel law," *Corrosion Science*, vol. 47, no. 12, pp. 3019–3033, 2005.

[4] C. LeRose, "The collapse of the silver bridge," *West Virginia Historical Society Quarterly*, vol. 15, no. 4, 2001, http://www.wvculture.org/history/wvhs/wvhs1504.html.

[5] CBC.ca, "Former Quebec premier to head probe into overpass collapse," 2006, http://www.cbc.ca/news/canada/story/2006/10/02/laval-montreal.html.

[6] R. Afifah and Latief, Struktur Wahana Atlantis Dikaji Ulang, 2011, http://megapolitan.kompas.com/read/2011/09/28/12185069/Struktur.Wahana.Atlantis.Dikaji.Ulang.

[7] B. Woodard, *Apartment Building Cited Several Times Before Porch Collapse Injured Man*, 2015, http://www.dnainfo.com/chicago/20150327/west-rogers-park/apartment-building-cited-several-times-before-porch-collapse-injured-man.

[8] K. Hornbostel, C. K. Larsen, and M. R. Geiker, "Relationship between concrete resistivity and corrosion rate—a literature review," *Cement and Concrete Composites*, vol. 39, pp. 60–72, 2013.

[9] J. Gao, J. Wu, J. Li, and X. Zhao, "Monitoring of corrosion in reinforced concrete structure using Bragg grating sensing," *NDT & E International*, vol. 44, no. 2, pp. 202–205, 2011.

[10] Z. W. Wang, M. Zhou, G. G. Slabaugh, J. Zhai, and T. Fang, "Automatic detection of bridge deck condition from ground penetrating radar images," *IEEE Transactions on Automation Science and Engineering*, vol. 8, no. 3, pp. 633–640, 2011.

[11] H.-W. Song and V. Saraswathy, "Corrosion monitoring of reinforced concrete structures-a review," *International Journal of Electrochemical Science*, vol. 2, no. 1, pp. 1–28, 2007.

[12] M. Ridha, S. Fonna, S. Huzni, and A. K. Ariffin, "Corrosion risk assessment of public buildings affected by the 2004 tsunami in Banda Aceh," *Journal of Earthquake and Tsunami*, vol. 7, no. 1, pp. 1–22, 2013.

[13] P. Marinier and O. B. Isgor, "Model-assisted non-destructive monitoring of reinforcement corrosion in concrete structures," in *Nondestructive Testing of Materials and Structures*, O. Büyüköztürk and M. A. Taşdemir, Eds., vol. 6 of *RILEM Bookseries*, pp. 719–724, Springer, New York, NY, USA, 2013.

[14] A. A. A. Hassan, K. M. A. Hossain, and M. Lachemi, "Corrosion resistance of self-consolidating concrete in full-scale reinforced beams," *Cement & Concrete Composites*, vol. 31, no. 1, pp. 29–38, 2009.

[15] M. Ridha, K. Amaya, and S. Aoki, "Boundary element simulation for identification of steel corrosion in concrete using magnetic field measurement," *Corrosion*, vol. 61, no. 8, pp. 784–791, 2005.

[16] M. Ridha, K. Amaya, and S. Aoki, "Multistep genetic algorithm for detecting corrosion of reinforcing steels in concrete," *Corrosion*, vol. 57, no. 9, pp. 794–801, 2001.

[17] K. Amaya and S. Aoki, "Effective boundary element methods in corrosion analysis," *Engineering Analysis with Boundary Elements*, vol. 27, no. 5, pp. 507–519, 2003.

[18] S. Aoki and K. Kishimoto, "Aplication of BEM to galvanic corrosion and cathodic protection," in *Electrical Engineering Applications*, C. A. Brebbia, Ed., vol. 7 of *Topics in Boundary Element Research*, pp. 65–86, Springer, New York, NY, USA, 1990.

[19] S. I. Kabanikhin, "Definitions and examples of inverse and ill-posed problems," *Journal of Inverse and Ill-Posed Problems*, vol. 16, no. 4, pp. 317–357, 2008.

[20] M. Pour-Ghaz, O. B. Isgor, and P. Ghods, "Quantitative interpretation of half-cell potential measurements in concrete structures," *Journal of Materials in Civil Engineering*, vol. 21, no. 9, pp. 467–475, 2009.

[21] S. Fonna, S. Huzni, M. Ridha, and A. K. Ariffin, "Inverse analysis using particle swarm optimization for detecting corrosion profile of rebar in concrete structure," *Engineering Analysis with Boundary Elements*, vol. 37, no. 3, pp. 585–593, 2013.

[22] S. Fonna, M. Ridha, S. Huzni, and A. K. Ariffin, "Comparison of GA and PSO in boundary element inverse analysis for rebar corrosion detection," *Applied Mechanics and Materials*, vol. 471, pp. 319–323, 2014.

[23] H. G. Wheat and Z. Eliezer, "Some electrochemical aspects of corrosion of steel in concrete," *Corrosion*, vol. 41, no. 11, pp. 640–645, 1985.

[24] D. Lesnic, J. R. Berger, and P. A. Martin, "A boundary element regularization method for the boundary determination in potential corrosion damage," *Inverse Problems in Engineering*, vol. 10, no. 2, pp. 163–182, 2002.

3

Effect of Stress Corrosion on Relaxation of Large Diameter BGFRP Bars

Guowei Li [iD],[1] **Sidi Kabba Bakarr** [iD],[1] **Jingqiu Wang**,[2] **Xue Liu**,[3] **and Chengyu Hong** [iD][4]

[1]*College of Civil and Transportation Engineering, Hohai University, Nanjing, 210098, China*
[2]*Key Laboratory of Ministry of Education for Geomechanics and Embankment Engineering, Hohai University, Nanjing, 210098, China*
[3]*Guangdong Nanyue Transportation Investment Construction Co., Ltd., Guangzhou, 510000, China*
[4]*Department of Civil Engineering, Shanghai University, Shanghai, 200444, China*

Correspondence should be addressed to Sidi Kabba Bakarr; skb2die4@yahoo.com

Academic Editor: Michael J. Schütze

Fibre reinforced polymer (FRP) rebars do not corrode like steel rebars when they are exposed to moisture such as water. Instead they have been shown to degrade when exposed to alkaline media and, in some cases, acids. It has especially demonstrated extensive deterioration when it has been simultaneously stressed and exposed to harsh environments. This combined effect has been termed as stress corrosion. The effect of stress corrosion on the stress relaxation of large sized prestressed basalt-glass fibre reinforced polymer (BGFRP) bars was analyzed by laboratory experiments. Two stressed bars were submerged in aqueous solutions of acid and alkaline in two separate plastic tanks under constant strain. Stress reduction values were observed over a period of about 7 months. Bars immersed in acid bath had an average stress relaxation of 9.2% and that in the alkali bath was observed to be about 13.4%. These results support earlier assertions that exposure of GFRP bars to alkali media is likely to be detrimental to the long-term durability of the reinforced structure.

1. Introduction

Large diameter steel rods have been used as reinforcement in geotechnical engineering as soil nails and ground and rock anchors for slopes, tunnels, excavations, etc. for years. Steel reinforcement is susceptible to corrosion which is a major reason for the deterioration of these structures. Fibre reinforced polymer (FRP) bars have recently been identified as an ideal replacement for steel reinforcement because of their advantages of being environmentally friendly, lightweight, high stiffness, and being manufactured according to specific purposes compared with their steel counterparts. Toxic corrosion inhibitors have been used in the past to curb corrosion and these methods have been scrutinized by environmentalists as they are significant biohazards and pollutants [1]. FRP bars being environmentally friendly could be a way to minimise the use of these toxic corrosion inhibitors. FRP bars are known for their resistance to conventional corrosion that arises from exposure to moisture such as water and deicing salts as in the case of steel. This advantage is the main reason why FRP bars have been viewed as a promising alternative for steel structural reinforcement. Soil nails and anchors are mainly used under tension. Using a GFRP (glass fibre reinforced polymer) bar as a soil nail takes full advantage of its high tensile strength and avoids the disadvantages that come from its low shear modulus as a brittle material [2].

Stress corrosion has become a major focus in the field research of GFRP bars. This condition arises due to long-term loading combined with the effects of exposure to harsh environmental conditions such as acids, alkalis, and deicing salts which are major factors impacting the durability of this material. The most significant obstacle preventing the extended use of FRP materials is a lack of long-term and durability performance data compared to the data available for traditional construction materials [3]. Because the use of GFRP bars as soil nails depends on prestressing the rod, the knowledge of the extent of its stress relaxation capabilities becomes very important. High values of stress relaxation can lead to loss of tensile capacity to adequately reinforce the structure which could be detrimental to the durability of the system. Stress relaxation is one of the widely accepted test methods for predicting the long-term mechanical performance of structural materials and is time and cost efficient [4]. This paper illustrates the effects of the combination of stress and an aqueous environment on the stress relaxation of the GFRP bars. In these studies two GFRP rods are prestressed and submerged in a pool of acid and alkaline solutions each. The long-term use and durability of the bars as soil anchors are analyzed by observing stress relaxation values which are the primary parameters of measurement used in this study.

Over the years, some studies have been conducted regarding the time-dependent behaviour and long-term durability of GFRP bars [5–8]. These studies have contributed significantly to this area of study. Some of them cover the important phenomena of creep and recently few have covered the relaxation behaviour of GFRP rods [9–12]. Another limitation of these previous works is that a significant amount of the tests conducted was on small diameter GFRP bars (less than 20mm). In this study, we will investigate GFRP bars to be used as soil nail elements which often adopts larger geometric size bar, as that of steel soil nail elements (more than 25 mm of diameter), for use as slope construction reinforcement [13].

Exposure of ground anchors to subsurface moisture like water, acids, and alkali contribute significantly to the durability of ground anchors. Corrosion of steel reinforcement is the major cause of deterioration of existing reinforced concrete (RC) structures, resulting in significant expenses for repair and maintenance and leading to shorter service life. To address the corrosion problem, fibre reinforced polymer (FRP) bars have recently emerged as a promising solution not only in the rehabilitation of existing structures but also for the construction of new and more durable RC structures. But due to their versatile applications in harsh environments and exposure to high alkalinity content of concrete, the durability performance of FRP bars and their bond with concrete are major concerns [14].

Glass fibres are damaged due to the combination of two processes: (1) chemical attack on the glass fibres by an alkaline cement environment and (2) concentration and growth of hydration products between individual filaments [15]. The embrittlement of fibres is due to the nucleation of calcium hydroxide on the fibre surface. The hydroxylation can cause fibre surface pitting and roughening. These act as flaws severely reducing fibre properties in the presence of moisture. In addition, calcium, sodium, and potassium hydroxides found in the concrete pore solutions aggressively affect glass fibres. Therefore, the degradation of glass fibres is not only due to high pH level, but also due to the combination of alkalis and moisture. During the service life of a geotechnical system it is possible for the reinforcement to come into contact with acids due to acid rain or fluids attaining acidity when seeping through soil and collecting minerals. Acids are also known to seep through reinforced concrete and cause negative effects on traditional steel reinforcement [16, 17]. Therefore, it is important to evaluate the effect an acidic environment might have on GFRP bars. The reaction rates of all these degradation phenomena increase with temperature [18].

Several authors have studied the effect of hazardous environments of the durability of fibre reinforced polymer composites [19–21], which included accelerated aging and combined effect of load and a simulated environment in acids, alkaline, salt water, etc. Nkurunziza et al. [22] critical review of the literature concerning the durability of GFRP reinforcing bars offers a substantial amount of useful information to the design engineer. The authors do a good job of explaining the degradation mechanism, listing the causes and advances in technology to combat the deterioration of GFRP bars used in reinforced concrete. In the concluding remarks, Nkurunziza et al. recognize that the durability tests cited in the review on the latest generation of GFRP bars subjected to stress higher than design limits, combined with aggressive mediums at elevated temperatures, have concluded that the strength reduction factors adopted by the current codes and guidelines are conservative. The factors adopted by the current codes and guidelines are based on few test results carried out on early generations of GFRP bars that have substantially evolved. Furthermore, accelerated testing techniques are very conservative and that tests more representative of actual field conditions are needed to accurately predict the long-term durability of the GFRP bars.

2. Materials

The following sections give descriptions of the materials used in the test.

2.1. Sand-Coated BGFRP Bars. The FRP bars used in this study are made of epoxy resin and two types of fibres including basalt and glass fibres. To improve the resistance of pure glass fibres composite to corrosion in alkaline environment, basalt-glass fibre hybrid composites with inner cores of glass fibre are covered by the basalt fibres with better alkali resistance. Physical properties of bars used are listed in Table 1.

As shown in Figure 1, the reinforcement materials used in the tests are the basalt-glass fibre hybrid composites with a diameter of 28mm produced by Zhongshan Pulwell Composites Co., Ltd., in Guangdong Province, China, also used by the authors in [23]; body consists of a thermosetting epoxy resin and the contents of each component (by weight) are resin, 19%, basalt fibre, 10%, glass fibre, 65%, and fine sand, 6%.

2.2. Stress Relaxation Test Deformation Devices. If an FRP bar is loaded using traditional wedge-shaped frictional grips, the combination of high compressive stresses and mechanical damage caused by the serrations on the wedge surface will lead to premature failure of the grip zone. Griping the FRP bar with a device which could undertake the tensile load for any measurement of mechanical properties is a key technique. In this study, the seamless steel pipe was used to grip the FRP bar by filling it with binding agent which could expand by itself to gradually create compressive stresses. A centralizer was designed to keep the bar at the centre of the steel pipe for

FIGURE 1: BGFRP bar specimen.

TABLE 1: Physical properties of FRP bars in study [23].

Average Diameter (mm)		Ratio of fibre weight of basalt to total fibre (%)	Density (g/cm^3)	Content (weight ratio %)				Fibre volume fraction (%)	
Basalt	Glass			Basalt	Glass	Resin	Fine sand	Basalt	Glass
2.10	25.35	13.16	2.07	10	65	19	6	58.95	58.76

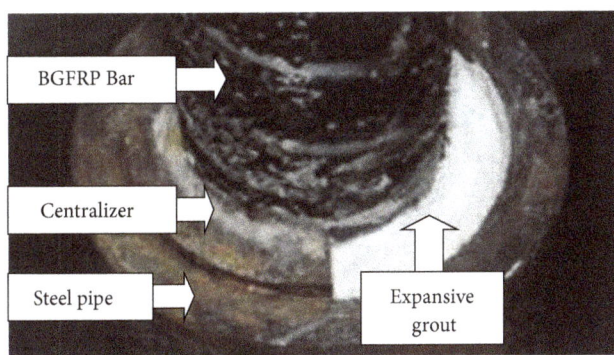

FIGURE 2: Diagram of centralizer keeping FRP bar central.

TABLE 2: FBG sensor positions.

Specimen Number	Sensor Position/cm
1 - Acidic PH=2	0, 19, 34, 49, 68
2 - Alkaline PH=13	0, 19, 34, 49, 69

the precise distribution of compressive stresses in the pipe as shown in Figure 2.

From the figure the white material shows the binding agent and half of the centralizer shown inserted into the gap. The first two half centralizers are fixed at one end of the pipe that is then inserted through the pipe until the other goes through the hole of the centralizers. The binding agent liquid made from cement, expansive material, and water is then poured into the pipe. The other two centralizers are fixed at the other end of the pipe and the specimen is left at room temperature and damp cured, by wrapping the sleeved ends with damp cloths and watering intermittently, for about 24 hours. The pipe at other end of the FRP rod was then filled with the binding material. Specimens were left to cure for another 24 hours and then fitted into the frames and left to properly cure at room temperature for not more than 15 days. The main advantage of this setup is that it can withstand a tensile stress that can reach 70% of its ultimate tensile strength for the large diameter FRP bar for the condition of constant deformation. This setup can satisfy the requirements

of the durability test of FRP bars simulated to the actual carry processes of prestressed structures reinforced with FRP bar [23].

The loading system for the FRP bar relaxation test consisted of a hollow jack, steel casing with threaded support rods connecting outer screw nuts and connection to two bearing plates and bed plates, as shown in Figure 3.

Optic fibre Bragg grating (FBG) sensors were installed at the centre of the FRP bar body through 2mm grooves and anchored by adhesive for the measurement of strain, as shown in Figure 5. Figure 4 shows the locations by distance of each sensor in the BGFRP bar.

Plastic rectangular immersion tanks were used to hold the acid and alkaline solutions. These tanks had holes through the smaller faces, through which the rods will pass as it is fixed on the frame and load applied as shown in Figure 6. Liquid pH values in the tanks were measured approximately, using litmus paper. Table 2 gives a precise description of the position of the sensors in the bar with the jack end of the rod being 0.

3. Experimental Procedure

The deformation equipment is assembled and the tank is put into the frame. The rod specimens are inserted through the holes in the plates and walls of the plastic tanks. The holes in the plastic tanks are sealed with rubber and adhesive tape and injected with Vaseline to block gap between rubber and channel. The solutions are then poured into the tanks and engulf the free end of the specimens, with anchored sections hanging out either end of the setup.

The specimens undergo cyclic loading in a stepwise manner, each stage of loading increments of about 10kN, with load cell values being recorded. At each stage of loading the regulatory load cell reading time rate of change is less than 2kN/h as is standard. Load levels are read by the load cell; dial indicators on the faces of the smaller faces of the load plate give deformation readings. After this process is completed, the load is then increased and the step is repeated, until the bolt is loaded to 90kN and then locked to maintain strain and tension. The stressed specimens are then locked for about 7 days to allow for prestress loss to approach zero.

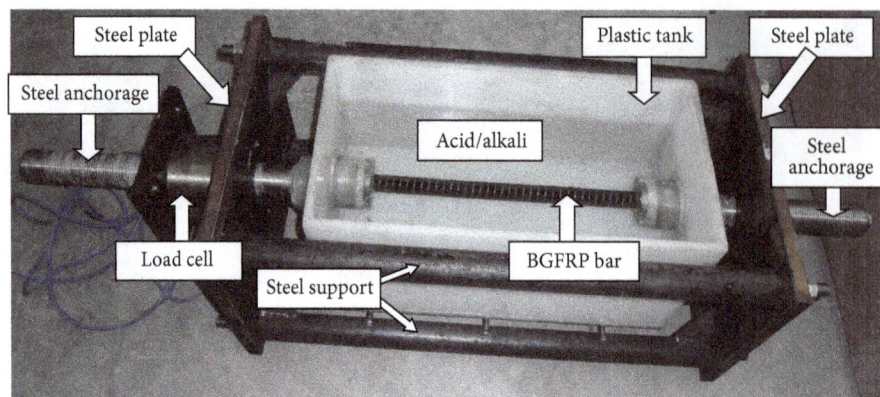

FIGURE 3: Photo of test apparatus.

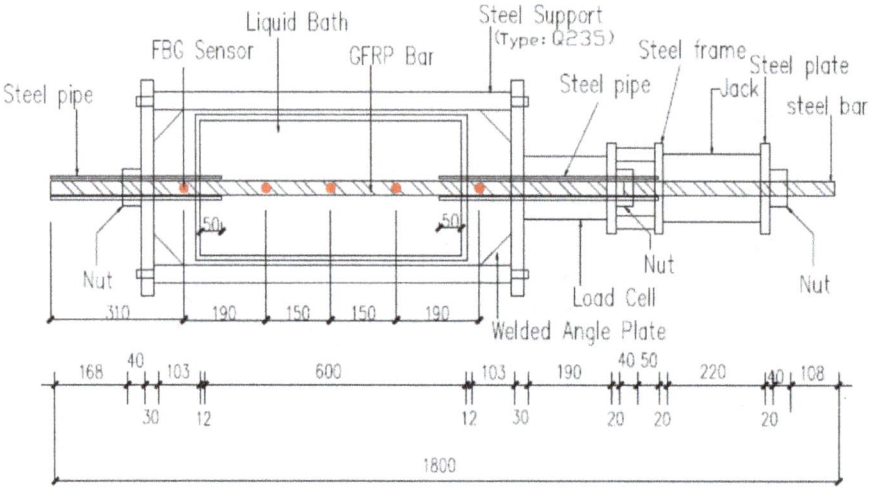

FIGURE 4: Detailed diagram of experimental setup.

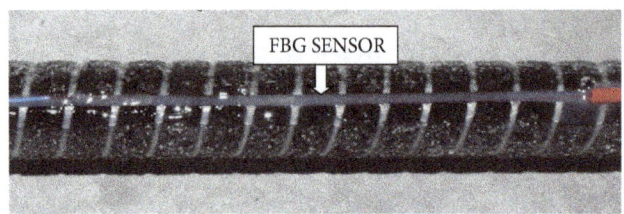

FIGURE 5: FBG sensor location on BGFRP bar.

Measurement and correlation of data starts after 10 days following the above procedure.

4. Experimental Results and Analysis

4.1. Stress Relaxation Analysis: Acid. The stress changes were recorded and analyzed for each sensor. From the trend shown in Figure 7, relaxation phenomenon is most prevalent in sensor 1 which relaxes from about 50 to 21MPa. Change in stress of sensors 2 and 3 was small from 135MPa and 148MPa to 122MPa and 134MPa, respectively. This can be attributed to

the distribution of stress in the specimen. Stress propagates from the loaded end of the rod to the other end and attempts to attain equilibrium. Hence the decrease in stress at the end anchorage points and that at the centre points are almost unchanged. Sensor 4 and sensor 5 were destroyed during stressing of the bar.

The average change in stress shown in Figure 7 is from FBG sensors 2 and 3. Individually they show relaxation percentages of 9.21 and 9.34%, respectively, which shows the similarity of the relaxation process in that region of the bar. From the graph, change in stress was calculated to be 9.3%, from 140 MPa to about 128 MPa.

4.2. Stress Relaxation Analysis: Alkaline. Stress changes were analyzed in the alkaline engulfed specimen and the figures for each sensor reading are shown in Figure 8.

The most relaxation by percentage is seen to occur at the anchorage ends of the specimen and sensors 1 and 5 (from 85 and 58MPa to 49 and 34MPa, respectively) as shown in Figure 8. Both curves showed a similar trend. Figure 8 shows the relaxation of the free central section of the rod

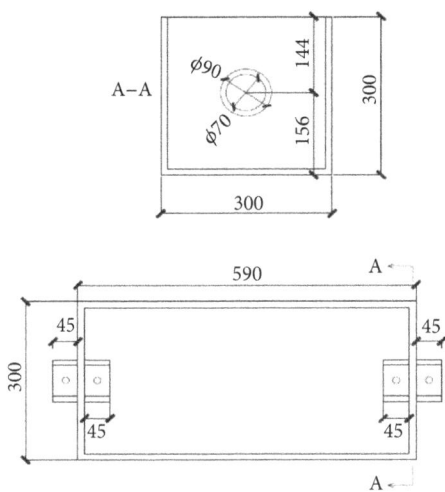

FIGURE 6: Diagram showing measurement of immersion tank.

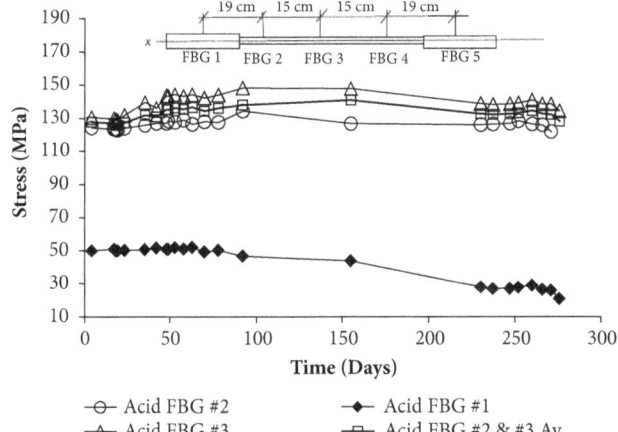

FIGURE 7: Stress versus time FBG sensors (acid).

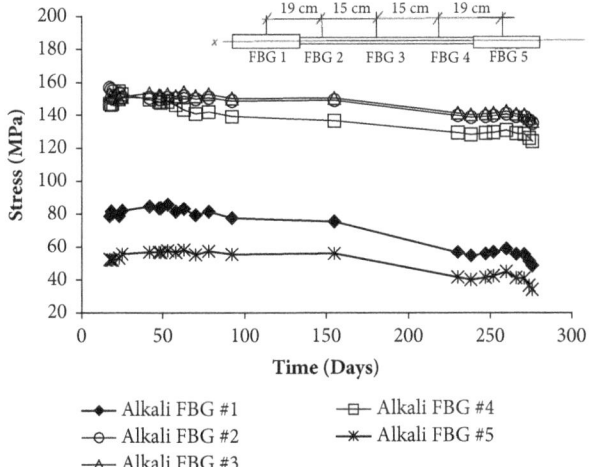

FIGURE 8: Stress versus time Alkali FBG sensors.

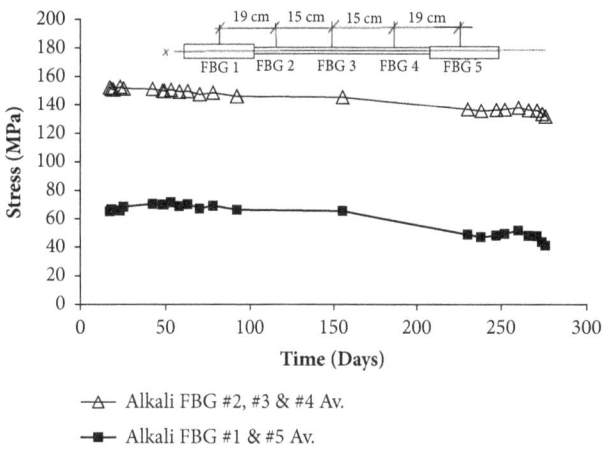

FIGURE 9: Stress versus time, average values (alkali).

monitored by sensors 2, 3, and 4 (from 156, 154, and 155MPa to 135, 136, and 124MPa, respectively) showing similar curve trends revealing a consistent pattern for both sections of the specimen. This trend is attributed to the distribution of stress in the specimen. The average change in stress in the anchorage and free section were also plotted as shown in Figure 9. Stress value averages were calculated from FBG sensors 1 and 5 and sensors 2, 3, and 4. Average relaxation percentages for the anchorage section and free section were 42% and 13.7%, respectively.

4.3. Combined Effect of Stress and Acidic/Alkaline Environment. Individually, a specimen exposed to a wet or corrosive environment and that exposed to a tensile force are both affected characteristically by the respective state. In this case, the specimens are both stressed and exposed to a hazardous environment. When these specimens are stressed, microcracks are expected to appear on the surface of the rod, thus facilitating the ingress of liquids into the bar, allowing it to penetrate the glass fibres. This only occurs if the axial force is high enough to cause microcracks. Nevertheless, the combined effect of sustained stress and a liquid or aqueous environment can lead to significant strength loss, interfacial degradation, and brittleness [24].

It is however important to note that the exposure of GFRP bars to acidic environments is not very prevalent in literature, as far as the author knows. Acids pose not as much a significant threat to the durability of the bars in itself. In combination with stress, the effects can lead to a penetration of the matrix by the acid, which will then unfavourably react with the fibres. Alkalis on their own have been known to have adverse effects on glass fibres and some types of matrices. In the case of our experiment, the synergistic effect of the acid/alkali and sustained stress must have been a factor in the stress relaxation observed in the test. Since the acid/alkali was in aqueous form and totally engulfed the specimen, this environment had an excess of highly mobile ions which is not usually the case in typical field conditions. Figure 10 shows the stress corrosion effect on relaxation, at room temperature, at a stress value of 90kN.

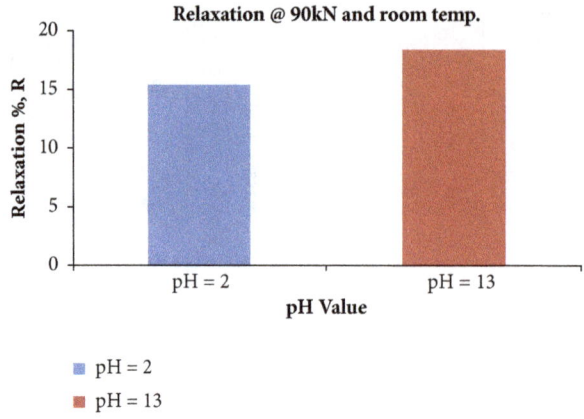

FIGURE 10: Combined effect of stress/load and environment on relaxation.

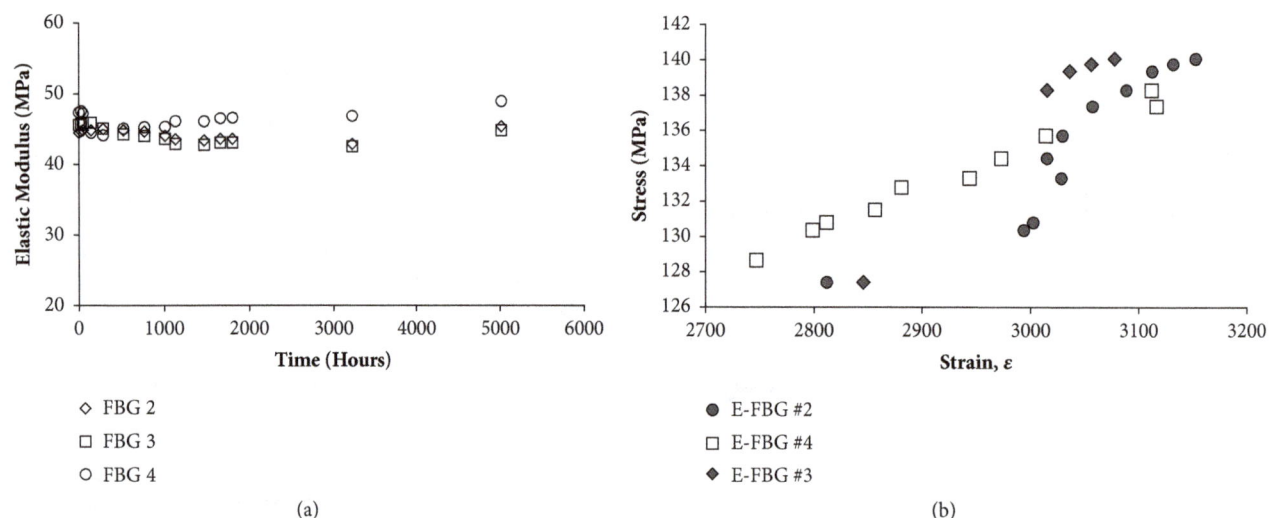

(a)

(b)

FIGURE 11: (a) Change in modulus versus time; (b) stress versus strain for FBG sensors.

Unfortunately other material parameters such as the residual tensile strength of the bar have not been conducted so all the conclusions arrived at are based on the stress relaxation. The tests were conducted at room temperature. More experiments of varying stress levels, pH, and temperature have to be conducted to provide more data which can be analyzed to give better understanding of these conditions.

4.4. Anchorage Effect and Bar Rheological Properties.
GFRP bars are part polymers and are expected to exhibit some amount of viscoelastic behaviour when subjected to long-term loading and various environmental conditions. Due to its rheological properties, the elastic modulus is a function of time, unlike in steel, which makes analysis of time-dependent deformation of FRPs more complex. An attempt was made at computing the elastic modulus during the experiment at the central regions of the acid specimen by obtaining stress values from load cell by dividing the load by the cross sectional

area of the bar (see (1)), and calculating the changing elastic modulus as shown in (2):

$$\text{Stress, } \sigma = \frac{\text{Load, F}}{\text{Cross sectional area, A}} \quad (1)$$

$$\text{Elastic modulus, E (t)} = \frac{\text{Stress, } \sigma}{\text{Strain, } \varepsilon} \quad (2)$$

E(t) was plotted for the 3 central FBG sensors (2, 3, and 4) as shown in Figure 11.

It is observed from Figure 11(a) that the change in elastic modulus over the entire period of the experiment was very small, all within values from 42 to 49 MPa. This may mean that a large part of the relaxation was likely caused by more of the anchorage device's inability to properly restrain the specimen, deformations in the tensile frame setup, and the synergistic effect of the load and aqueous solutions, rather than the rheological properties of the bar itself. This phenomenon is also exhibited by the relatively linear stress versus strain curves as shown in Figure 11(b).

5. Conclusions

Two large diameter BGFRP bars were instrumented with FBG sensors, stressed to about 90 kN, each exposed to aqueous acid and alkali in plastic tanks. The aim was to observe the combined effect of tension and the simulated environment on the relaxation of the bars. From the analysis of data and test results and in consideration of experiment duration and conditions, the following conclusions were arrived at:

(a) As expected, relaxation is higher in the specimen submerged in alkali than that in the acid

(b) Results also show that even though acids are not as corrosive to FRP as alkalis are, it may still pose problems because of the high relaxation value obtained from the tests

(c) The test may also have exhibited high relaxation due to the ingress of the aqueous acid/alkali, especially in the case of the acid which may have needed help with entering the bars due to the cracks that may have propagated on the surface of the rod

(d) There was little change in the elastic modulus of the acid specimen which showed that a large part of the relaxation was due to the anchorage device's inability to properly restrain the bar, tensile frame deformation, and the simulated environment, rather than its rheological properties

(e) Further analysis of stress relaxation process is needed in the anchorage area to better understand the effect of the pressure exerted by the grout on the bar specimen

(f) Although the steel sleeve anchorage method has been proved to be effective, more tests should be done especially in the choice of grout used in anchorage for different applications. This may be done by testing different expansive agents within the steel sleeve and come up with better alternatives

The tests lasted for only 7 months. This time is not long enough to obtain the desired results since it takes some time for the liquids to diffuse into the specimen. Therefore, a longer test period would be recommended to get more accurate results. The tests also demonstrate relaxation with an excess amount of mobile ions in the aqueous environments and this overcompensates for actual field conditions. It would be useful for further tests that mimic actual field conditions to be conducted.

Conflicts of Interest

The authors declare that they have no conflicts of interest.

Acknowledgments

The authors would like to acknowledge the financial support provided by the National Natural Science Foundation of China (Project nos. 41472240 and 41602352), the Fundamental Research Funds for the Central Universities (Grant nos. 2015B25514 and 2015B17214), and the Government of Guangdong Province and Ministry of Education of China (Project no. 2009B09060011).

References

[1] A. A. Torres-Acosta, W. Martínez-Molina, and E. M. Alonso-Guzmán, "State of the Art on Cactus Additions in Alkaline Media as Corrosion Inhibitors," *International Journal of Corrosion*, vol. 2012, Article ID 646142, 9 pages, 2012.

[2] Z. Chen, L. Zheng, Q. Jin, and X. Li, "Durability study on glass fiber reinforced polymer soil nail via accelerated aging test and long-term field test," *Polymer Composites*, vol. 38, no. 12, pp. 2863–2873, 2017.

[3] A. Micelli and F. Nanni, "Mechanical Properties and Durability of FRP Rods," Center for Infrastructure Engineering Studies (CIES, University Missouri-Rolla, Rep. No.00-22 CIES), 2001.

[4] Y. J. Tong, L. H. Xu, and C. Q. Li, "The Stress Relaxation of Glass Fibre Composites With Low Velocity Impact Damage," in *Proceedings of the Asia-Pacific Conference on FRP in Structures (APFIS 2007)*, S. T. Smith, Ed., International Institute for FRP in Construction, 2007.

[5] G. Nkurunziza, B. Benmokrane, A. S. Debaiky, and R. Masmoudi, "Effect of sustained load and environment on long-term tensile properties of glass fiber-reinforced polymer reinforcing bars," *ACI Structural Journal*, vol. 102, no. 4, pp. 615–621, 2005.

[6] F. Micelli, A. Nanni, and A. La Tegola, "Effects of Conditioning Environment on GFRP Bars," 22nd SAMPE Europe International Conference, CNIT Paris, March 27-29, 2001.

[7] R. Masmoudi, G. Nkurunziza, B. Benmokrane, and P. Cousin, "Durability of glass FRP composite bars for concrete structure reinforcement under tensile sustained load in wet and alkaline environments," in *Proceedings of the Canadian Society for Civil Engineering - 31st Annual Conference: 2003 Building our Civilization*, pp. 916–924, Canada, June 2003.

[8] Y. A. Al-Salloum and T. H. Almusallam, "Creep effect on the behavior of concrete beams reinforced with GFRP bars subjected to different environments," *Construction and Building Materials*, vol. 21, no. 7, pp. 1510–1519, 2007.

[9] I. Sasaki and I. Nishizaki, "Tensile Load Relaxation of Frp Cable System During Long-Term Exposure Tests," in *Proceedings of theProceedings of the 6th International Conference on FRP Composites in Civil Engineering, CICE 2012*, pp. 1–8, 2012.

[10] R. Sovjak, J. Fornusek, P. Konvalinka, and J. L. Vitek, "Creep and Stress Relaxation of Concrete Slab with Pre-stressed GFRP," in *Proceedings of the 4th Asia-Pacific Conference on FRP in Structures*, pp. 299–304, Seoul, Korea, 2009.

[11] G.-W. Li, C. Ni, H.-F. Pei, W.-M. Ge, and C. W. W. Ng, "Stress relaxation of grouted entirely large diameter B-GFRP soil nail," *China Ocean Engineering*, vol. 27, no. 4, pp. 495–508, 2013.

[12] G.-W. Li, C.-Y. Hong, J. Dai et al., "FBG-Based Creep Analysis of GFRP Materials Embedded in Concrete," *Mathematical*

Problems in Engineering, vol. 2013, Article ID 631216, 9 pages, 2013.

[13] H.-H. Zhu, J.-H. Yin, A. T. Yeung, and W. Jin, "Field pullout testing and performance evaluation of GFRP soil nails," *Journal of Geotechnical and Geoenvironmental Engineering*, vol. 137, no. 7, pp. 633–642, 2011.

[14] Y. Chen, "Accelerated ageing tests and long-term prediction models for durability of FRP bars in concrete," Dissertation submitted to the College of Engineering and Mineral Resources at West Virginia University 2007.

[15] V. M. Karbhari, K. Murphy, and S. Zhang, "Effect of concrete based alkali solutions on short-term durability of E-glass/vinylester composites," *Journal of Composite Materials*, vol. 36, no. 17, pp. 2101–2121, 2002.

[16] M. Criado, S. Fajardo, and J. M. Bastidas, "Corrosion Behaviour of a New Low-Nickel Stainless Steel Reinforcement: A Study in Simulated Pore Solutions and in Fly Ash Mortars," *International Journal of Corrosion*, vol. 2012, Article ID 847323, 8 pages, 2012.

[17] F. Yingfang, H. Zhiqiang, and L. Jianglin, "Ultrasonic Measurement of Corrosion Depth Development in Concrete Exposed to Acidic Environment," *International Journal of Corrosion*, vol. 2012, Article ID 749185, 8 pages, 2012.

[18] M. Robert, P. Wang, P. Cousin, and B. Benmokrane, "Temperature as an accelerating factor for long-term durability testing of FRPs: Should there be any limitations?" *Journal of Composites for Construction*, vol. 14, no. 4, pp. 361–367, 2010.

[19] S. Putic, M. Stamenovic, J. Petrovic, M. Rakin, and B. Medjo, "Effect of alkaline solutions on the tensile properties of glass-polyester pipes," *Acta Periodica Technologica*, no. 42, pp. 185–195, 2011.

[20] F. E. Tannous and H. Saadatmanesh, "Environmental Effects on the Mechanical Properties of E-Glass FRP Rebars," *ACI Materials Journal*, vol. 95, no. 2, pp. 87–100, 1998.

[21] B. Benmokrane, P. Wang, T. M. Ton-That, H. Rahman, and J.-F. Robert, "Durability of glass fiber-reinforced polymer reinforcing bars in concrete environment," *Journal of Composites for Construction*, vol. 6, no. 3, pp. 143–153, 2002.

[22] G. Nkurunziza, A. Debaiky, P. Cousin, and B. Benmokrane, "Durability of GFRP bars: A critical review of the literature," *Progress in Structural Engineering and Materials*, vol. 7, no. 4, pp. 194–209, 2005.

[23] L. Guo-wei, P. Hua-Fu, and H. Cheng-yu, "Study on the Stress Relaxation Behavior of Large Diameter B-GFRP Bars Using FBG Sensing Technology," *International Journal of Distributed Sensor Networks*, vol. 2013, Article ID 201767, 12 pages, 2013.

[24] G. Carra and V. Carvelli, "Ageing of pultruded glass fibre reinforced polymer composites exposed to combined environmental agents," *Composite Structures*, vol. 108, no. 1, pp. 1019–1026, 2014.

4

Hot Corrosion of $SrTiO_3$ Perovskite in Na_2SO_4 + 50 wt.% V_2O_5 and Na_2SO_4 + 10 wt.% NaCl Environments at $900\,^{\circ}C$

M. Krishna Prasad,[1] **K. Srinivasa Rao,**[2] **Madhusudhan Reddy,**[3] **and Gosipathala Sreedhar** ⓘ[4]

[1]*GMR Institute of Technology, Rajam 532127, India*
[2]*Andhra University College of Engineering, Visakhapatnam 530003, India*
[3]*Defence Metallurgical Research Laboratory, Hyderabad 500066, India*
[4]*CSIR-Central Electrochemical Research Institute, Karaikudi 630 006, India*

Correspondence should be addressed to Gosipathala Sreedhar; gsreedhar@cecri.res.in

Academic Editor: Francisco Javier Perez Trujillo

This study examines the phase stability of perovskite $SrTiO_3$ in Na_2SO_4 + 50 wt.% V_2O_5 and Na_2SO_4 + 10 wt.% NaCl environments at $900\,^{\circ}C$. Hot corrosion results show the formation of Sr_2VO_4, SrV_2O_6, and $SrTiV_5O_{11}$ phases in Na_2SO_4 + 50 wt.% V_2O_5 environment and $Sr_3Ti_2O_7$, Na_4TiO_4, and TiO_2 phases in Na_2SO_4 + 10 wt.% NaCl environment. Morphological observations revealed the austerity of hot corrosion attack on $SrTiO_3$. The Sr^{2+} ions leached out from $SrTiO_3$ and reacted with corrosive environments. These observations clearly indicate the destabilization of $SrTiO_3$ in both environments.

1. Introduction

Thermal barrier coatings (TBCs) are generally applied as thermal insulating layers on the hot metallic surfaces such as combustion engine parts, gas turbine blades, and aero-engine parts. TBCs can enhance the operating temperature and reduce the cooling requirements ensuing higher engine efficiency and part life of an engine [1–5]. In order to gratify the requirements of TBC materials, the properties such as low thermal conductivity, higher melting point, and matching thermal expansion coefficient with bond coat are needed. Another predominant factor for determining the durability of TBCs is the stability against hot corrosion. Hot corrosion occurs when the engine runs with low quality fuel which contains sodium, vanadium, and sulfur as impurities in it and also in chloride (marine) environments. The sodium and vanadium salts are having lower melting point ($700–880\,^{\circ}C$) than gas turbine operating temperature ($>1300\,^{\circ}C$). Hence, their salts can easily condensate in the form of Na_2SO_4, V_2O_5, and NaCl on the top coat of TBCs and undergo hot corrosion

in two types (Types I and II). Type I generally occurs above the melting point of sodium sulphate (Tm ~ $884\,^{\circ}C$) with Type II vice versa. In both the types, these salts react with ceramic top coat and show the way for destabilization and degradation of TBCs.

In recent years, 6–8 wt.% yttria stabilized zirconia (YSZ) has been widely used as a top coat material for TBC and much research has been involved in understanding the hot corrosion behavior of YSZ in corrosive environments. The authors reported that these molten salts could cause several microcracks and spallation of the coating [6–9]. Particularly, in vanadium containing environments, YSZ is found to be unstable and forms YVO_4 and m'-ZrO_2 (destabilization). Destabilization causes the volume change and leads to spallation of the coating.

Therefore, there is a need to develop alternate materials to enhance the hot corrosion stability of the coating. The candidate materials, such as garnets ($Y_3Al_5O_{12}$), La, Sm, Nd, Yb, and Gd doped ZrO_2, YSZ-Ta_2O_5, monazite ($LaPO_4$), and lanthanum aluminates ($LaMgAl_{11}O_{19}$) were developed in recent years and their stability was examined [10–14]. Perovskite (ABO$_3$) based ceramics (SrZrO$_3$, SrCeO$_3$, and BaZrO$_3$) have also drawn much attention towards TBCs due to their low thermal conductivity, higher melting point, and matching thermal expansion coefficient with bond coat material. Particularly, the SrTiO$_3$ has higher melting point of 2080°C, thermal expansion coefficient ~10^{-5} K^{-1} (1727°C), and thermal conductivity in the range of 11 to 2.5 W·m^{-1} K^{-1} (25°C to 1100°C). But the review of literature showed that the published work on SrTiO$_3$ is mostly devoted to understanding the thermophysical properties [15–20]. However, the literature on hot corrosion behavior of SrTiO$_3$ is sparse and reactions between SrTiO$_3$ and molten melts remain unclear. Therefore, further studies are needed to understand the hot corrosion behavior of SrTiO$_3$.

Hence, the present work is focused on studying the phase stability of perovskite type SrTiO$_3$ in Na$_2$SO$_4$ + 50 wt.% V$_2$O$_5$ and Na$_2$SO$_4$ + 10 wt.% NaCl environments at 900°C.

2. Materials and Methods

2.1. Sample Preparation. Strontium titanate powders were synthesized via hydrothermal route [21] using strontium hydroxide (99%, Alfa Aesar) and titanium tetrachloride (99.9%, Fisher Scientific) as precursor materials. To make the pellets, obtained strontium titanate powders were pressed with 370 MPa pressure using a uniaxial die (1.3 cm internal diameter). Polyvinyl alcohol (PVA) was used as a binder (1 : 10 ratio) in the preparation of strontium titanate pellets. Compacted pellets were kept in a furnace at 500°C for 48 h to remove the binder. Then these pellets were heat treated at 900°C for 48 h.

2.2. Hot Corrosion Tests. A mixture of Na$_2$SO$_4$ and V$_2$O$_5$ salts in 1 : 1 weight ratio and another mixture of Na$_2$SO$_4$ and NaCl salts in 9 : 1 weight ratio were prepared and dissolved separately in distilled water for getting 50 wt.% of each salt solution. Samples were preheated at 200°C for 30 min to achieve good adhesion of salts. Salt solution was applied on all sides of surface using camel hair brush. Salt coated samples were kept in a furnace and heated at 120°C for 2 h to remove the moisture. The samples were weighed to determine the salt coverage and found in the range of 20–30 mg/cm^2. Then these salt coated samples were kept in an alumina boat and heated at 900°C for 100 h in air. After hot corrosion, samples were taken out and washed with running water and dried.

2.3. X-Ray Diffraction Analysis. Before and after hot corrosion, the nature of phases was investigated by X-ray diffractometer (Bruker, D8 advance) with Cu Kα radiation with a scan rate of 6°/min. All the phases were identified using the JCPDS-PCPDFWIN software package.

2.4. Morphological Analysis. The surface morphologies and elemental analysis of samples were examined by Field Emission Scanning Electron Microscope (FESEM-ZEISS-SUPRA 55VP) attached with EDX (OXFORD instruments).

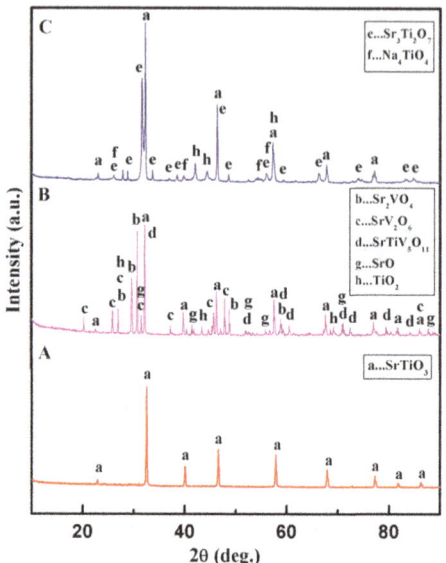

FIGURE 1: XRD patterns of the SrTiO$_3$, (A) before exposure to hot corrosion and after exposure to hot corrosion at 900°C in (B) Na$_2$SO$_4$ + 50 wt.% V$_2$O$_5$ environment and (C) Na$_2$SO$_4$ + 10 wt.% NaCl environment.

3. Results

3.1. X-Ray Diffraction Analysis. Figure 1 shows the X-ray diffraction patterns of the SrTiO$_3$ sample (A) before hot corrosion and after subjecting to hot corrosion at 900°C in (B) Na$_2$SO$_4$ + 50 wt.% V$_2$O$_5$ environment and (C) Na$_2$SO$_4$ + 10 wt.% NaCl environment. Before hot corrosion, it is observed that SrTiO$_3$ is in cubic phase (JCPDS-79-0176). After exposure to hot corrosion in Na$_2$SO$_4$ + 50 wt.% V$_2$O$_5$ environment (Figure 1(B)), additional phases such as Sr$_2$VO$_4$, SrV$_2$O$_6$, and SrTiV$_5$O$_{11}$ (JCPDS-81-0854, 28-1267, and 48-0540, resp.) were identified. After exposure to hot corrosion in Na$_2$SO$_4$ + 10 wt.% NaCl environment (Figure 1(C)), additional phases such as Sr$_3$Ti$_2$O$_7$ (JCPDS-78-2479), Na$_4$TiO$_4$ (JCPDS-42-0513), and TiO$_2$ were identified. It is worthwhile to mention that SrTiO$_3$ phase is found to be unstable in both Na$_2$SO$_4$ + 50 wt.% V$_2$O$_5$ and Na$_2$SO$_4$ + 10 wt.% NaCl environments at 900°C. Succinctly, this indicates destabilization of SrTiO$_3$.

3.2. Morphological and Microchemical Analysis. Surface morphology of SrTiO$_3$ sample before and after exposure to hot corrosion in both Na$_2$SO$_4$ + 50 wt.% V$_2$O$_5$ and Na$_2$SO$_4$ + 10 wt.% NaCl environments at 900°C was examined to identify the severity of hot corrosion attack. Figure 2 shows the macroscopic view of the SrTiO$_3$ before and after exposure to hot corrosion. Figure 2(a) (before hot corrosion) reveals that surface is free from cracks and scales. Nonuniform and severe hot corrosion attack was noticed on the surfaces which were subjected to hot corrosion in both Na$_2$SO$_4$ + 50 wt.% V$_2$O$_5$ (Figure 2(b)) and Na$_2$SO$_4$ + 10 wt.% NaCl environments (Figure 2(c)) at 900°C.

Figure 3 shows the surface morphology of SrTiO$_3$ before subjecting to hot corrosion. Figures 3(a) and 3(b) show the

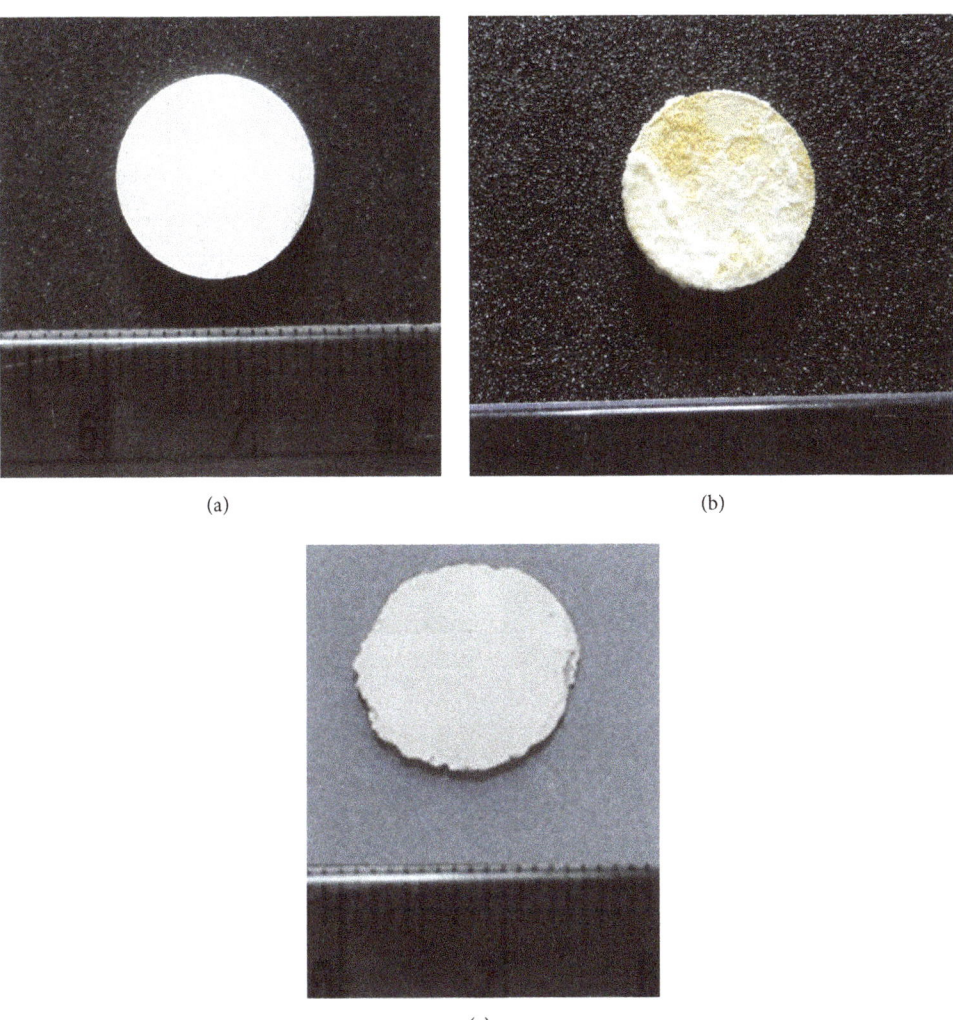

(a)

(b)

(c)

FIGURE 2: Macroscopic view of SrTiO$_3$ surface, (a) before exposure to hot corrosion and after exposure to (b) Na$_2$SO$_4$ + 50 wt.% V$_2$O$_5$ environment and (c) Na$_2$SO$_4$ + 10 wt.% NaCl environment at 900˚C.

(a)

(b)

FIGURE 3: (a) Lower magnification image of SrTiO$_3$ sample before exposure to hot corrosion and (b) higher magnification image of (a). EDX analysis was done on the surface of (b) at region 1 (corresponding to SrTiO$_3$).

TABLE 1: The elemental compositions of SrTiO$_3$ samples before and after being subjected to hot corrosion in both Na$_2$SO$_4$ + 50 wt.% V$_2$O$_5$ and Na$_2$SO$_4$ + 10 wt.% NaCl environments at 900°C.

S. No	Figures	Environments	Regions	Elements (wt.%)			
				O	Ti	V	Sr
1	Figure 3(b)	Before hot corrosion	1	24.80	22.82	-	52.38
2	Figure 4(b)	Na$_2$SO$_4$ + 50 wt.% V$_2$O$_5$	1	24.94	23.14	-	51.92
3	Figure 4(c)	Na$_2$SO$_4$ + 50 wt.% V$_2$O$_5$	1	25.03	-	18.83	56.14
4	Figure 6(b)	Na$_2$SO$_4$ + 10 wt.% NaCl	1	26.90	27.92	-	45.18

(a)   (b)

(c)

FIGURE 4: Surface morphology of SrTiO$_3$ after being subjected to hot corrosion in Na$_2$SO$_4$ + 50 wt.% V$_2$O$_5$ environment at 900°C, (a and b) lower and higher magnification images of SrTiO$_3$, shows octahedron shaped morphology. EDX analysis was demonstrated on (b) at region 1 (corresponding to strontium titanate); (c) higher magnification image at other regions of (a) shows irregular shape morphology. EDX analysis was done on (c) at region 1 (corresponding to strontium vanadates).

lower and higher magnification FESEM images of SrTiO$_3$. Figure 3(b) shows the spherical shape morphology. EDX analysis was performed to identify the chemical compositions at region 1 (Figure 3(b)) and the data is presented in Table 1. Region 1 is composed of Sr, Ti, and O (corresponding to SrTiO$_3$). These results are in good agreement with the XRD results (Figure 1(A)). The surface morphology of SrTiO$_3$ after subjecting to hot corrosion in Na$_2$SO$_4$ + 50 wt.% V$_2$O$_5$ environment at 900°C reveals the octahedron shape morphology (Figure 4(a)). Higher magnification image of Figure 4(a) is shown in Figure 4(b) and EDX analysis was done at region 1 (data is presented in Table 1). Region 1 is composed of Sr, Ti, and O (corresponding to SrTiO$_3$). Spherical type of morphology of strontium titanate changes onto octahedral shape when treated in sulphate solution of venadate at 900°C.

Figure 4(c) shows irregular shape morphology. EDX point analysis was carried out at region 1 (data is presented in Table 1) and this location is composed of Sr, V, and O (corresponding to strontium vanadates). Figure 5 shows the severity of hot corrosion on SrTiO$_3$ after exposure to Na$_2$SO$_4$ + 50 wt.% V$_2$O$_5$ environment and it shows the porous layer and corrosion products on the surface of SrTiO$_3$.

The surface morphology of SrTiO$_3$ after exposure to Na$_2$SO$_4$ + 10 wt.% NaCl environments at 900°C is shown in Figure 6. Lower magnification image (Figure 6(a)) reveals that nonuniform, severe hot corrosion and groove attack were observed in Na$_2$SO$_4$ + 10 wt.% NaCl environment. Higher magnification image (Figure 6(b)) reveals the plate type of morphology. EDX analysis was carried out at region 1 (data is presented in Table 1) in Figure 6(b) and this location is

FIGURE 5: Surface morphology of $SrTiO_3$ after being subjected to hot corrosion in Na_2SO_4 + 50 wt.% V_2O_5 environment at 900°C. A porous layer and corrosion products were observed on the surface.

composed of Sr, Ti, and O (corresponding to $Sr_3Ti_2O_7$). Similar type of morphology for $Sr_3Ti_2O_7$ has also been observed by previous author [22].

4. Discussion

The phase stability of $SrTiO_3$ against hot corrosion is being studied in Na_2SO_4 + 50 wt.% V_2O_5 and Na_2SO_4 + 10 wt.% NaCl environments at 900°C. Though $SrTiO_3$ is an oxide, it underwent destabilization in both environments. Massive hot corrosion attack is being observed on surface of $SrTiO_3$ after being subjected to both environments at 900°C. Significant literature is available for $SrO-V_2O_5$ binary systems but not for $SrO-NaVO_3$. It is reasonable to predict the chemical reaction between $SrTiO_3$ and $NaVO_3$ using $SrO-V_2O_5$ [23, 24] and $SrO-VO_2-V_2O_5$ phase diagrams [25]. First, $SrTiO_3$ decomposed into SrO and TiO_2 phases by leaching of Sr^{2+} ions from $SrTiO_3$ due to the presence of corrosive environments (see (1)). When the temperature reaches 630°C in furnace, $NaVO_3$ forms and acts as good solvent medium for oxides. At this stage, the reactions between sodium vanadate and oxides are faster.

$$SrTiO_3 (s) \longrightarrow SrO (s) + TiO_2 (s) \qquad (1)$$

$$SrO (s) + 2NaVO_3 (l) \longrightarrow SrV_2O_6 (s) + Na_2O (s) \qquad (2)$$

$$4SrO (s) + 2NaVO_3 (l)$$
$$\longrightarrow 2Sr_2VO_4 (s) + Na_2O (s) + \frac{1}{2}O_2 (g) \qquad (3)$$

Habibi et al. [14] have also reported that the liquid $NaVO_3$ will increase the atom mobility of oxide elements. SrO reacts with $NaVO_3$ liquid melt and forms SrV_2O_6 (see (2)). Again Na_2O reacts separately with V_2O_5 and SO_3 and forms $NaVO_3$ and Na_2SO_4, respectively. This reaction is cyclic. SrO and TiO_2 can also react with $NaVO_3$ and form $SrTiV_5O_{11}$. At 900°C, SrO reacts with $NaVO_3$ and forms the Sr_2VO_4 phase (see (3)). This phase is identified as predominant corrosion product phase in XRD (Figure 1(B)). The above sequence of reactions describes the formation mechanism of corrosion products. Higher magnification image of $SrTiO_3$

(Na_2SO_4 + 50 wt.% V_2O_5 at 900°C) (Figure 5) shows the severe hot corrosion and corrosion products on it. Due to damage caused by Na_2SO_4 + 50 wt.% V_2O_5, a porous layer and corrosion products were formed and observed in the Figure 5. This reveals the severity of hot corrosion on $SrTiO_3$ in Na_2SO_4 + 50 wt.% V_2O_5 environment.

The Na_2SO_4 + 10 wt.% NaCl environment also caused the leaching Sr^{2+} ions and subsequently formation of SrO and reacts with chloride to form $SrCl_2$ (see (4)). In general thermodynamic stability of oxides is higher than the chlorides. Hence formed chlorides are further transformed into oxides (see (5)). Further they reacted with themselves and formed $Sr_3Ti_2O_7$ at 900°C (see (6)). Tilley [26] has studied the Sr-Ti-O system at 1100–1473°C and reported the formation of three phases such as Sr_2TiO_4, $Sr_3Ti_2O_7$, and $SrTiO_3$. Jacob and Rajitha [27] have studied the pseudo-binary system (SrO + TiO_2) at 775 to 977°C and reported the thermodynamics properties of Sr_2TiO_4, $Sr_3Ti_2O_7$, $Sr_4Ti_3O_{10}$, and $SrTiO_3$ using solid state galvanic cells (see (6)). Based on the previous works [22, 27] the reaction between SrO and TiO_2 can be expressed as follows:

$$SrO (s) + Cl_2 (g) \longrightarrow SrCl_2 (s) + \frac{1}{2}O_2 (g) \uparrow \qquad (4)$$

$$SrCl_2 (s) + \frac{1}{2}O_2 (g) \longrightarrow SrO (s) + Cl_2 (g) \uparrow \qquad (5)$$

$$3SrO (s) + 2TiO_2 (s) \longrightarrow Sr_3Ti_2O_7 (s) \qquad (6)$$

$$TiO_2 (s) + 2Na_2O (s) \longrightarrow Na_4TiO_4 (s) \qquad (7)$$

The TiO_2 phase can react with Na_2O to form Na_4TiO_4 (see (7)). There has been no previous work on hot corrosion behavior of $SrTiO_3$ in NaCl medium. The possible role of Na_2O against SrO and TiO_2 phases is discussed here. According to Coulomb's law (lattice energy depends on the ionic radii and charges of the ions) the compounds with +2 ions have more lattice energy than the compounds with +1 ion. In addition to this, Ti atoms are having smaller ionic radius (94 pm) than that of Sr atoms (112 pm) [28]. Hence, the Ti-O structured atoms would be expected to have larger lattice energy which gives higher attractive force towards Na_2O. Based on our above discussions the order of lattice energy can be expressed as

$$TiO_2 > SrO > Na_2O. \qquad (8)$$

The lattice energy of TiO_2 (12150 kJ/mole) is more than that of SrO (3223 kJ/mole) [29]. As per theory, the lattice energy is directly proportional to the phase stability of the compound, which indicates the less stability for SrO at 900°C. But, in the case of TiO_2, it needs higher energy to separate the Ti^{2+} and O^{2-} ions. As we discussed earlier, the strong electrical attraction between the Na^+ (Na_2O) and TiO_2 ions might cause the formation of Na_4TiO_4 phase (see (7)). These discussions are well in agreement with XRD (Figure 1(C)) results. This is the possible reason why the reaction between TiO_2 and NaCl was highly favored compared with that with SrO. According to the binary phase diagrams of Na_2O-TiO_2 [30, 31], Na_4TiO_4 melts at 850°C and this was the reason

(a)   (b)

Figure 6: (a) Surface morphology of SrTiO$_3$ after exposure to hot corrosion in Na$_2$SO$_4$ + 10 wt.% NaCl environment at 900°C and (b) higher magnification image of (a) shows the plate type morphology. EDX analysis was performed on (b) at region 1 (corresponding to Sr$_3$Ti$_2$O$_7$).

for Na$_4$TiO$_4$ being identified as minor corrosion product compared to that of Sr$_3$Ti$_2$O$_7$ in XRD (Figure 1(C)).

5. Conclusions

The hot corrosion behavior of perovskite type SrTiO$_3$ was studied in Na$_2$SO$_4$ + 50 wt.% V$_2$O$_5$ and Na$_2$SO$_4$ + 10 wt.% NaCl environments at 900°C for 100 h.

(1) SrTiO$_3$ has been found to be unstable in both Na$_2$SO$_4$ + 50 wt.% V$_2$O$_5$ and Na$_2$SO$_4$ + 10 wt.% NaCl environments at 900°C.

(2) The Sr^{2+} ion leaches from its own phase and reacts with NaVO$_3$ to form Sr$_2$VO$_4$, SrV$_2$O$_6$ and SrTiV$_5$O$_{11}$ in Na$_2$SO$_4$ + 50 wt.% V$_2$O$_5$ environment.

(3) The Na$_2$SO$_4$ + 10 wt.% NaCl environment also caused the leaching of Sr^{2+} ions. The SrO reacted with TiO$_2$ and formed Sr$_3$Ti$_2$O$_7$. The reaction between Na$_2$O and TiO$_2$ caused the formation of Na$_4$TiO$_4$.

(4) This study clearly indicates the destabilization of SrTiO$_3$ in both mediums.

Conflicts of Interest

The authors declare that they have no conflicts of interest.

Acknowledgments

The author Dr. G. Sreedhar would like to acknowledge the DST, India, for sponsoring Project SB/EMEQ-036/2014 (GAP 09/15).

References

[1] N. P. Padture, M. Gell, and E. H. Jordan, "Thermal barrier coatings for gas-turbine engine applications," *Science*, vol. 296, no. 5566, pp. 280–284, 2002.

[2] D. J. Wortman, B. A. Nagaraj, and E. C. Duderstadt, "Thermal barrier coatings for gas turbine use," *Materials Science and Engineering: A Structural Materials: Properties, Microstructure and Processing*, vol. 120-121, no. 2, pp. 433–440, 1989.

[3] D. R. Clarke and S. R. Phillpot, "Thermal barrier coating materials," *Materials Today*, vol. 8, no. 6, pp. 22–29, 2005.

[4] R. A. Miller, "Current status of thermal barrier coatings - An overview," *Surface and Coatings Technology*, vol. 30, no. 1, pp. 1–11, 1987.

[5] A. Uzun, I. Çevik, and M. Akçil, "Effects of thermal barrier coating on a turbocharged diesel engine performance," *Surface and Coatings Technology*, vol. 116-119, pp. 505–507, 1999.

[6] E. H. Kisi and C. J. Howard, "Crystal structures of zirconia phases and their inter-relation," *Key Engineering Materials*, no. 153-154, pp. 1–36, 1998.

[7] G. Sreedhar and V. S. Raja, "Hot corrosion behavior of YSZ/Al$_2$O$_3$ dispersed NiCrAlY plasma-sprayed coatings in Na$_2$SO$_4$-10 wt.% NaCl melt," *Corrosion Science*, vol. 52, no. 8, pp. 2592–2602, 2010.

[8] R. L. Jones and C. E. Williams, "Hot corrosion studies of zirconia ceramics," *Surface and Coatings Technology*, vol. 32, no. 1-4, pp. 349–358, 1987.

[9] S. Park, J. Kim, M. Kim, H. Song, and C. Park, "Microscopic observations of degradations of yttria and ceria stabilized zirconia thermal barrier coatings under hot corrosion," *Surface and Coatings Technology*, vol. 190, pp. 357–365, 2005.

[10] C. Xialong, Z. Yu, G. Lijion, Z. Binglin, W. Ying, and C. Xueqiong, "Hot corrosion behaviour of plasma sprayed YSZ/LaMgAl$_{11}$O$_{15}$ composite coatings in molten sulfate-vanadate salt," *Corrosion Science*, vol. 53, pp. 2335–2343, 2011.

[11] J. Wu, X. Wei, N. P. Padture et al., "Low thermal conductivity rare earth zirconates for potential thermal barrier coating applications," *Journal of the American Ceramic Society*, vol. 85, no. 12, pp. 3031–3035, 2002.

[12] R. Vassen, X. Cao, F. Tietz, D. Basu, and D. Stöver, "Zircoantes as new material for thermal barrier coatings," *Journal of the American Ceramic Society*, vol. 83, no. 8, pp. 2023–2028, 2000.

[13] Y. J. Su, R. W. Trice, K. T. Faber, H. Wang, and W. D. Porter, "Thermal conductivity, phase stability, and oxidation resistance of Y$_3$Al$_5$O$_{12}$(YAG)/Y$_2$O$_3$-ZrO$_2$ (YSZ) thermal barrier coatings," *Oxidation of Metals*, vol. 61, no. 3-4, pp. 253–271, 2004.

[14] M. H. Habibi, L. Wang, J. Liang, and S. M. Guo, "An investigation on hot corrosion behavior of YSZ-Ta$_2$O$_5$ in Na$_2$SO$_4$+V$_2$O$_5$ salt at 1100°C," *Corrosion Science*, vol. 75, pp. 409–414, 2013.

[15] S. Yamanaka, K. Kurosaki, T. Oyama et al., "Thermophysical properties of perovskite-type strontium cerate and zirconate,"

Journal of the American Ceramic Society, vol. 88, no. 6, pp. 1496–1499, 2005.

[16] W. Ma, D. E. Mack, R. Vaben, and D. Stover, "Perovskite-type strontium zircoante as a new material for thermal barrier coatings," *Journal of the American Ceramic Society*, vol. 91, no. 8, pp. 2630–2635, 2008.

[17] B. Heimburg, W. Beele, K. Kempter et al., "Process for producing a ceramic thermal barrier layer for gas turbine engine component," U.S. Patent 6602553 B2, 2003.

[18] C. J. Howard and H. T. Stokes, "Structures and phase transformations in perovskites-A group theoretical approach," *Acta Crystallographica Section A*, vol. 61, pp. 93–111, 2005.

[19] B. J. Kennedy, C. J. Howard, and B. C. Chakoumakos, "hakoumakos, High temperature phase transitions in $SrZrO_3$," *Physical Review B: Condensed Matter and Materials Physics*, vol. 59, no. 6, pp. 4023–4027, 1999.

[20] Y. Zhao and D. J. Weidner, "Thermal expansion of $SrZrO_3$ and $BaZrO_3$ perovskites," *Physics and Chemistry of Minerals*, vol. 18, no. 5, pp. 294–301, 1991.

[21] C. Chen, X. Jiao, D. Chen, and Y. Zhao, "Effects of precursors on hydrothermally synthesized $SrTiO_3$ powders," *Materials Research Bulletin*, vol. 36, no. 12, pp. 2119–2126, 2001.

[22] Z. F. Wei, X. L. Chen, F. M. Wang, W. C. Li, M. He, and Y. Zhang, "Phase relations in the ternary system $SrO-TiO_2-B_2O_3$," *Journal of Alloys and Compounds*, vol. 327, no. 1-2, pp. L10–L13, 2001.

[23] G. Sreedhar, M. M. Alam, and V. S. Raja, "Hot corrosion behavior of plasma sprayed YSZ/Al_2O_3 dispersed NiCrAlY coatings on Inconel-718 super alloy," *Surface and Coatings Technology*, vol. 204, no. 3, pp. 291–299, 2009.

[24] J.-S. Park, J. Luo, L. Adijanto, J. M. Vohs, and R. J. Gorte, "The stability of lanthanum strontium vanadate for solid oxide fuel cells," *Journal of Power Sources*, vol. 222, pp. 123–128, 2013.

[25] E. E. Kaul, *Experimental investigation of new low-dimensional spin systems in vanadium oxides*, Technische Universität Dresden, Dresden, Germany, 2005.

[26] R. J. D. Tilley, "An electron microscope study of perovskite-related oxides in the SrTiO system," *Journal of Solid State Chemistry*, vol. 21, no. 4, pp. 293–301, 1977.

[27] K. T. Jacob and G. Rajitha, "Thermodynamic properties of strontium titanates: Sr_2TiO_4, $Sr_3Ti_2O_7$, $Sr_4Ti_3O_{10}$ and $SrTiO_3$," *The Journal of Chemical Thermodynamics*, vol. 43, no. 1, pp. 51–57, 2011.

[28] J. A. Dean, *Lange's Handbook of Chemistry*, New York, USA, 15th edition, 1972.

[29] H. D. B. Jenkins, *CRC Handbook of Chemistry and Physics*, Florida, USA, 79th edition, 1998.

[30] R. S. Roth, T. Negas, and L. P. Cook, "Phase Diagrams for Ceramists, Columbus," *Journal of the American Ceramic Society*, vol. 5, article 88, 1983.

[31] V. Tathavadkar and A. Jha, "The effect of molten sodium titanate and carbonate salt mixture on the alkali roasting of ilmenite and rutile minerals," in *Proceedings of VII International conference on molten slags fluxes and salts*, pp. 255–262, SAIMM, 2004.

Experimental Assessment of Rebar Corrosion in Concrete Slab Using Ground Penetrating Radar (GPR)

Ahmad Zaki [ID],[1] Megat Azmi Megat Johari,[2]
Wan Muhd Aminuddin Wan Hussin,[2] and Yessi Jusman [ID][3]

[1]*Department of Civil Engineering, Universitas Abdurrab, Pekanbaru, 28291 Riau, Indonesia*
[2]*School of Civil Engineering, Engineering Campus, Universiti Sains Malaysia, Nibong Tebal, 14300 Penang, Malaysia*
[3]*Department of Electrical Engineering, Faculty of Engineering, Universitas Muhammadiyah Yogyakarta Kasihan, Bantul 55183, Yogyakarta, Indonesia*

Correspondence should be addressed to Ahmad Zaki; ahmad.zaki@univrab.ac.id

Academic Editor: Michael I. Ojovan

Corrosion of steel reinforcement is a major cause of structural damage that requires repair or replacement. Early detection of steel corrosion can limit the extent of necessary repairs or replacements and costs associated with the rehabilitation works. The ground penetrating radar (GPR) method has been found to be a useful method for evaluating reinforcement corrosion in existing concrete structures. In this paper, GPR was utilized to assess corrosion of steel reinforcement in a concrete slab. A technique for accelerating reinforcement bar corrosion using direct current (DC) power supply with 5% sodium chloride (NaCl) solution was used to induce corrosion to embedded reinforcement bars (rebars) in this concrete slab. A 2 GHz GPR was used to assess the corrosion of the rebars. The analysis of the results of the GPR data obtained shows that corrosion of the rebars could be effectively localized and assessed.

1. Introduction

Corrosion of steel reinforcement is a worldwide problem of concrete structures [1]. Corrosion has been recognized as a major deterioration phenomenon which leads to the structural concrete degradation due to environmental actions [2]. Many reports have highlighted that all over the world concrete structures are damaged by corrosion and the costs associated with repair and maintenance required worldwide have exceeded billions of dollars [3]. Successful repair to concrete structures requires reliable information on the concrete structural conditions, including the cause of damage, degree of damage, and effect of damage on the actual structural behaviour.

Visual inspection of the whole structure to assess the condition of corrosion in concrete structures regarding reinforcement corrosion is a common regular inspection method but it is highly dependent on the expertise of the operator as unseen corrosion is not easily detected [2]. Meanwhile, the half-cell potential (HCP) technique is the most widely used nondestructive test to detect and localize rebar corrosion. The half-cell potential technique is based on electrochemical principles which provides information pertaining to the probability of reinforcement corrosion in concrete structures [2, 4]. On the other hand, ground penetrating radar (GPR), an alternative nondestructive testing (NDT) method, has become a valuable tool for inspection of concrete structures over the past few years [5]. GPR is capable of early detection of reinforcement corrosion in concrete structures [6–11]. Based on the capability of the GPR method to investigate corrosion, this paper proposes an assessment of rebar corrosion using GPR on a concrete slab with embedded corroded steel bars.

2. Literature Review

2.1. Ground Penetrating Radar (GPR). GPR, an electromagnetic (EM) investigative method, is mostly used in the reflection mode where a signal is emitted via an antenna into the structure under investigation. The reflected energy caused by changes in material properties is recorded and analyzed [12, 13]. Development of GPR has evolved over the past 35 years. Early research related to GPR focused on the possibilities and applicability of GPR. The first application of GPR in 1904 was to detect metal objects [14]. The use of GPR for civil engineering applications commenced in the 1980s [15]. Cantor [16] developed the basis of analysis and interpretation techniques in these areas. Furthermore, Clemena [17] utilized GPR for testing of concrete structures. GPR is a possible method for periodic inspection and maintenance of concrete structures [18]. Generally, GPR technology may be employed for localization of reinforcing bars, localization of cracking, localization and assessment of voids, localization of honeycombed or cracking, corrosion detection, estimation of bar size, concrete mix proportions, and environmental conditions [10, 19–31].

The principles of nondestructive testing (NDT) by GPR involve the transmission of EM waves into a structure under investigation. The propagation of EM waves of GPR depends on the corresponding dielectric properties of the materials [32–34]. Permittivity in turn depends on the EM properties which are influenced by temperature, moisture content, salt content, pore structure, and pulse frequency [35]. Whenever EM waves encounter an interface of two media having different dielectric constants, part of the wave is reflected back to the receiving antenna. The amount of scattering depends upon the contrast in dielectric properties of the two media [36]. The values of the dielectric constant for a variety of materials range from 1 to 81 (1 = air, 81 = water) [35]. EM waves propagate through the medium and are reflected at interfaces of materials with different dielectric properties. The reflected waves are recorded by a receiving antenna. These received waves are then converted into a voltage wave, which is called a trace (a-scan), as shown in Figure 1. GPR systems generate rapid successions of traces, which are displayed as a so-called radargram (b-scan) by investigating a material along a line, as shown in Figure 2(a). The other waves are displayed as time slice (c-scan) by investigating a material along a grid pattern, as shown in Figure 2(b) [37].

2.2. Accelerated Corrosion Process.
In practice, corrosion of steel reinforcement in concrete structures is a long-term process. It takes quite a real long time for the initiation and propagation of corrosion. For a research study, it is not easy to achieve different degrees (i.e., levels of mass loss) of corrosion within a short period of time. Therefore, various methods for accelerating reinforcement corrosion in a concrete structure have been used by several researchers [39]. In many studies the impressed current technique has been used to study the mechanical behaviour of corroded concrete structures [40], bond behaviour of corroded rebars [41], structural behaviour of corroded structural concrete elements [42, 43], and the performance of concrete structures [44]. The advantages of using this technique are that a high degree of corrosion can be obtained within a short period of time and the easy control of the desired degree of corrosion.

The anodic reaction releases electrons, while the cathodic reaction consumes electrons. Figure 3 shows schematically the process of corrosion of steel in concrete [3].

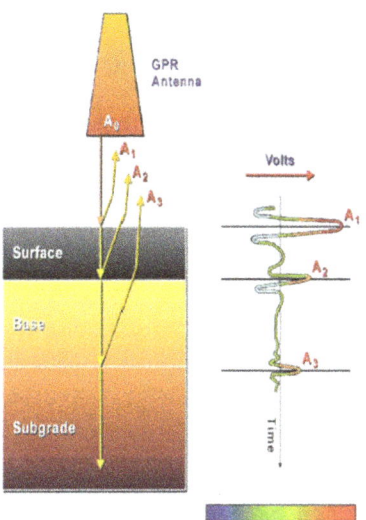

FIGURE 1: Single trace (a-scan) [38].

The anodic reaction depends on the pH of the electrolyte and the presence of anions. The anodic reaction for rebar corrosion is given by

$$\text{Fe} \longrightarrow \text{Fe}^{2+} + 2\text{e}^- \quad \text{(for iron and steel)} \tag{1}$$

The impressed current technique is used for accelerating corrosion of steel reinforcement in concrete by applying direct current (DC) through an external power supply. The positive terminal of this DC power supply is connected to the rebar making it act as anode, while the negative terminal is connected to a counter electrode becoming a cathode. The counter electrode could be in the form of an internal bar [45], external mesh [46], or external plate [41, 47]. The concrete specimens should be partially immersed in sodium chloride (NaCl) solution in a suitable tank, where the NaCl solution is in direct contact with the bottom of the structure. Here, an NaCl solution functioned as the electrolyte. The amount of corrosion is related to the energy consumed, which is a function of voltage (V), amperage (A), and time interval. Usually, a constant voltage between the anode and cathode is applied through the DC power supply. The principle of the method is described in ASTM G1-03 [48] and ASTM G31-72 [49].

Corrosion is often reported in terms of weight loss per unit time or thickness loss per unit time and steel surface area. Corrosion rate is directly proportional to the measured electrical current density and total mass loss is related to the measured electrical current by Faraday's Law [50] as given by

$$\Delta m = \frac{MIt}{zF} \tag{2}$$

where

Δm = mass of steel consumed (g);

M = atomic or molecular weight of metal (56 g for Fe);

I = current (amperes), t = time that current or potential is applied (seconds);

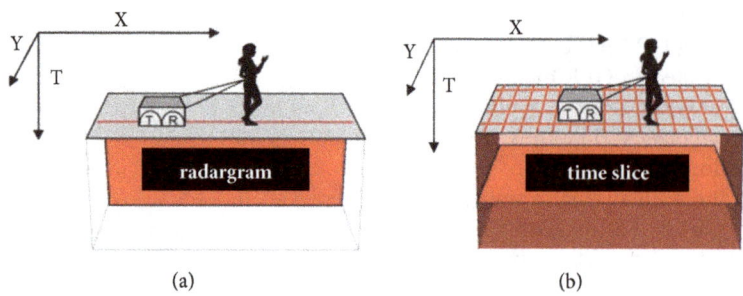

FIGURE 2: (a) Radargram (*b*-scan) and (b) time slice (*c*-scan) [37].

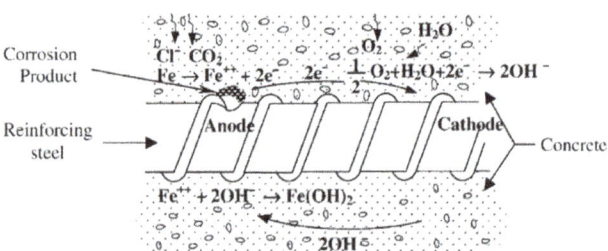

FIGURE 3: Process of corrosion of steel in concrete [3].

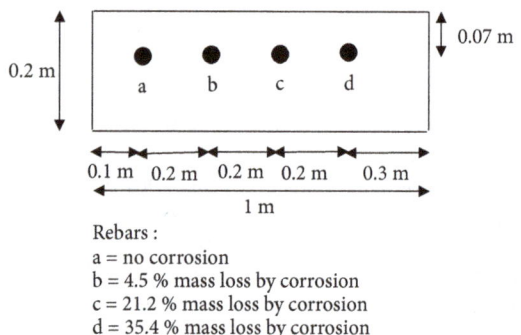

Rebars:
a = no corrosion
b = 4.5 % mass loss by corrosion
c = 21.2 % mass loss by corrosion
d = 35.4 % mass loss by corrosion

FIGURE 4: Schematic representation of concrete slab with rebars at different degrees of corrosion.

z = ionic charge or electrons transferred in the half-cell reaction (2 for Fe);

F = Faraday's constant (96,500 amp/sec).

Furthermore, the actual degree of corrosion or level of mass loss for the corroded steel specimens could be measured as the difference in mass of the bars before and after testing through equation [51]:

$$Ml = \frac{Mi - Mf}{Mi} \times 100 \qquad (3)$$

where

Ml = mass loss (%);

Mi = mass of noncorroded specimen weighed before the corrosion test (g)

Mf = mass of corroded specimen weighed after the corrosion test (g).

3. Materials and Methods

3.1. Specimens. Normal concrete with C30 grade was used in the experiment. The mix proportion includes 380 kg/m³ Portland cement, 190 kg/m³ water (w/c = 0.5), 780 kg/m³ uncrushed sand (fine aggregate), and 1080 kg/m³ crushed granite (coarse aggregate) with a maximum aggregate size of 20 mm. Average 28-day and 150-day compressive strengths are 44.43 MPa and 58.88 MPa, respectively. Type deformed steel reinforcement is selected for the rebars having a diameter of 20 mm. The exposed length of the rebars is 0.6 m.

A concrete slab was prepared having dimensions of length (l) = 1 m, width (w) = 0.5 m, and height (h) = 0.2 m. Figure 4

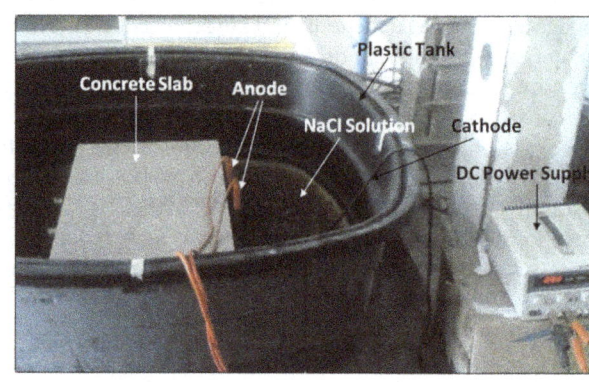

FIGURE 5: A photo of the impressed current technique.

shows a schematic representation of this concrete slab. After 28 days of moist curing, a DC current was impressed on three embedded rebars (i.e., rebars b, c, and d) embedded in this concrete slab. Corrosion of the individual rebars was accelerated by varying the duration of the applied current in order to achieve the intended degree of corrosion. The positive terminal of a DC power supply was connected to a rebar as anode, while the negative terminal was connected to a cathode. Here, a copper plate was used as the external cathode completely submerged in an NaCl solution. The slab was partially immersed in a solution of 5% NaCl and 95% distilled water stored in a plastic tank. The NaCl solution was in direct contact with the bottom of the slab. A photo of the test setup is shown in Figure 5.

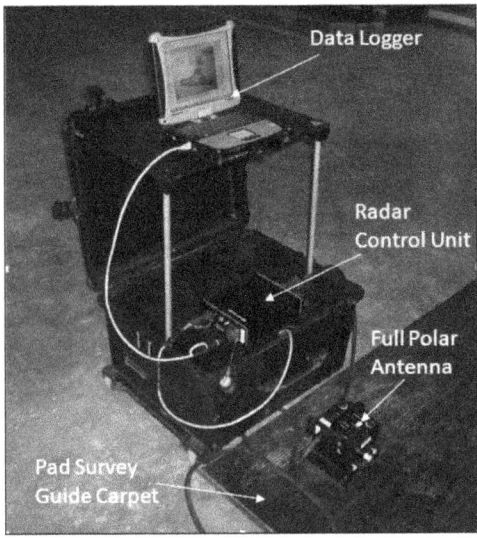

FIGURE 6: GPR ALADDIN SK2 kit.

TABLE 1: The duration of current applied for each rebar.

Rebar	Current applied (days)	Corrosion Degrees (%)	Current Density (mA/cm^2)
a	0	0.359	0
b	20.85	4.527	3.973
c	54.53	21.263	19.034
d	84.91	35.359	31.158

During the full duration of the test, a total anodic current of 0.4 A was applied to accelerate the corrosion process for a set of 3 rebars (rebar b-c-d). The process was continued until the different rebars were corroded to the required degree by varying the exposure times. The required duration of the applied current is calculated from the different degrees of corrosion. Faraday's law was applied to the accelerated technique. Firstly, the mass loss (Δm) was calculated using Faraday's law as illustrated in (2) [50]. Secondly, the corrosion levels of the rebars were based on the percentage of the difference between weight of initial and final mass using (2) [52]. Table 1 shows the duration of current applied for each of the 3 rebars.

3.2. GPR Acquisitions. The GPR equipment used in this research was manufactured by IDS (Ingegneria Dei Sistemi S.p.A) Pisa, Italy, as shown in Figure 6. The special full-polar high-frequency antenna (2 GHz) was combined with the patented Pad Survey Guide (PSG). In this work, the previously described reinforced concrete slab was scanned using the GPR equipment. The 2 GHz GPR antenna with two polarizations (transversal and longitudinal) was able to identify both types of scanned targets (shallow and deep) in just one scan.

The slab was scanned in line on a PSG carpet. A line of scan step on the PSG carpet was 0.78 cm, as shown in Figure 7(a). The PSG carpet held alternate upper and lower guides in order to make the slide perfectly fit and for the antenna to realize straight scans. Also, the slide held a special film on its external surface removing minimum friction during the drag of the antenna on the PSG, as shown in Figure 7(b). Finally, the PSG held on its edge a graduated horizontal pole with the indication of numbers and letters in a progressive sequence, in order to give the user a reference point at the beginning of the scan.

The scanned GPR results were stored in a computer completed by IDS in the GPR set. After obtaining the images from GPR, GRED software was used to obtain data images which were subsequently processed using image processing techniques. Data image could be presented as *a*-scan, *b*-scan, *c*-scan, and 3D images.

4. Results and Discussions

4.1. Corrosion Levels. The corrosion levels of the 4 rebars embedded in the concrete slab are presented as qualitative and quantitative results. The qualitative results are shown in Figure 9. It shows the postcorrosion slab after the completion of the corrosion process (rebars a = no corrosion, b = 4.5% mass loss, c = 21.2% mass loss, and d = 35.4% mass loss) based on the calculation of (3). There are cracks in the concrete cover above rebar c and rebar d. The concrete cover of rebar d shows a clear macrocrack along the full length of the rebar. This type of crack could affect the integrity of the concrete slab.

After that, the concrete slab was broken to retrieve the corroded rebars. The rebars were cleaned according to ASTM G1-2003 standard to remove all corrosion products [48]. The cleaned rebars with different degree of corrosion after cleaning are shown in Figure 10. Finally, the cleaned rebars were weighed to quantify the mass loss due to corrosion according to (3).

4.2. Assessing Rebar Corrosion Using GPR. Figure 11 shows the *a*-scan of line 25 before the accelerated corrosion process. The line is used as a reference because line 25 is in the middle of the rebar and also represents the condition of the rebars. Rebars a, b, c, and d are conditions of rebars which agree with information contained in Table 1. In Figures 11–14, S_d refers to the direct wave signal and S_r is the reflected wave signal from the rebar-concrete interface. There are no significant differences in the waves peak (i.e., direct wave and reflected wave) in terms of amplitude and travel time for all four rebars, which indicates that the rebars are in good condition (no corrosion). The waves are very important for reinforced concrete durability as it can provide information on concrete cover, i.e., the thickness and quality (presence of defects) of the concrete cover [53]. Information on the wave peaks is shown in Table 2.

Figure 12 shows the *a*-scan of the concrete slab at line 25 after the completion of the corrosion process of rebar b at the 21st day. Table 3 provides information on the quantification of the *a*-scan shown in Figure 12. The amplitude of direct wave and reflected wave peak was increased because the frequency was attenuated due to the accelerated corrosion [34, 53, 54]. The amplitudes of rebars b, c, and d were higher than that of rebar a. These conditions could be caused by the

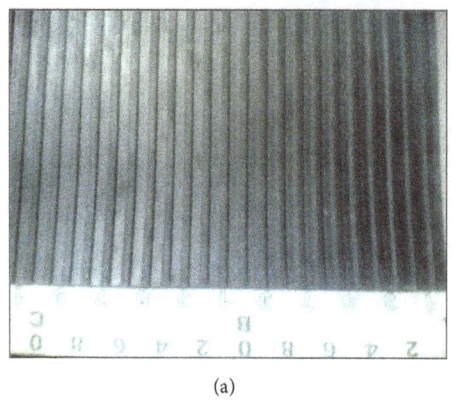

(a)

Slide of GPR

(b)

FIGURE 7: Features: (a) PSG carpet with lines of scan and (b) slide of GPR.

TABLE 2: Quantification of a-scan derived from Figure 8.

	Rebar a		Rebar b		Rebar c		Rebar d	
	Amplitude Peak	Travel Time	Amplitude Peak	Travel Time	Amplitude Peak	Travel Time	Amplitude Peak	Travel Time
S_d	0.410	-0.19	0.41	-0.19	0.410	-0.19	0.410	-0.19
S_r	0.321	1.127	0.295	1.170	0.321	1.127	0.321	1.127

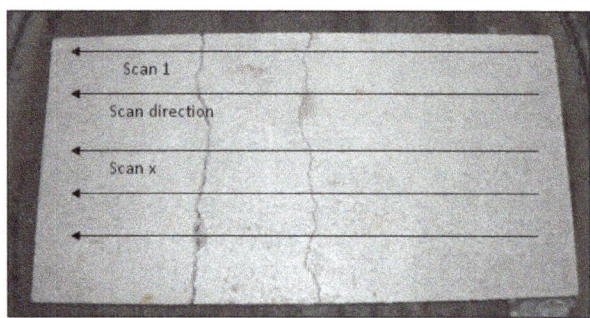

FIGURE 8: Scan direction of GPR.

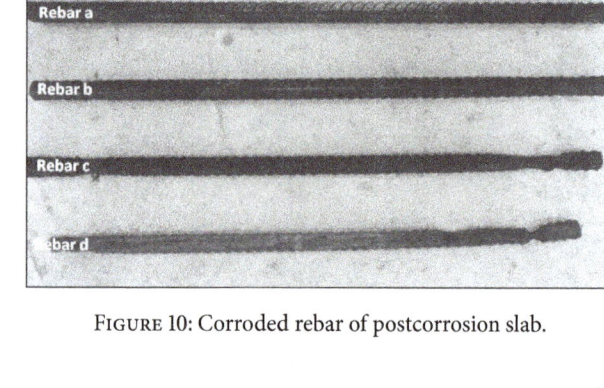

FIGURE 10: Corroded rebar of postcorrosion slab.

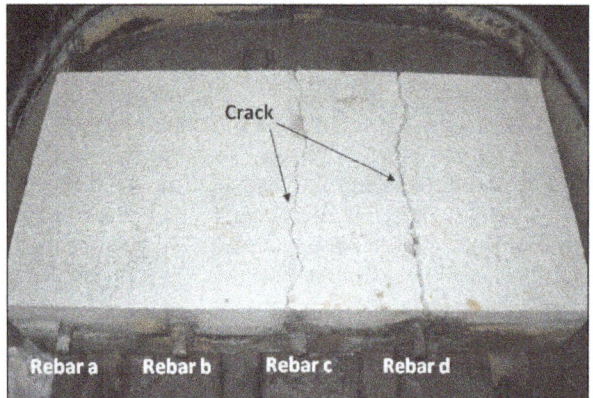

FIGURE 9: Concrete slab image after accelerated corrosion.

high content of chloride [35]. This may happen after cracking of the concrete cover, allowing faster ingress of chloride ions to the rebar; however prior to cracking hardly any chloride ions were present near the rebar. In this condition, chloride ions were forced to accumulate around the rebar. The accumulation of ions attenuated and absorbed the reflected wave in both time and frequency.

Figure 13 shows the a-scan of the concrete slab at line 25 after the completion of the corrosion process of rebar c at the 55th day. The frequencies of the waves shift to lower regime due to the effect of the corrosion process. The reflected waves of rebar c have lower amplitude than that of rebars a, b, and d because it is likely that the corrosion process of rebar c was affected by the accumulation of chloride ions in the concrete cover zone of the rebar which is more than for the other rebars. The higher content of chloride ions retarded and absorbed the waves in both time and frequency [55, 56]. The quantification values of the waves are shown in Table 4.

Figure 14 shows the a-scan of the concrete slab at line 25 after the completion of the corrosion process of rebar d at the 85th day. The direct wave peaks of rebar d are higher than that of rebars a, b, and c. This condition may have happened due to the cracked concrete cover which affects rebar c. The presence of voids in the crack zones could have increased the amplitude

TABLE 3: Quantification of a-scan derived from Figure 9.

	Rebar a		Rebar b		Rebar c		Rebar d	
	Amplitude Peak	Travel Time	Amplitude Peak	Travel Time	Amplitude Peak	Travel Time	Amplitude Peak	Travel Time
S_d	0.410	0.310	0.505	0.311	0.466	0.307	0.570	0.311
S_r	0.357	1.059	0.460	1.062	0.416	1.062	0.407	1.050

TABLE 4: Quantification of a-scan derived from Figure 10.

	Rebar a		Rebar b		Rebar c		Rebar d	
	Amplitude Peak	Travel Time	Amplitude Peak	Travel Time	Amplitude Peak	Travel Time	Amplitude Peak	Travel Time
S_d	0.480	0.496	0.490	0.496	0.500	0.496	0.510	0.496
S_r	0.330	1.620	0.350	1.620	0.220	1.860	0.320	1.620

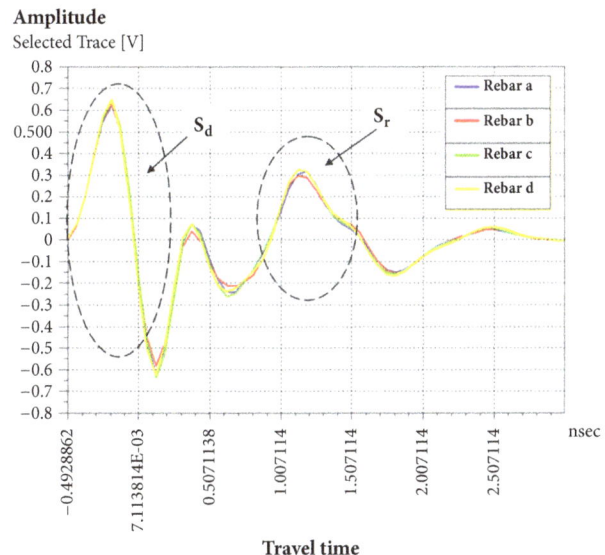

FIGURE 11: a-scan of the postcorrosion slab before accelerated corrosion.

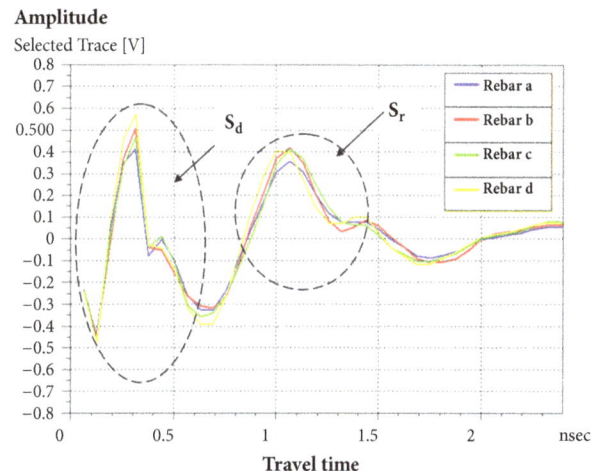

FIGURE 12: a-scan of the postcorrosion slab after accelerated corrosion of rebar b.

[34]. However, these phenomena did not happen for rebar d. As shown in Table 5, the higher content of moisture and chloride from the induced corrosion affects the amplitude of the reflected waves of GPR [35]. The moisture content and chloride content increase due to corrosion. Even though rebar a is supposed to have no corrosion, while rebar b was induced to have a low level of corrosion, rebars b and d show higher amplitudes (0.370 and 0.321, respectively) of reflected waves than rebar a (0.289). Rebar c has lower amplitude (0.213) and higher travel time (1.964 ns). It is likely that this has happened due to the high contents of chloride ions which cause an attenuating influence of waves and has stronger impact on rebar c than the other rebars [56, 57].

Figure 15 shows the b-scan of the rebars of the concrete slab in line 25. In Figure 15(a), there are no differences between the hyperbolic images and the respective depth of the rebars because the rebars are in good condition (no corrosion). In the case of Figure 15(b), the corrosion process has reduced the frequency signal of the waves whose effect is to reduce the hyperbolic image of the rebars. Meanwhile, in Figure 15(c), the image is fuzzy because of the formation and

transport of corrosion products which cause an attenuating influence of waves and has strong impact to b-scan image [57]. The corrosion products of rebar c have diffused to a shallower cover depth, as shown in Figure 16(a). This may be due to the fact that the types of corrosion products when accelerating the corrosion process by an impressed current are different and less voluminous than the products formed under normal free corrosion conditions.

Finally, Figure 15(d) shows that the hyperbolic image of rebar c is significantly reduced due to corrosion. This can be explained by the b-scan image of rebar c which is fuzzier due to corrosion products that have diffused to closer proximity within the concrete surface [58], as shown in Figure 16(a). However, different phenomena could be observed in the b-scan image of rebar d. Rebar d is supposed to have a higher level of corrosion than rebar c, but the opposite condition is displayed from the b-scan. In theory, the image becomes fuzzy due to an increase of the corrosion level [59]. However, the image of rebar d shown is much clearer. The depth of the rebar image becomes shallower than the reference (rebar a). The probable explanation is that the energy of the wave was less attenuated due to corrosion products that diffused to a much shallower concrete cover [58], as shown in Figure 16(b). The amount of corrosion products of rebar d is more than that of rebar c; however the corrosion products of rebar d

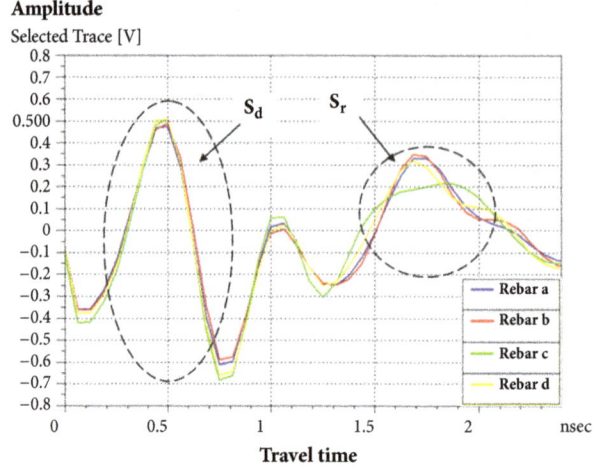

FIGURE 13: *a*-scan of the postcorrosion slab after accelerated corrosion of rebar c.

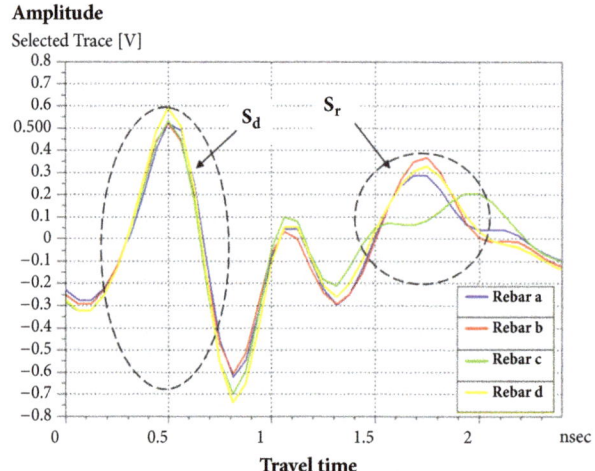

FIGURE 14: *a*-scan of the postcorrosion slab after accelerated corrosion of rebar d.

TABLE 5: Quantification of *a*-scan derived from Figure 11.

	Rebar a		Rebar b		Rebar c		Rebar d	
	Amplitude Peak	Travel Time	Amplitude Peak	Travel Time	Amplitude Peak	Travel Time	Amplitude Peak	Travel Time
S_d	0.528	0.502	0.528	0.502	0.528	0.502	0.590	0.502
S_r	0.289	1.750	0.370	1.750	0.213	1.964	0.321	1.750

(Figure 16(b)) have not diffused significantly to the concrete cover surface like the corrosion products of rebar c, as shown in Figure 16(a).

5. Conclusions

In this paper, it has been demonstrated that the presence of rebar corrosion in a concrete slab has been successfully assessed using GPR. The processes start with concrete slab fabrication followed by the results of imposed current technique and data collection by scanning a reinforced concrete slab using 2 GHz of the GPR. Several results can be obtained by using the GPR technique for assessing rebar corrosion in a concrete slab. The results are presented by the *a*-scan

and the *b*-scan. The results show that rebar corrosion can be detected and identified at an early stage before visual damage or other signs of corrosion have appeared. The results of the *a*-scan, i.e., lower amplitude and larger travel times of the waves for corroded rebars, are probably due to increased chloride contents and presence of corrosion products. For the *b*-scan, the form of blurring and dimming of rebar image features shows a distinct indication of rebar corrosion due to chloride contents and corrosion product which have influenced the threshold level of image. Generally, the results of an a-scan can represent the condition of rebar corrosion damage. The impressed current results in uniform corrosion attack whereas in practice (real structures) chloride-induced corrosion normally will be characterised by strongly localised

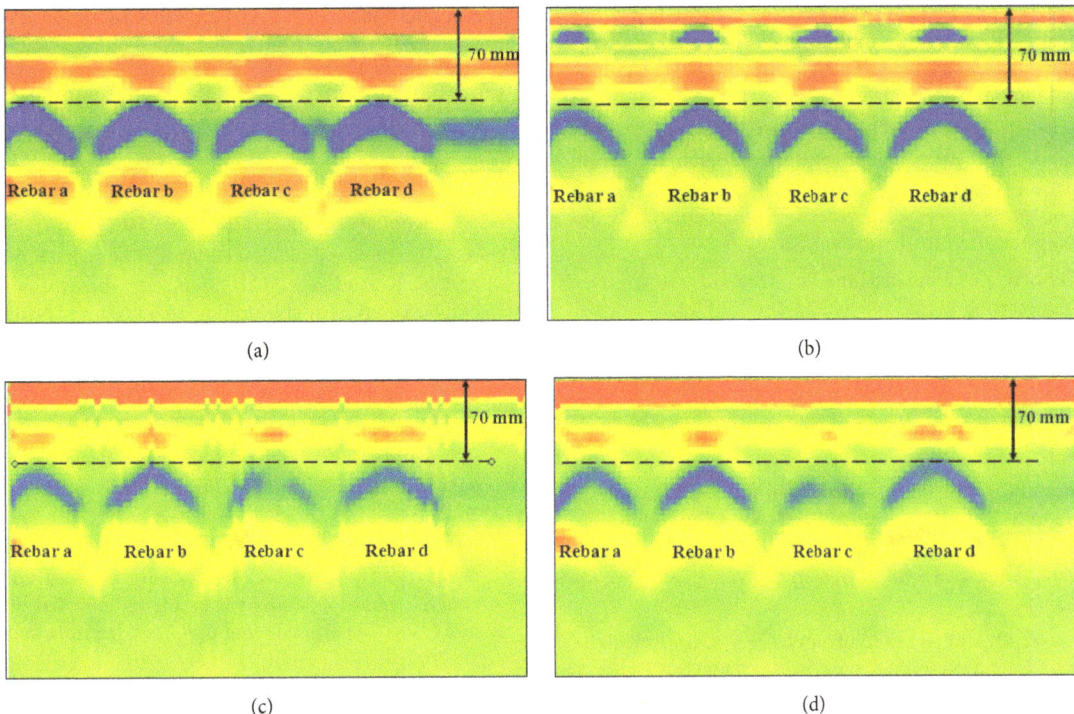

FIGURE 15: *b*-scan of the postcorrosion slab: (a) before accelerated corrosion, (b) after accelerated corrosion of rebar b, (c) after accelerated corrosion of rebar c, and (d) after accelerated corrosion of rebar d.

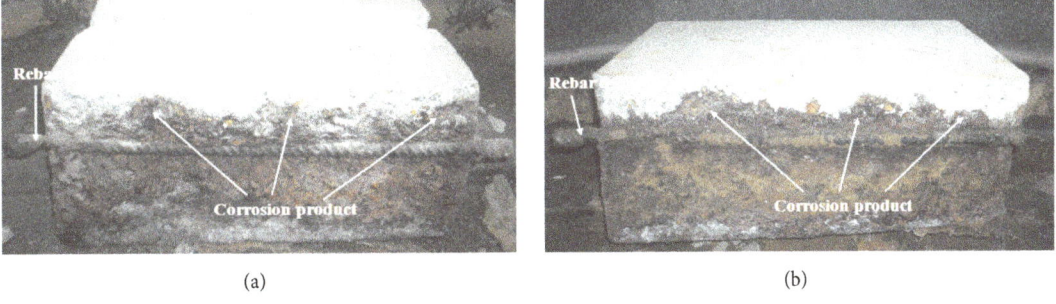

FIGURE 16: Cut sections of the postcorrosion slab: (a) rebar c and (b) rebar d.

corrosion attack. In summary the results obtained by GPR indicate that GPR can also effectively be used for assessment of localised corrosion damage of concrete slab in terms of chloride contents and corrosion product.

Conflicts of Interest

There are no conflicts of interest regarding the publication of this paper.

Acknowledgments

The authors would like to acknowledge the financial support provided by Universiti Sains Malaysia for providing RU-PGRS Grant Scheme entitled "Concrete Health Monitoring Using Non-Destructive Testing (NDT) Methods" and the financial support provided by Universitas Abdurrab, Pekanbaru, Indonesia.

References

[1] M. Badawi and K. Soudki, "Control of corrosion-induced damage in reinforced concrete beams using carbon fiber-reinforced polymer laminates," *Journal of Composites for Construction*, vol. 9, no. 2, pp. 195–201, 2005.

[2] B. Elsener, C. Andrade, J. Gulikers, R. Polder, and M. Raupach, "Hall-cell potential measurements—potential mapping on reinforced concrete structures," *Materials and Structures*, vol. 36, no. 261, pp. 461–471, 2003.

[3] S. Ahmad, "Reinforcement corrosion in concrete structures, its monitoring and service life prediction—a review," *Cement and Concrete Composites*, vol. 25, no. 4-5, pp. 459–471, 2003.

[4] A. Zaki, H. K. Chai, D. G. Aggelis, and N. Alver, "Non-destructive evaluation for corrosion monitoring in concrete: A review and capability of acoustic emission technique," *Sensors*, vol. 15, no. 8, pp. 19069–19101, 2015.

[5] H. Wiggenhauser, H. W. Reinhardt, D. O. Thompson, and D. E. Chimenti, "NDT in civil engineering: experience and results of the for 384 research group," in *Proceedings of the Review of Progress in Quantitative Nondestructive Evaluation*, vol. 29, pp. 47–54, Kingston, Rhode Island, RI, USA, 2010.

[6] G. R. Olhoeft, "Maximizing the information return from ground penetrating radar," *Journal of Applied Geophysics*, vol. 43, no. 2-4, pp. 175–187, 2000.

[7] C. W. Chang, C. H. Lin, and H. S. Lien, "Measurement radius of reinforcing steel bar in concrete using digital image GPR," *Construction and Building Materials*, vol. 23, no. 2, pp. 1057–1063, 2009.

[8] W. Al-Nuaimy, Y. Huang, M. Nakhkash, M. T. C. Fang, V. T. Nguyen, and A. Eriksen, "Automatic detection of buried utilities and solid objects with GPR using neural networks and pattern recognition," *Journal of Applied Geophysics*, vol. 43, no. 2–4, pp. 157–165, 2000.

[9] X.-Q. He, Z.-Q. Zhu, Q.-Y. Liu, and G.-Y. Lu, "Review of GPR Rebar Detection. Progress In Electromagnetics Research Symposium," in *Proceedings of the Review of GPR Rebar Detection. Progress In Electromagnetics Research Symposium*, Beijing, China, 2009.

[10] M. I. Hasan and N. Yazdani, "An experimental study for quantitative estimation of rebar corrosion in concrete using ground penetrating radar," *Journal of Engineering (United States)*, vol. 2016, 2016.

[11] S. Hong, H. Wiggenhauser, R. Helmerich, B. Dong, P. Dong, and F. Xing, "Long-term monitoring of reinforcement corrosion in concrete using ground penetrating radar," *Corrosion Science*, vol. 114, pp. 123–132, 2017.

[12] J. Hugenschmidt and R. Mastrangelo, "GPR inspection of concrete bridges," *Cement and Concrete Composites*, vol. 28, no. 4, pp. 384–392, 2006.

[13] W. Wai-Lok Lai, X. Dérobert, and P. Annan, "A review of Ground Penetrating Radar application in civil engineering: A 30-year journey from Locating and Testing to Imaging and Diagnosis," *NDT & E International*, vol. 96, pp. 58–78, 2018.

[14] C. P. F. Ulricksen, "Application of impulse radar to civil engineering," in *Application of impulse radar to civil engineering*, Lund University of Technology. Sweden: Lund, 1982.

[15] M. W. Juranty and S. B. Chase, "Ground-penetrating radar evaluation of bridge decks," in *Proceedings of the Nondestructive Evaluation of Aging Infrastructure*, pp. 59–67, Oakland, CA, USA, 2004.

[16] T. R. Cantor, "Review of penetrating radar as applied to non-destructive testing of concrete," in *ACI SP-82*, V. M. Malhotra and T. R. Cantor, Eds., vol. 20, Insituy Nondestructive Testing of Concrete, American Concrete Institute, 1984.

[17] G. G. Clemena, "Short pulse radar methods," in *Handbook on Non-destructive Testing of Concrete*, V. M. Malhotra and N. J. Carino, Eds., vol. 11, CRC Press, Boston, Mass, USA, 1991.

[18] S. Laurens, J. P. Balayssac, J. Rhazi, G. Klysz, and G. Arliguie, "Non-destructive evaluation of concrete moisture by GPR: Experimental study and direct modeling," *Materials and Structures/Materiaux et Constructions*, vol. 38, no. 283, pp. 827–832, 2005.

[19] C. Maierhofer and S. Leipold, "Radar investigation of masonry structures," *NDT & E International*, vol. 34, no. 2, pp. 139–147, 2001.

[20] J. H. Bungey, "Sub-surface radar testing of concrete: A review," *Construction and Building Materials*, vol. 18, no. 1, pp. 1–8, 2004.

[21] V. Barrile and R. Pucinotti, "Application of radar technology to reinforced concrete structures: A case study," *NDT & E International*, vol. 38, no. 7, pp. 596–604, 2005.

[22] A. V. Varnavina, L. H. Sneed, A. K. Khamzin, E. V. Torgashov, and N. L. Anderson, "An attempt to describe a relationship between concrete deterioration quantities and bridge deck condition assessment techniques," *Journal of Applied Geophysics*, vol. 142, pp. 38–48, 2017.

[23] P. Wiwatrojanagul, R. Sahamitmongkol, S. Tangtermsirikul, and N. Khamsemanan, "A new method to determine locations of rebars and estimate cover thickness of RC structures using GPR data," *Construction and Building Materials*, vol. 140, pp. 257–273, 2017.

[24] A. V. Varnavina, A. K. Khamzin, L. H. Sneed et al., "Concrete bridge deck assessment: Relationship between GPR data and concrete removal depth measurements collected after hydrodemolition," *Construction and Building Materials*, vol. 99, pp. 26–38, 2015.

[25] S. Yehia, N. Qaddoumi, S. Farrag, and L. Hamzeh, "Investigation of concrete mix variations and environmental conditions on defect detection ability using GPR," *NDT & E International*, vol. 65, pp. 35–46, 2014.

[26] M. Torres-Luque, E. Bastidas-Arteaga, F. Schoefs, M. Sánchez-Silva, and J. F. Osma, "Non-destructive methods for measuring chloride ingress into concrete: State-of-the-art and future challenges," *Construction and Building Materials*, vol. 68, pp. 68–81, 2014.

[27] X. Dérobert, J. F. Lataste, J.-P. Balayssac, and S. Laurens, "Evaluation of chloride contamination in concrete using electromagnetic non-destructive testing methods," *NDT & E International*, vol. 89, pp. 19–29, 2017.

[28] A. Zaki, S. Kabir, B. H. Abu Bakar, M. A. Megat Johari, and Y. Jusman, "Application of Image Processing for Detection of Corrosion Using Ground Penetrating Radar," in *Proceedings of the The 6th International Conference on Information Communication Technology and Systems*, Surabaya, 2010.

[29] A. Zaki and S. Kabir, "Radar-based quantification of corrosion damage in concrete structures," in *Proceedings of the Progress in Electromagnetics Research Symposium, PIERS 2011 Marrakesh*, pp. 794–798, Morocco, March 2011.

[30] S. Kabir and A. Zaki, "Detection and quantification of corrosion damage using ground penetrating radar (GPR)," in *Proceedings of the Progress in Electromagnetics Research Symposium, PIERS 2011 Marrakesh*, pp. 790–793, Morocco, March 2011.

[31] S. N. A. Mohd Kanafiah, N. D. M. Kamal, A. Z. Ahmad Firdaus et al., "Recognition system of Underground Object Shape using ground penetrating radar datagram," in *Proceedings of the 5th IEEE International Conference on Control System, Computing and Engineering, ICCSCE 2015*, pp. 488–491, Malaysia, November 2015.

[32] A. Robert, "Dielectric permittivity of concrete between 50 Mhz and 1 GHz and GPR measurements for building materials evaluation," *Journal of Applied Geophysics*, vol. 40, no. 1-3, pp. 89–94, 1998.

[33] M. N. Soutsos, J. H. Bungey, S. G. Millard, M. R. Shaw, and A. Patterson, "Dielectric properties of concrete and their influence

on radar testing," *NDT & E International*, vol. 34, no. 6, pp. 419–425, 2001.

[34] Z. M. Sbartaï, S. Laurens, J.-P. Balayssac, G. Ballivy, and G. Arliguie, "Effect of concrete moisture on radar signal amplitude," *ACI Materials Journal*, vol. 103, no. 6, pp. 419–426, 2006.

[35] C. Maierhofer, "Nondestructive evaluation of concrete infrastructure with ground penetrating radar," *Journal of Materials in Civil Engineering*, vol. 15, no. 3, pp. 287–297, 2003.

[36] U. B. Halabe, *Condition Assessment of Reinforced Concrete Structures Using Electromagnetic Waves*, Massachusetts Institute of Technology, Massachusetts, Mass, USA, 1990.

[37] A. Taffe and C. Maierhofer, "Guidelines for NDT methods in civil engineering," in *Proceedings of the The International Symposium on Non-Destructive Testing in Civil Engineering, NDT-CE*, Berlin, Germany, 2003.

[38] X. He, Z. Zhu, G. Lu, and Q. Lu, "Bridge management with GPR," in *Proceedings of the 2009 International Conference on Information Management, Innovation Management and Industrial Engineering, ICIII 2009*, pp. 329–332, China, December 2009.

[39] S. Ahmad, "Techniques for inducing accelerated corrosion of steel in concrete," *Arabian Journal for Science and Engineering*, vol. 34, no. 2 C, pp. 95–104, 2009.

[40] A. A. Almusallam, "Effect of degree of corrosion on the properties of reinforcing steel bars," *Construction and Building Materials*, vol. 15, no. 8, pp. 361–368, 2001.

[41] C. Fang, K. Lundgren, L. Chen, and C. Zhu, "Corrosion influence on bond in reinforced concrete," *Cement and Concrete Research*, vol. 34, no. 11, pp. 2159–2167, 2004.

[42] A. H. Al-Saidy, A. S. Al-Harthy, K. S. Al-Jabri, M. Abdul-Halim, and N. M. Al-Shidi, "Structural performance of corroded RC beams repaired with CFRP sheets," *Composite Structures*, vol. 92, no. 8, pp. 1931–1938, 2010.

[43] A. Zaki, H. K. Chai, A. Behnia, D. G. Aggelis, J. Y. Tan, and Z. Ibrahim, "Monitoring fracture of steel corroded reinforced concrete members under flexure by acoustic emission technique," *Construction and Building Materials*, vol. 136, pp. 609–618, 2017.

[44] K. Sakr, "Effect of cement type on the corrosion of reinforcing steel bars exposed to acidic media using electrochemical techniques," *Cement and Concrete Research*, vol. 35, no. 9, pp. 1820–1826, 2005.

[45] Y. Yuan, Y. Ji, and S. P. Shah, "Comparison of two accelerated corrosion techniques for concrete structures," *ACI Structural Journal*, vol. 104, no. 3, pp. 344–347, 2007.

[46] G. Nounu and Z.-U. Chaudhary, "Reinforced concrete repairs in beams," *Construction and Building Materials*, vol. 13, no. 4, pp. 195–212, 1999.

[47] A. A. Almusallam, A. S. Al-Gahtani, and A. R. Aziz, "Effect of reinforcement corrosion on bond strength," *Construction and Building Materials*, vol. 10, no. 2, pp. 123–129, 1996.

[48] ASTM, *G1-Standard Practice for Preparing, Cleaning, and Evaluating Corrosion Test Specimens*, 2003.

[49] ASTM, *G3 -72-Standard Practice for Laboratory Immersion Corrosion Testing of Metals*, 2004.

[50] T. A. El Maaddawy and K. A. Soudki, "Effectiveness of impressed current technique to simulate corrosion of steel reinforcement in concrete," *Journal of Materials in Civil Engineering*, vol. 15, no. 1, pp. 41–47, 2003.

[51] C. A. Apostolopoulos and D. Michalopoulos, "Impact of corrosion on mass loss, fatigue and hardness of BSt500 s steel," *Journal of Materials Engineering and Performance*, vol. 16, no. 1, pp. 63–67, 2007.

[52] C. A. Apostolopoulos, D. Michalopoulos, and P. Koutsoukos, "The corrosion effects on the structural integrity of reinforcing steel," *Journal of Materials Engineering and Performance*, vol. 17, no. 4, pp. 506–516, 2008.

[53] G. Klysz, J.-P. Balayssac, and S. Laurens, "Spectral analysis of radar surface waves for non-destructive evaluation of cover concrete," *NDT & E International*, vol. 37, no. 3, pp. 221–227, 2004.

[54] W. L. Lai, S. C. Kou, W. F. Tsang, and C. S. Poon, "Characterization of concrete properties from dielectric properties using ground penetrating radar," *Cement and Concrete Research*, vol. 39, no. 8, pp. 687–695, 2009.

[55] C. L. Barnes, J.-F. Trottier, and D. Forgeron, "Improved concrete bridge deck evaluation using GPR by accounting for signal depth-amplitude effects," *NDT & E International*, vol. 41, no. 6, pp. 427–433, 2008.

[56] W.-L. Lai, T. Kind, M. Stoppel, and H. Wiggenhauser, "Measurement of accelerated steel corrosion in concrete using ground-penetrating radar and a modified half-cell potential method," *Journal of Infrastructure Systems*, vol. 19, no. 2, pp. 205–220, 2013.

[57] C. Maierhofer and A. Taffe, "Guidelines for NDT methods in civil engineering," in *Nondestructive Testing in Civil Engineering (NDTCE)*, Berlin, Germany, 2003.

[58] W. L. Lai, T. Kind, and H. Wiggenhauser, "Detection of accelerated reinforcement corrosion in concrete by ground penetrating radar," in *Proceedings of the 13th Internarional Conference on Ground Penetrating Radar, GPR 2010*, Italy, June 2010.

[59] R. W. Arndt, J. Cui, D. R. Huston, D. O. Thompson, and D. E. Chimenti, "Monitoring of reinforced concrete corrosion and deterioration by periodic multi-sensor non-destructive evaluation," in *Proceedings of 37th Annual Review of Progress in Quantitative Nondestructive Evaluation, QNDE 2010*, vol. 20A, pp. 1371–1378, San Diego, Ca, USA.

A Comparative Study of Hydrogen-Induced Cracking Resistances of API 5L B and X52MS Carbon Steels

Rodrigo Monzon Figueredo, Mariana Cristina de Oliveira, Leandro Jesus de Paula, Heloisa Andréa Acciari ⓘ, and Eduardo Norberto Codaro ⓘ

School of Engineering, São Paulo State University (UNESP), Guaratinguetá, SP, Brazil

Correspondence should be addressed to Heloisa Andréa Acciari; heloisa@feg.unesp.br

Academic Editor: Ramana M. Pidaparti

Susceptibility to hydrogen-induced cracking of API 5L B and X52MS low-carbon steels in NACE 177-A, 177-B, and 284-B solutions has been investigated by the present work. A metallographic analysis of these steels was performed before and after NACE TM0284 standard testing. Corrosion products were characterized by scanning electron microscopy and X-ray dispersive energy spectrometry, which were subsequently identified by X-ray diffraction. Thus it was found that pH directly affects the solubility of corrosion products and hydrogen permeation. Both steels showed generalized corrosion in solution 177-A, and a discontinuous film was formed on their surfaces in solution 177-B; however, only the API 5L B steel failed the HIC test and exhibited greater crack length ratio in solution 177-A. In solution 284-B whose pH is higher, the steels exhibited thick mackinawite films with no internal cracking.

1. Introduction

In the latest decades, the world has been experiencing a continuous growth in the demand for oil and its derivatives. Despite all efforts to diversify the energy matrix world, current projections still reveal a considerable increase in the demand for oil and natural gas. This growing demand and the fact that many countries aim to obtain strategic autonomy from their energy matrix have led to increased exploitation complex oil and gas sources [1]. These sources present greater geological (depth, pressure, and temperature) and geographical (remote regions and marine currents) difficulties, as well as products with higher concentrations of hydrogen sulfide, commonly known as sour gas. The corrosion of steel pipelines by hydrogen sulfide has been a recurrent problem for oil and natural gas exploitation and production industry, which has aroused special attention after the discovery of the Brazilian pre-salt in 2006. Hydrogen sulfide together with other characteristics of the medium, such as temperature and pH condition, can cause general corrosion, localized corrosion, embrittlement, and cracking of pipelines; moreover, there is a large history of such occurrences in valves and welded joints. Pipeline rupture or its parts also have great environmental and economic impacts due to leaks and explosions, which lead to a halt in production [2–5]. Although corrosion consequences are well known, causes and mechanisms by which each phenomenon occurs are still not well understood. General corrosion is attributed to the preferential dissolution of the ferritic phase and it often manifests by a cementite scale (Fe_3C) formation. Localized corrosion arises from the galvanic couples between the ferritic phase and nonmetallic (e.g., MnS) or intermetallic (e.g., Fe_3C) inclusions and it manifests as pitting [6, 7].

The steps of a possible mechanism of hydrogen-induced cracking (HIC) can be, first, interstitial diffusion of hydrogen atoms under the steel surface. It occurs at a rate that depends on the packaging of the iron unit cell, that is, greater diffusibility in the ferrite phase (bcc) than in the more compact austenite phase (fcc). Then, hydrogen concentration at trapping sites: punctiform defects (gaps), linear defects (dislocations), two-dimensional defects (grain contours, mainly triple grain junction), and discontinuities (pores and contours of intermetallic particles such as Fe_3C, nonmetallic inclusions such as MnS, and impurities as oxides). The next step can be

TABLE 1: Chemical composition of API 5L X52MS and API 5L B steels (wt.%).

Steel	C	Si	Mn	P	S	V	Nb	Ti
X52MS	0.03	0.21	1.18	0.01	0.000	0.004	0.03	0.01
B	0.10	0.32	0.86	0.02	0.002	0.002	0.02	0.02

TABLE 2: Mechanical properties of steels.

Steel	FL (MPa)	RL (MPa)	EL (%)	Hardness (HV_{10})	Tenacity at $-20°C$ (J)
X52MS	440	485	44	173	403
B	373	458	41	170	205

hydrogen atoms interaction with other atoms or cations in these sites of the iron crystalline network and hence decrease of the reticular energy (decohesion). After that, the recombination of hydrogen atoms and pressurization of these sites causes internal stress. Finally, this pressurization is manifested mainly in the form of blisters near the steel surface and internal cracks aligned in the direction of material rolling (HIC) [2, 8].

Since the chemical compositions of petroleum, natural gas, and produced water vary widely, and laboratory tests cannot reproduce the internal conditions of oil and gas pipelines, standard tests have been developed for qualifying of carbon steels. Corrosion standard tests of carbon steels to be used in oil and gas production establish specific conditions for pH, temperature, H_2S partial pressure, and time of exposure [8, 9]. It is important to note that the test solutions described by these standards do not simulate a sour environment but rather provide a reproducible test environment capable of evaluating the susceptibility to cracking of different steels in a relatively short period of time. In the pipe manufacturing industry, it is common to classify steels according to their corrosion resistance in one of these test solutions. However, there is always uncertainty in trying to extrapolate results based on production and test history, that is, how a product would behave in a low pH medium when it was originally designed and qualified for being used in a high pH medium.

Thus, the present work aims to study the resistance to hydrogen-induced cracking of API 5L B and X52MS steels in 177-A, 177-B, and 284-B NACE solutions. The corrosion products were characterized by electron microscopy (SEM) and X-ray dispersive energy spectrometry (EDS) and identified by X-ray diffraction (XRD).

2. Materials and Methods

API 5L X52MS and API 5L B carbon steels pipes are commonly manufactured in Brazil. The first one is produced to resist harsh sour conditions, whereas the last one is used in sweet or slightly acidic environments. The pipes were manufactured according to standard [10] through the cold forming process by three-stage pressing (edge pressing, U-shaped pressing, and O-shaped pressing), followed by longitudinal submerged arc welding, and expansion in order to calibrate the final geometry of pipes.

TABLE 3: Chemical composition of solutions 177-A and 177-B.

Compounds	Concentration (wt.%)	
	177-A	177-B
NaCl	5.0	5.0
HCH_3COO	0.50	2.5
$NaCH_3COO$	-	0.41

Chemical composition of these steels (Table 1) was determined by optical emission spectrometry with a Thermo ARL 3460 spectrometer and the ARL WinOE software.

The API 5L X52MS steel is quite able to support operation in H_2S-containing environments, being qualified in the manufacturing plant through tests in solution 177-A (Table 2). It should have maxima CLR (crack length ratio), CTR (crack thickness ratio), and CSR (crack sensitivity ratio) at 15%, 5.0%, and 2.0%, respectively, to be considered qualified [3]. The API 5L B steel used herein has not been heat-treated, thus not being appropriate for all the sour environments. It was chosen so as to compare hydrogen-induced cracking resistances of the two steels. Samples of $100 \times 20 \times 10$ mm were taken from the pipes by plasma cutting at approximately 90 degrees from the longitudinal weld.

To evaluate their resistance to HIC, two flasks were used for each test, one containing 10 L of test solution (Tables 3 and 4) and the other one with 10 $100 \times 20 \times 10$ mm samples, being five of each steel (Figure 1). These samples were placed on a plastic grid at 25 mm from the bottom of the vessel being at least 15 mm apart from each other. A screwed cap and a rubber gasket ring closed the vessels. An inlet valve with a delivery tube was used for gas injection and solution test transferring. An outlet valve was used for internal pressure relief and gas washing. The vessels and gas-washing bottle were placed inside a laboratory fume hood equipped with a hydrogen sulfide detector. The test solution was purged of air with $N_{2(g)}$ 99.999% for 1 h at 100 mL/min per liter of test solution prior to transferring the test solution into the test vessel. Then, 8.0 L of test solutions was transferred by positive pressure of $N_{2(g)}$. The test vessel containing the test solution was purged with $N_{2(g)}$ for 1 h at 100 mL/min per liter of test solution and then saturated with $H_2S_{(g)}$ 99.9% at the same flow rate so as to ensure that the test solution

TABLE 4: Composition of solution 284-B adjusted to pH 8.2 with NaOH.

Compounds (g L^{-1})	NaCl	Na$_2$SO$_4$	MgCl$_2$·6H$_2$O	CaCl$_2$	SrCl$_2$·6H$_2$O
	24.53	4.09	11.11	1.16	0.042
	KCl	NaHCO$_3$	KBr	H$_3$BO$_3$	NaF
	0.695	0.201	0.101	0.027	0.003

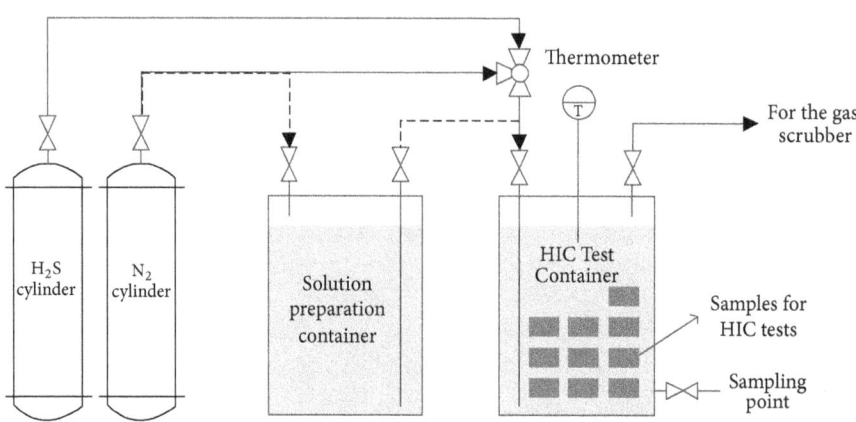

FIGURE 1: Hydrogen-induced cracking test apparatus.

remains saturated with H$_2$S$_{(g)}$ for 96 hours of testing. The tests were performed at a temperature of 24°C ± 3°C. At every 24 h, an aliquot of the test solution was collected in order to measure pH and H$_2$S$_{(g)}$ concentration, with the former being by iodometric titration. According to standard test method [8], the standard pH range for 177-A solution was 2.6 to 2.8 at the beginning of the test and <4.0 at the end of the test; for 177-B, the range was set at 3.4 to 3.6 at the beginning and <4.0 at the end; for 284-B the range is from 8.1 to 8.3 at the beginning and 4.8 to 5.4 at the end of the test. In this work, pH values and H$_2$S$_{(g)}$ concentrations (≥2,300 ppm) were in good agreement with the aforementioned NACE standard. After 96 h, the desulfurization process was initiated by bubbling with N$_{2(g)}$. The test vessel was drained and opened, and the samples were carefully washed in deionized water, dried under stream of N$_{2(g)}$, and introduced into a desiccator containing silica, which was connected to a vacuum pump.

Scanning electron microscopy (SEM) analyses were carried out on surfaces and cross sections of samples with a Jeol 6350 scanning electron microscope. Local chemical composition was determined by X-ray dispersive energy spectrometry (EDS) using a Thermo C10015 probe controlled by the Noran System SIX software. X-ray diffraction (XRD) analyses were carried out on corroded surfaces using the Bruker D8 Advance Eco diffractometer equipped with the DIFFRAC.EVA software. In order to study the temporal evolution of corrosion products in a longer immersion time, a further experiment was performed, thus exposing a new sample to solution 284-B for 35 days at room temperature. In this case, by following the preparation and transfer stages of solution 284-B, the test vessel containing the test solution was purged with N$_{2(g)}$ for 1 h at 100 mL/min per liter of 284-B

and then saturated with H$_2$S$_{(g)}$ 99.9% for 1 h at the same flow rate (i.e., nonstationary conditions).

3. Results and Discussion

API 5L X52MS carbon steel shows the ferritic-bainitic matrix with high grain refinement and random inclusion distribution (Figure 2(a)). On the other hand, API 5L B carbon steel shows a ferrite-pearlite microstructure and presence of banding (Figure 2(b)).

Cross-sectional analyses of the API 5L X52MS samples have not revealed the occurrence of cracks in any of the test solutions (Table 5), which is probably due to a low concentration of alloying elements with subsequent low segregation and banding of the microstructure. The API 5L B samples obtained high values of CLR and CTR for solution 177-A, that is, 40% higher than those observed in solution 177-B on average, and no cracks were found in solution 284-B, as shown in Table 6.

These results are directly associated with hydrogen permeation throughout steel. The concentration of hydrogen ions decreases as the pH and deposited film thickness increase, which can hinder hydrogen diffusion and penetration into steel [6, 11]. Cracks can be observed in the cross sections of samples exposed to solutions 177-A and 177-B, respectively, in Figures 3 and 4.

SEM in Figure 5(a) shows cracks towards the same direction, that is, parallel to the sample surface. Figure 5(b) shows details of crack nucleation from the inclusion of MnS into steel B. Figure 5(c) shows the results of the EDS analysis of a region indicated by an arrow in Figure 5(b). Cracks were mainly developed during the elongated inclusions of MnS, as observed in Figure 5(b). The composition of the

TABLE 5: HIC test results for the API 5L X52MS steel.

	177-A				177-B				284-B		
CP	CLR	CTR	CSR	CP	CLR	CTR	CSR	CP	CLR	CTR	CSR
1	0	0	0	7	0	0	0	12	0	0	0
2	0	0	0	8	0	0	0	13	0	0	0
3	0	0	0	9	0	0	0	14	0	0	0
4	0	0	0	10	0	0	0	15	0	0	0
5	0	0	0	11	0	0	0	16	0	0	0
Med	0	0	0	Med	0	0	0	Med	0	0	0

TABLE 6: HIC test results for the API 5L B steel.

	177-A				177-B				284-B		
CP	CLR	CTR	CSR	CP	CLR	CTR	CSR	CP	CLR	CTR	CSR
16	26%	3%	1%	21	29%	3%	1%	26	0	0	0
17	35%	4%	1%	22	27%	3%	1%	27	0	0	0
18	48%	4%	2%	23	32%	4%	1%	28	0	0	0
19	35%	5%	2%	24	16%	2%	0%	29	0	0	0
20	46%	7%	3%	25	32%	1%	0%	30	0	0	0
Med	38%	5%	2%	Med	27%	2%	1%	Med	0	0	0

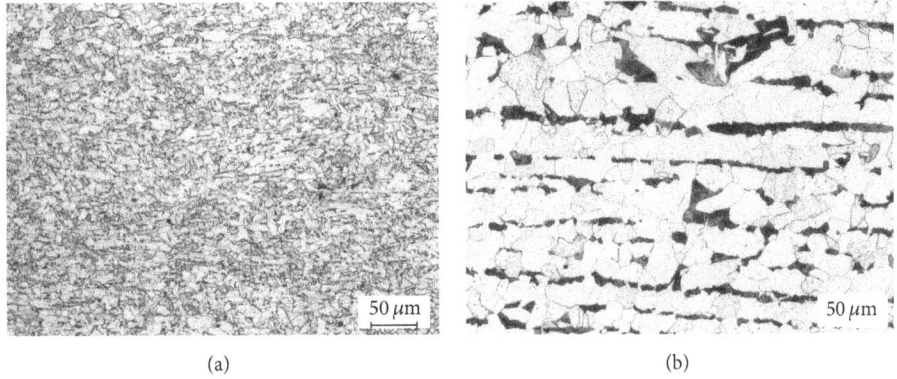

(a)   (b)

FIGURE 2: Microstructure of the central region of the API 5L steel plates: (a) X52MS and (b) B (metallographic etchant: 3.0% nital solution).

FIGURE 3: Stepwise cracks in the cross section of the API 5L B sample exposed to solution 177-A.

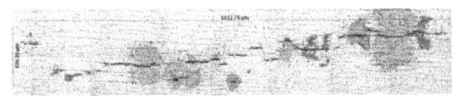

FIGURE 4: Stepwise cracks in the cross section of the API 5L B sample exposed to solution 177-B.

inclusion at the crack tip in Figure 5(b) is presented in the EDS spectrum of Figure 5(c), in which S and Mn peaks can be clearly observed.

It is well known that the cracking in H_2S environments is associated with the presence of nonmetallic inclusions, such as MnS and Ti or V carbonitrides, and a banded structure. The heterogeneous distribution of these inclusions also favors cracking due to the higher concentration of hydrogen traps in some regions. Segregations associated with high levels of P and S in steel contribute to the accumulation of hydrogen in the material [7, 12–16].

The surface analysis has revealed that, for the lowest pH solution (177-A solution), there was no film formation on the sample, as shown in Figures 6(a) and 6(b) and confirmed by the EDS analysis of region 3, Figure 6(c), in which only peaks associated with Fe and Mn are observed. The carbon steel exhibited generalized corrosion at low pH, probably due to high solubility of iron sulfide and preferential attack of ferritic phase [6, 7]. In solution 177-B, a discontinuous layer with some degree of crystallinity was observed on a region of the surface shown in Figures 6(d) and 6(e). The EDS analysis

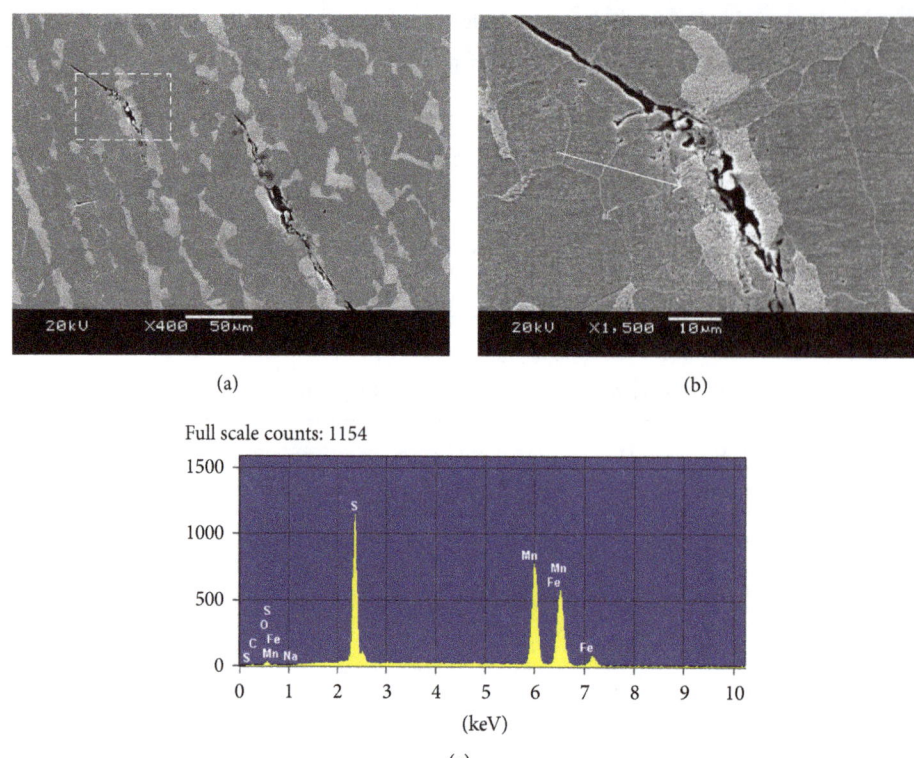

FIGURE 5: Micrographs showing details of crack nucleation from the inclusion of MnS into the API 5L B steel: (a) 400x and (b) 1500x; (c) EDS analysis of the region indicated by an arrow at (b) micrograph.

indicates sulfur-rich sulfides (nonstoichiometric FeS) in the crystalline region depicted in Figure 6(f). Similar surface characteristics were reported in another work on low-carbon steels in 5.0% NaCl solution saturated with H_2S and pH 4.2 [17]. It was not possible to identify any iron sulfides by the XRD analysis due to a small amount of corrosion products.

In solution 284-B at a higher pH, there was a considerable amount of insoluble sulfides resulting from the saturation step with H_2S, as shown in Figures 6(g) and 6(h) and confirmed by the EDS of region 3 in Figure 6(i). This thick film probably provides greater resistance to diffusion of ionic species, thus hindering the corrosion process in comparison with discontinuous films formed at a lower pH. The XRD surface analysis has only identified corrosion products formed in solution 284-B. The diffractogram of corrosion products obtained after 96 h of exposure (Figure 7) shows peaks at approximately 18°, 30°, 35°, 38°, 50°, and 54° which are the same as those of mackinawite (M), that is, a tetragonal crystalline form of iron sulfide [18]. A similar diffractogram has been reported by Zheng et al. [19] and Zhou et al. [20].

Several iron sulfide polymorphs have been identified within oil and gas pipelines. The most cited monosulfides are tetragonal mackinawite, hexagonal troilite, monoclinic pyrrhotite, and hexagonal pyrrhotite, all which are relatively stable in oxygen-free acid media [21–23]. Corrosion tests at room temperature using artificial sea water or any other salt solutions saturated with hydrogen sulfide often lead to one or two types of iron sulfides, that is, mackinawite (kinetically favored product) and pyrrhotite (thermodynamically favored product).

In order to study the temporal evolution of corrosion products, another corrosion test was performed in a longer immersion time in solution 284-B. The diffractogram of corrosion products obtained after 35 days of exposure only showed mackinawite peaks (Figure 8). When compared to Figure 7, mackinawite peaks increase and iron peaks decrease, which is probably due to surface coverage and the sulfide layer thickening during immersion time.

4. Conclusion

Cracking resistance seems to be directly associated with the steel microstructure and medium acidity. API 5L B and X52MS steels showed generalized corrosion in solution 177-A. Both steels formed a discontinuous layer when exposed to solution 177-B. Only the API 5L B steel failed the HIC test and exhibited a larger crack length ratio in solution 177-A. The SEM-EDS analysis reveals that cracks were initiated in the MnS inclusions. In solution 284-B, a thick mackinawite film was formed on both steels, and no crack in their microstructures was observed. When the API 5L X52MS steel was exposed to a longer immersion time, mackinawite was the only corrosion product obtained from the process.

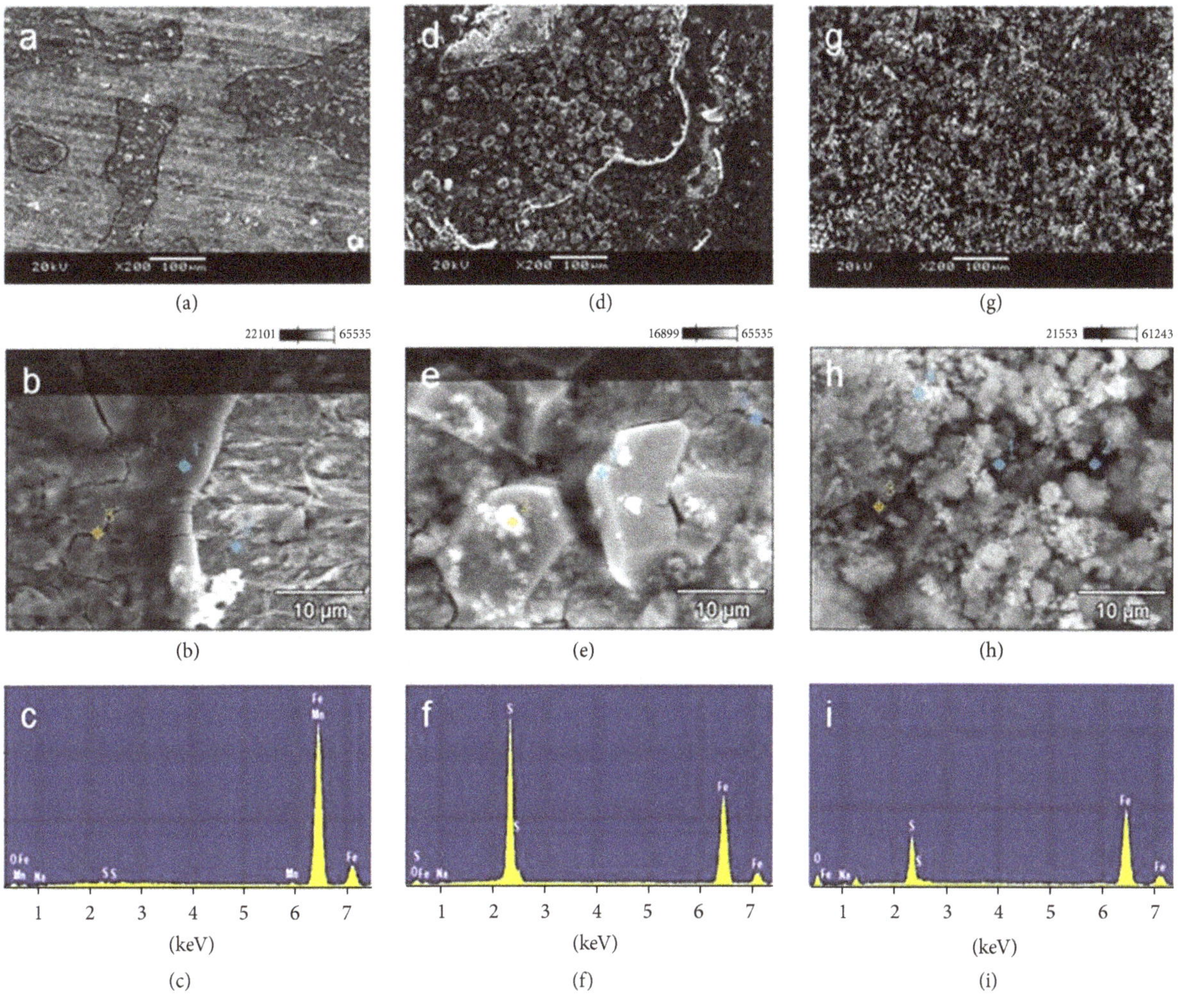

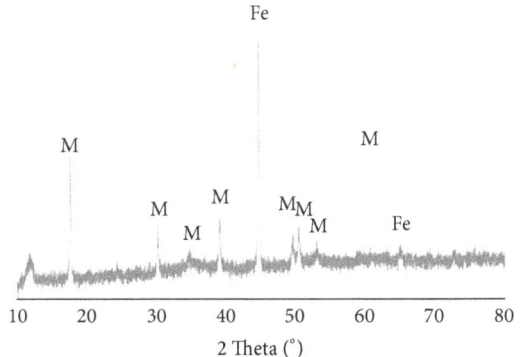

FIGURE 6: SEM and EDS analyses of corrosion products in the API 5L X52MS steel. This steel exposed to 177-A solution: (a) 200x, (b) 3000x, and (c) EDS. This one exposed to 177-B solution: (d) 200x, (e) 3000x, and (f) EDS. This one exposed to 284-B solution: (g) 200x, (h) 3000x, and (i) EDS.

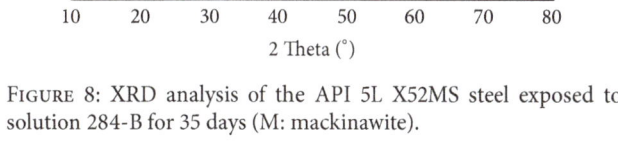

FIGURE 7: XRD analysis of the API 5L X52MS steel exposed to solution 284-B for 96 h (M: mackinawite).

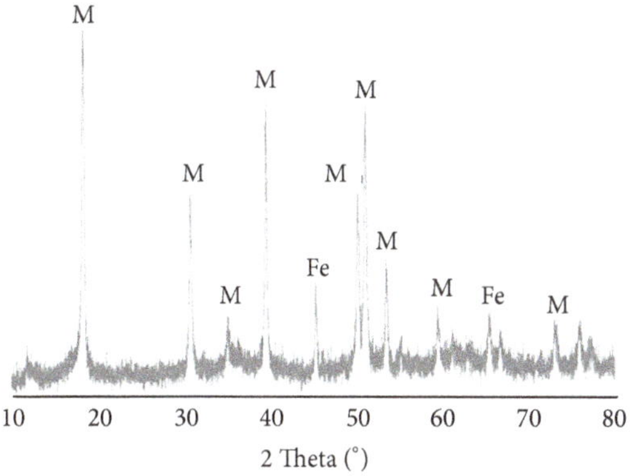

FIGURE 8: XRD analysis of the API 5L X52MS steel exposed to solution 284-B for 35 days (M: mackinawite).

Conflicts of Interest

The authors declare that they have no conflicts of interest.

Acknowledgments

The authors would like to thank PROEX/UNESP, CAPES, and FAPESP (Process 2017/11361-5) for the financial support.

References

[1] IEA, *Resources to Reserves 2013 - Oil, Gas and Coal Technologies for the Energy Markets of the Future*, International Energy Agency, Paris, France.

[2] M. A. Mohtadi-Bonab, J. A. Szpunar, R. Basu, and M. Eskandari, "The mechanism of failure by hydrogen induced cracking in an acidic environment for API 5L X70 pipeline steel," *International Journal of Hydrogen Energy*, vol. 40, no. 2, pp. 1096–1107, 2015.

[3] ISO, "Petroleum and Natural Gas Industries - Materials for use in H_2S-Containing Environments in Oil and Gas Production - Part 1," Tech. Rep. ISO 15156-1, International Organization for Standardization, Switzerland, 2009.

[4] D. Talbot, *Corrosion Science and Technology*, CRC Press, New York, NY, USA, 2007.

[5] B. P. Tissot and D. H. Welte, *Petroleum Formation and Occurrence*, Springer, Berlin, Germany, 2nd edition, 1984.

[6] C. Zhou, S. Zheng, C. Chen, and G. Lu, "The effect of the partial pressure of H2S on the permeation of hydrogen in low carbon pipeline steel," *Corrosion Science*, vol. 67, pp. 184–192, 2013.

[7] J. Sojka, M. Jérôme, M. Sozańska, P. Váňová, L. Rytířová, and P. Jonšta, "Role of microstructure and testing conditions in sulphide stress cracking of X52 and X60 API steels," *Materials Science and Engineering: A Structural Materials: Properties, Microstructure and Processing*, vol. 480, no. 1-2, pp. 237–243, 2008.

[8] NACE International, "Evaluation of Pipeline and pressure vessel steel for resistance to hydrogen-induced crack," Tech. Rep. ANSI/NACE TM0284, NACE International, Houston, Tex, USA, 2017.

[9] NACE International, "Standard test method laboratory testing of metals for resistance to sulfide stress cracking and stress corrosion cracking in H_2S environments," Tech. Rep. ANSI/NACE TM0177, NACE International, Houston, Tex, USA, 2016.

[10] API 5L, "Specification for Line Pipe," Tech. Rep., API Publishing Services, Washington, Wash, USA, 2012.

[11] K. Kobayashi, T. Omura, M. Okatsu, T. Hara, and N. Ishikawa, "Proposal of HIC test solution with buffer capacity in NACE TM0284," in *Proceedings of the NACE - International Corrosion Conference Series*, Orlando, Fla, USA, 2013.

[12] F. Huang, J. Liu, Z. Deng, J. Cheng, Z. Lu, and X. Li, "Effect of microstructure and inclusions on hydrogen induced cracking susceptibility and hydrogen trapping efficiency of X120 pipeline steel," *Materials Science and Engineering: A Structural Materials: Properties, Microstructure and Processing*, vol. 527, no. 26, pp. 6997–7001, 2010.

[13] L. Gan, F. Huang, X. Zhao, J. Liu, and Y. F. Cheng, "Hydrogen trapping and hydrogen induced cracking of welded X100 pipeline steel in H_2S environments," *International Journal of Hydrogen Energy*, vol. 43, no. 4, pp. 2293–2306, 2018.

[14] M. Elboujdaini and R. W. Revie, "Metallurgical factors in stress corrosion cracking (SCC) and hydrogen-induced cracking (HIC)," *Journal of Solid State Electrochemistry*, vol. 13, no. 7, pp. 1091–1099, 2009.

[15] A. C. Palmer and R. A. King, *Subsea Pipeline Engineering*, PennWell, Tulsa, Okla, USA, 2008, Subsea Pipeline Engineering.

[16] M. A. Mohtadi-Bonab and M. Eskandari, "A focus on different factors affecting hydrogen induced cracking in oil and natural gas pipeline steel," *Engineering Failure Analysis*, vol. 79, pp. 351–360, 2017.

[17] P. Bai, S. Zheng, and C. Chen, "Electrochemical characteristics of the early corrosion stages of API X52 steel exposed to H_2S environments," *Materials Chemistry and Physics*, vol. 149, pp. 295–301, 2015.

[18] J. Ning, Y. Zheng, B. Brown, D. Young, and S. Nesic, "Construction and verification of pourbaix diagrams for hydrogen sulfide corrosion of mild steel," in *Proceedings of the NACE - International Corrosion Conference Series*, Dallas, Tex, USA, 2015.

[19] S. Zheng, C. Zhou, X. Chen, L. Zhang, J. Zheng, and Y. Zhao, "Dependence of the abnormal protective property on the corrosion product film formed on H2S-adjacent API-X52 pipeline steel," *International Journal of Hydrogen Energy*, vol. 39, no. 25, pp. 13919–13925, 2014.

[20] C. Zhou, X. Chen, Z. Wang, S. Zheng, X. Li, and L. Zhang, "Effects of environmental conditions on hydrogen permeation of X52 pipeline steel exposed to high H_2S-containing solutions," *Corrosion Science*, vol. 89, no. C, pp. 30–37, 2014.

[21] S. N. Smith, B. Brown, and W. Sun, "Paper 11081 of NACE International," in *Proceedings of the Paper 11081 of NACE International*, Houston, Tex, USA, 2011.

[22] B. Craig, "Corrosion product analysis - a road map to corrosion in oil and gas production," *Materials Performance*, vol. 41, no. 8, pp. 56–58, 2002.

[23] S. N. Smith, "Corrosion product analysis in oil and gas pipelines," *Materials Performance*, vol. 42, no. 8, pp. 44–47, 2003.

7

Experimental Investigation into Corrosion Effect on Mechanical Properties of High Strength Steel Bars under Dynamic Loadings

Hui Chen[ID],[1,2] Jinjin Zhang[ID],[1] Jin Yang,[1] and Feilong Ye[1]

[1]*Department of Building Engineering, Oujiang College, Wenzhou University, Wenzhou 325035, China*
[2]*Department of Structural Engineering, Tongji University, Shanghai 200092, China*

Correspondence should be addressed to Hui Chen; chenhui0306@wzu.edu.cn

Academic Editor: Ramazan Solmaz

The tensile behaviors of corroded steel bars are important in the capacity evaluation of corroded reinforced concrete structures. The present paper studies the mechanical behavior of the corroded high strength reinforcing steel bars under static and dynamic loading. High strength reinforcing steel bars were corroded by using accelerated corrosion methods and the tensile tests were carried out under different strain rates. The results showed that the mechanical properties of corroded high strength steel bars were strain rate dependent, and the strain rate effect decreased with the increase of corrosion degree. The decreased nominal yield and ultimate strengths were mainly caused by the reduction of cross-sectional areas, and the decreased ultimate deformation and the shortened yield plateau resulted from the intensified stress concentration at the nonuniform reduction. Based on the test results, reduction factors were proposed to relate the tensile behaviors with the corrosion degree and strain rate for corroded bars. A modified Johnson-Cook strength model of corroded high strength steel bars under dynamic loading was proposed by taking into account the influence of corrosion degree. Comparison between the model and test results showed that proposed model properly describes the dynamic response of the corroded high strength rebars.

1. Introduction

Structure deterioration induced by corrosion of reinforcing bars is one of the major problems in civil engineering. The corrosion of reinforcing bars (rebars) not only leads to the cracking of the concrete cover, but also causes the serious damage of Reinforced concrete (RC) structures. Therefore, investigation of the deterioration of mechanical properties of corroded steel bars is crucial for predicting the serviceability and durability of RC structures [1]. The static tensile test results of corroded rebars have shown that, with the development of the corrosion degree, the nominal yield and ultimate strengths and the ultimate strain of a corroded rebar decrease. Meanwhile, the yield plateau shortens or even disappears [2, 3]. Empirical formulas have been proposed to evaluate the yield and ultimate strengths of corroded reinforcing bars [4, 5], and the mathematical models of stress-strain relationship for corroded rebars in different environment condition have also been established [6–10].

RC structures may suffer impact or explosion loads during their service life, such as bridge and underground protective structures [11]. The strain rate of rebars in concrete structures may reach $10 \, s^{-1}$ under impact loading. Evident effects of strain rates on the mechanical properties of rebars are found in their tensile tests under a high strain rate [12]. Both their yield and ultimate strengths increase when the strain rate is increased, but the yield strength has a more obvious increment than the ultimate strength [13, 14]. The test results of corroded medium and low strength rebars under dynamic loadings have indicated that the strain rate effect decreased with the increase of corrosion degree [15]. In recent years, more and more high strength steel bars are being used in modern RC structure because of their good strength and ductility combination. However, the effect of corrosion and strain rate on the mechanical properties of high strength rebars has not been discussed.

With the development of computer technology, finite element simulation has played a key role, especially in the

TABLE 1: Chemical composition (by mass[a]) of steel bar HRB500 (%).

C	Si	Mn	P	S	Cr	Mo	V	Ni	Cu
0.24	0.27	1.44	0.03	0.02	0.08	<0.02	0.10	0.022	0.02

[a]The balance was Fe.

engineering area. An ideal constitutive model should be able to precisely predict the behaviors of materials under different loading conditions, such as static and dynamic modes and low and high strains, as many engineering materials behave significantly different under different loading conditions. A variety of rate dependent constitutive models have been proposed for high strain rate metal materials, such as Johnson-Cook (J-C) model and Cowper-Symonds model. Among them, The J-C constitutive model is one of the most widely used models because it performs in a simple yet effective manner [16]. As a semiempirical model, it describes the behaviors of plastic materials at high strains, high strain rates, and high temperatures. However, the existing J-C constitutive model does not consider the coupled effect of strain rate and corrosion degree on flow stress, causing limited capability of predicting material properties.

The intention of this paper is to experimentally investigate the effect of corrosion on mechanical properties of corroded high strength steel. A comprehensive test program is designed to observe, monitor, and evaluate corrosion behavior of corroded high steel bars and its effect on their mechanical properties in concrete under static and dynamic loading. The high strength rebars with different corrosion degrees were got by the impressed current method in concrete. To investigate the impact of corrosion degree on its mechanical properties under dynamic loading, tensile tests were conducted on the corroded high strength rebars by using a high strain rate test system. The mechanical performance characteristics (yield strength, ultimate strength and ultimate strain, etc.) were examined. Based on the test results, reduction factors were proposed to relate the tensile behaviors with the corrosion degree and strain rate for corroded bars and a modified model of Johnson-Cook strength for corroded high strength steel bars under dynamic loading was proposed by taking into account the influence of corrosion. The results produced from the tests can contribute to the body of knowledge of corrosion behavior and its effect on mechanical properties of corroded rebars under dynamic loading, which can be used to predict mechanical behavior of corroded RC structures under impact loading.

2. Experimental Program

2.1. Materials and Specimens. The high strength rebars HRB500 commonly used in China were used to investigate the impact of strain rate and corrosion on the mechanical behavior of rebars under static and dynamic loading. HRB500 rebars refer to the hot rolled ribbed rebar with the yield strength not less than 500 MPa. Their original diameters were 25 mm.

As is well known, the mechanical properties of metal are affected by its chemical composition, morphology, and

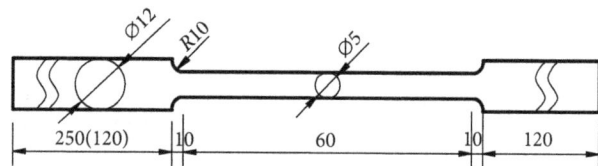

FIGURE 1: Dimension of specimens (Unit: mm, not to scale).

FIGURE 2: Preparation of RC slabs for the accelerated corrosion process.

microstructure which vary significantly. The chemical composition of HRB500 steel bar was tested before accelerated corrosion test, and the results are shown in Table 1. The standard specimens were made according to the ASTM-E8-09 standard [17] as shown in Figure 1, the reduced section was 5 mm in diameter and 60 mm in length, and the section of grip part was 12 mm in diameter. The length of grip part for strain rates ranging from $0.0002\,s^{-1}$ to $0.1\,s^{-1}$ was 70 mm and for strain rates ranging from $2\,s^{-1} \sim 50\,s^{-1}$ was 250 mm and 120 mm at two ends, respectively. Both ends of each rebar were extended 50 mm beyond the desired length for the connection with electric wires.

The grip parts of the specimens, which would not be expected to corroded, were wrapped by epoxy. Then, rebar specimens were embedded in concrete as shown in Figure 2, and the reduced sections of specimens were corroded by the impressed current method. The constituents of the concrete used for the preparation of the corroded high strength specimens consisted of ordinary Portland cement, sand with a maximum diameter of 5 mm, coarse aggregate with a maximum diameter of 15 mm, and tap water. Cubic samples with the dimension of $150 \times 150 \times 150\,mm^3$ were prepared. Mix proportion of the concrete mixtures is shown in Table 2. The 28-day axial compressive strength of concrete was found to be 29.5 MPa on an average.

2.2. Accelerated Corrosion Test. The concrete slab specimens of high strength rebars were cured in a natural indoor

TABLE 2: Mix proportion of the concrete mixtures.

w/c	Cement/kg/m³	Sand/kg/m³	Coarse aggregate/kg/m³	Water/kg/m³
0.50	674	506	1031	337

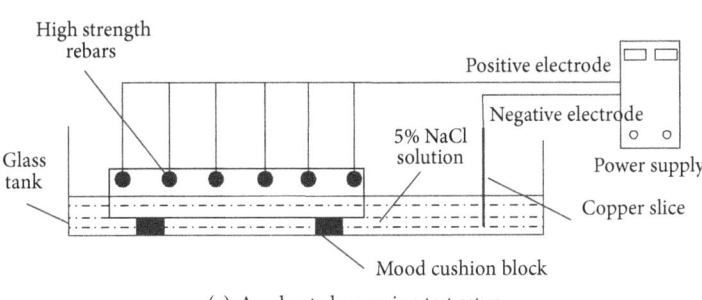

(a) Accelerated corrosion test setup

(b) Accelerated corrosion test site

FIGURE 3: Accelerated corrosion test.

environment and room temperature for 28 days after casting. Then they were placed into glass tanks and partially immersed in 5% NaCl solution. The solution level in the tanks was controlled to be about 30 mm lower than the lower surface of the rebars, as shown in Figure 3. In the previous studies on mechanical properties of corroded rebars, accelerated corrosion by means of the impressed current technique is widely used to obtain corroded rebars because of its advantages in terms of time and cost, and its feasibility to control the rate of corrosion [18, 19]. There are some differences between accelerated corrosion and natural corrosion in corrosion products, distribution of corrosion expansion force, and distribution of bond stress [20]. However, when the implied current density is less than $200\,\mu A/cm^2$, the differences between naturally corroded rebars and artificially corroded rebars could be neglected [15]. The rebars subject to both natural corrosion and accelerated corrosion showed the similar degradation of mechanical properties [3]. A constant current was applied during the corrosion process to generate a constant current density of $100\,\mu A/cm^2$. Based on Faraday's law, the total current required was calculated based on the respective steel surface area and was checked regularly to adjust for any drift. The targeted average corrosion degrees of the specimens were listed in Table 3. Based on Faraday's law of induction [21], as given in (1), the required time for the prescribed corrosion degrees of 0.05, 0.10, 0.15, 0.20, and 0.25 was 25 days, 50 days, 75 days, 100 days, and 125 days, respectively. Corrosion current was measured every day in the first week of the test and then weekly until the end of tests:

$$t_c = \frac{ZF \cdot r \cdot \rho \eta_{sp}}{86400 \cdot 2A_{Fe} \cdot i},\qquad(1)$$

where A_{Fe} is the atomic mass of iron ($A = 56\,g$), t_c is corrosion duration (days), Z is the valence of iron, which is 2 in this case (iron), F is Faraday's constant (96500 A·s), i is the density of impressed current (A/cm²), ρ is the mass density of iron (g/cm³), r is the radius of to-be-corroded rebars (mm), and η_{sp} is the targeted average corrosion degree.

TABLE 3: Test results of corroded high strength steel bars.

Specimen	η_s	$\eta_{s,max}$	f_{yc} (MPa)	f_{uc} (MPa)
HS00-00	0	0	499.11	639.16
HS05-00	0.05	0.064	469.06	597.91
HS15-00	0.14	0.171	412.53	516.93
HS20-00	0.19	0.231	395.21	496.05
HS25-00	0.25	0.288	361.60	463.97
HS00-0.01	0	0	528.65	666.67
HS00-0.1	0	0	541.89	676.85
HS00-2	0	0	573.47	696.21
HS00-10	0	0	592.31	706.90
HS00-50	0	0	611.66	727.78
HS05-10	0.054	0.064	562.26	664.63
HS15-10	0.145	0.174	503.18	580.09
HS20-10	0.201	0.241	479.25	557.17
HS25-10	0.247	0.296	444.10	519.99
HS10-00	0.098	0.118	439.01	564.81
HS10-0.01	0.095	0.114	470.59	589.76
HS10-0.1	0.097	0.116	472.12	609.12
HS10-2	0.104	0.124	496.56	611.66
HS10-10	0.096	0.115	522.54	625.92
HS10-50	0.105	0.126	528.14	644.77

Notes. (1) The number of a specimen, HSAA-BB, represents high strength steel rebar HRB500 with the targeted amount of corrosion, $\eta_{sp} = 0$. AA under strain rate of BBs^{-1}; (2) η_s is the average corrosion degree. $\eta_{s,max}$ is the maximum corrosion degree. f_{yc} and f_{uc} are the corresponding yielding and ultimate loads measured during the tensile tests of the corroded high strength rebars (MPa), respectively.

After the completion of each corrosion test, the slab was broken, and the corresponding rebars were then collected. The accelerated corroded rebars were cleaned using a hydrochloric acid solution and dried according to ASTM G1-03 standard [22], as shown in Figure 4.

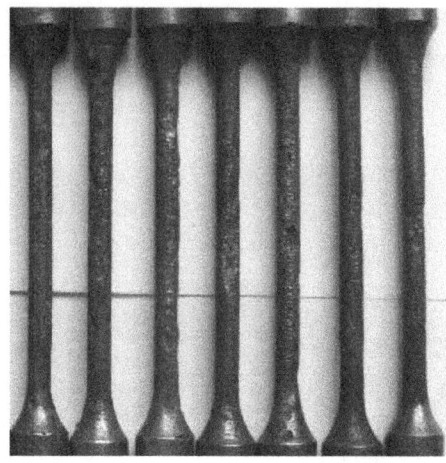

FIGURE 4: Corroded rebars after the accelerated corrosion process.

FIGURE 5: Three-dimensional laser scanner.

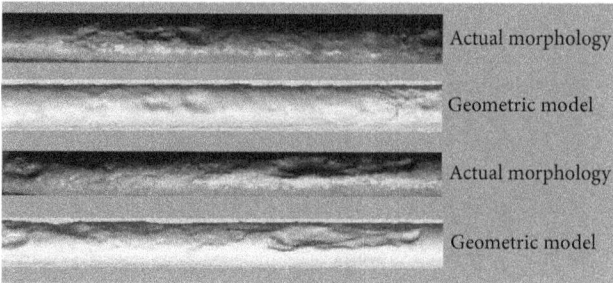

FIGURE 6: Actual corroded rebar and its 3D geometric model.

2.3. Measurement of Corrosion Degree. After acid cleaning and drying, geometric models of all corroded rebars were built using a 3D laser scanner (Creaform) with high measuring accuracy, convenient operation, automatic reconstruction of 3D objective shapes, and high repeatability, as shown in Figure 5. Figure 6 shows a comparison between a photo of a corroded rebar after acid cleaning and the corresponding 3D geometric model, which indicates that the accuracy of the 3D laser scanning (i.e., 0.04 mm) was satisfactory.

Through observation of the virtual models, it was clear that the corrosion was not evenly distributed circumferentially and longitudinally, as seen in Figure 6. ProEngineer software was used to obtain the areas of the discrete cross sections along the longitudinal axis of each corroded rebar at intervals of 1 mm. Accordingly, the average corrosion degree and the maximum corrosion degree were shown in Table 3.

Figures 7(a) and 7(b) indicate some typical profiles at different corrosion levels. The profiles of these cross sections were obtained from the orthographic projections of the intersection line between the cutting plane and the rebar surface. The inner irregular curve represents the profile of corroded cross sections, while the external thin circle indicates the outline of the sound steel bars. The corrosion degree was marked in the figures. The numbers on the top and below in Figures 7(a) and 7(b) represent the maximum corrosion

degree and the average corrosion degree, respectively. The maximum corrosion degree ranged from 0.064 to 0.296 for corrosion pit. From Figure 7, it can be seen that the corrosion penetration is far from circumferentially uniform. Through visual observation, the different pit shapes could be wide and shallow, elliptical, or even undercutting.

2.4. Tensile Tests of Corroded High Strength Rebars under Static and Dynamic Loading. After evaluating corrosion levels, the corroded bars were subjected to displacement-controlled tensile testing. The tensile tests for specimens were completed on an Instron VHS160/100-20 testing machine with strain rates of $2\,s^{-1} \sim 50\,s^{-1}$, in which the deformation of a rebar was measured by using laser displacement meters with a data acquisition frequency of 1000 Hz, as shown in Figure 8(a). The tensile tests for specimens were completed on a Zwick electronic testing machine with strain rates of $0.0002\,s^{-1} \sim 0.1\,s^{-1}$. An extensometer with a data acquisition frequency of 100 Hz was installed in the middle of a rebar specimen, with the gauge length of 50 mm, as shown in Figure 8(b). The static and dynamic loading material properties (elastic modulus, nominal yield stress, and ultimate tensile stress) were measured according to the ASTM E8 standard on tensile testing equipment. The yield and ultimate loads of the specimens are shown in Table 3.

3. Experimental Results and Discussion

3.1. Failure Modes. The typical failure modes of the specimens are shown in Figure 9; by comparing the broken section of corroded rebars, it was found that there were no obvious differences in the necking zones among the specimens with the same corrosion degree under the static and dynamic loading. However, most of the corroded rebars broke in the minimum cross section. The necking phenomenon disappeared gradually with the increase of the corrosion degree, which means that the deformation ability of the high strength rebar decreased.

3.2. Stress-Strain Curves. The strain of the test bar was determined by dividing the increase of the gauge length by the initial gauge length. Three methods were mainly used in earlier studies to calculate the stress of a corroded steel bar from the measured force, which are based on the average reduced cross-sectional area [3, 5], the original

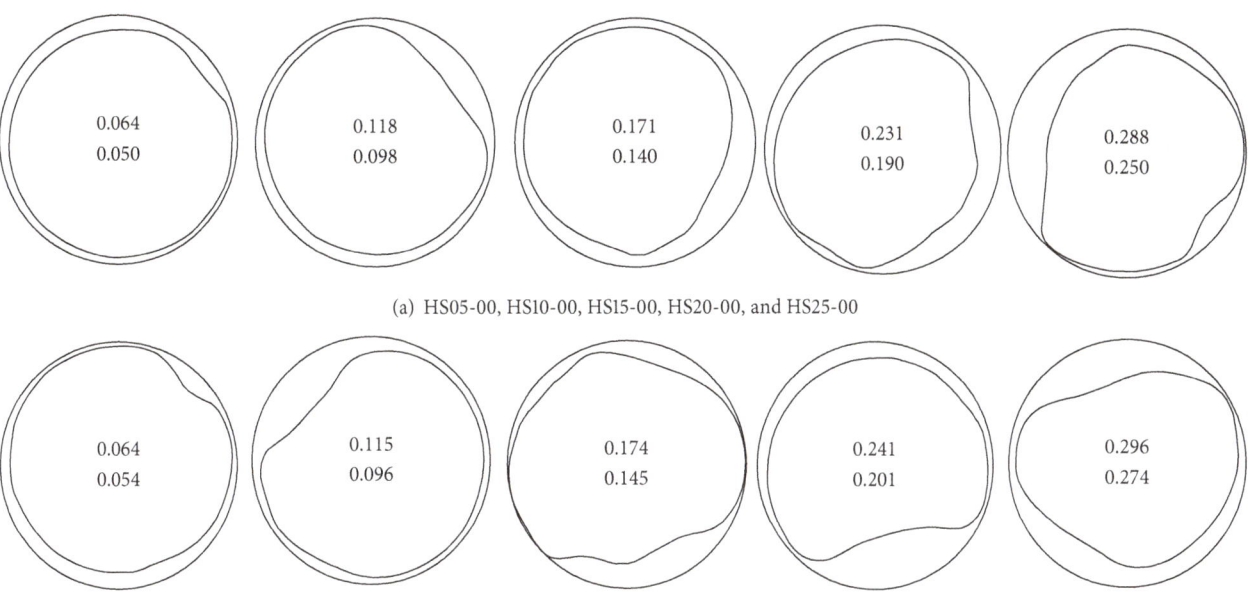

(a) HS05-00, HS10-00, HS15-00, HS20-00, and HS25-00

(b) HS05-10, HS10-10, HS15-10, HS20-10, and HS25-10

FIGURE 7: Typical cross-sectional profiles of bars under accelerated corrosion.

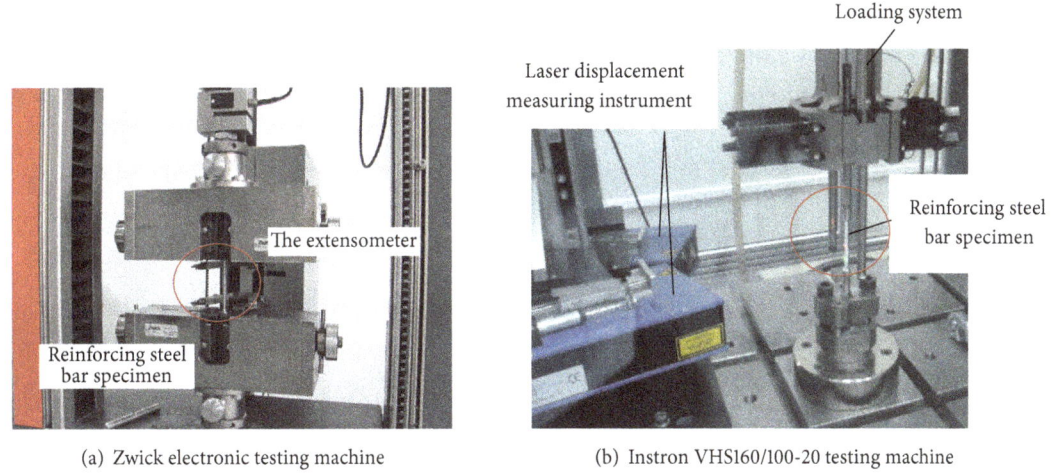

(a) Zwick electronic testing machine

(b) Instron VHS160/100-20 testing machine

FIGURE 8: Tensile test stand and dimension of specimens.

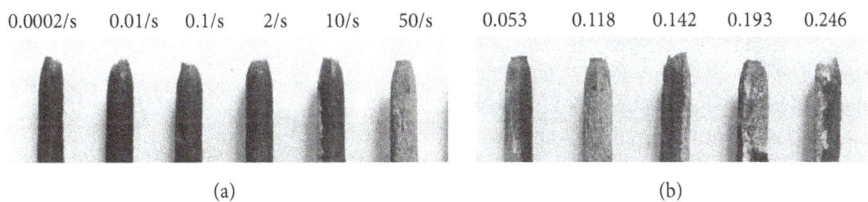

(a) (b)

FIGURE 9: Failure modes of corroded rebars (a) Specimens of HRB 500 with corrosion degree of 0.10 under different strain rates. (b) Specimens of HRB 500 with different corrosion degree under static loads.

cross-sectional area [6], and the smallest cross-sectional area of the bar, respectively [15]. There is no consensus among researchers regarding which method is more appropriate than the others. However, the method based on the original cross-sectional area was used in most of the previous studies. This research adopted it to calculate the stress of corroded bars.

The measured engineering nominal stress-strain curves are presented in Figure 10 and the resulting average true stress-strain curve was got and determined with (2) and (3):

$$\tilde{\varepsilon} = \ln(1 + \varepsilon),$$
$$\tilde{\sigma} = \ln(1 + \sigma), \tag{2}$$

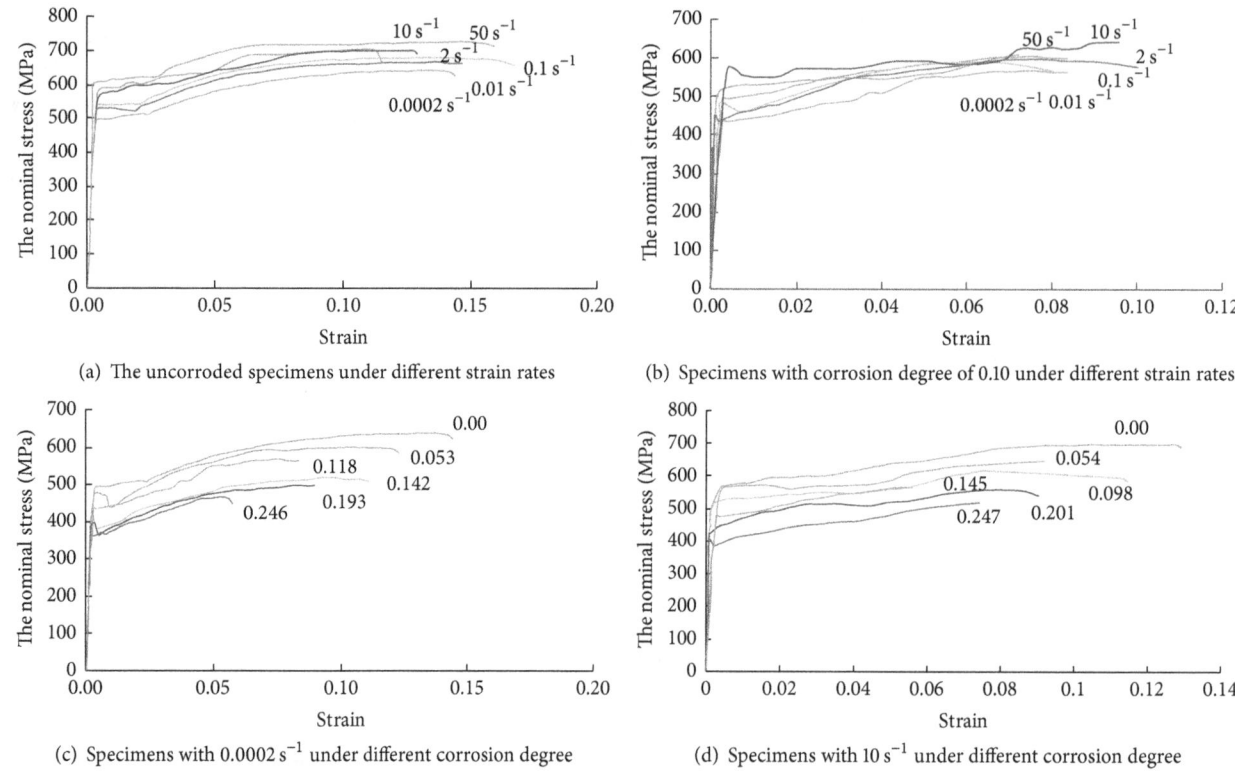

(a) The uncorroded specimens under different strain rates

(b) Specimens with corrosion degree of 0.10 under different strain rates

(c) Specimens with 0.0002 s⁻¹ under different corrosion degree

(d) Specimens with 10 s⁻¹ under different corrosion degree

FIGURE 10: The nominal stress-strain curves.

where ε and σ were the corresponding average engineering strain and stress (MPa), respectively. $\tilde{\varepsilon}$ and $\tilde{\sigma}$ were the true strain and stress (MPa), respectively.

The typical nominal stress-strain curves for specimens with different corrosion degrees are shown in Figures 10(a)–10(d). The number near a curve in Figures 10(a) and 10(b) stands for the strain rate; and the numbers near a curve in Figures 10(c) and 10(d) represent the average corrosion degree, respectively. The yield and ultimate strengths of the specimens are given in Table 3. The results showed that the mechanical properties of corroded high strength steel bars were strain rate dependent. Both the yield and ultimate load increased as the strain rate increased. It can be seen that, with the development of corrosion degree, the nominal yield strength, ultimate strength, and ultimate deformation decreased under dynamic loading, and the yield plateau shortened or even disappeared completely. Both the nominal strength and deformation capacities of corroded high strength bars tended to decrease with increasing corrosion degree. However, the trend was less clear for the deformation capacity than for the nominal strength. This is due to the fact that the nominal yield and strength capacity was mainly related to the minimum cross-sectional area while the deformation capacity was not only related to the minimum cross-sectional area but also related to the shape change along the bar, as shown in Figure 11, which shows the distribution of cross-sectional areas along the length of the bar obtained from the 3D scan for corroded rebars HS15-00 and HS20-00. This is consistent with the findings from Zhang et al. [3, 15], that the decreased nominal yield and ultimate strengths of

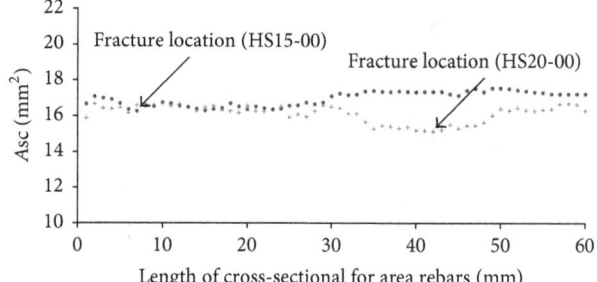

FIGURE 11: Distribution of cross-sectional areas for corroded rebars HS15-00 and HS20-00.

the corroded rebars were mainly caused by the reduction of cross-sectional areas for rebars, and the decreased ultimate deformation and the shortened yield plateau resulted from the intensified stress concentration at the corrosion pitting.

4. Development of the Modified J-C Model for Corroded High Strength Rebars

4.1. Original J-C Model. The normal equation of J-C model is as follows [16]:

$$\sigma_s = \left[A + B\left(\varepsilon^p\right)^n \right] \left[1 + C \ln \frac{\dot{\varepsilon}^p}{\dot{\varepsilon}_1} \right] \left[1 - T^{*m} \right], \qquad (3)$$

where σ_s is the Von Mises equivalent flow stress (MPa), A is the yield stress at a given reference temperature and a given reference strain rate, B is the coefficient of strain hardening, C is the coefficient of strain rate hardening, n is the strain hardening exponent, ε^p is the equivalent plastic strain, and $\dot{\varepsilon}^* = \dot{\varepsilon}/\dot{\varepsilon}_0$ is the dimensionless strain rate with $\dot{\varepsilon}$ being the strain rate and $\dot{\varepsilon}_0$ the reference strain rate. $T^* = (T - T_r)/(T - T_r)(T_m - T_r)$. $(T_m - T_r)$ is the dimensionless temperature, T is the experimental temperature (°C), T_r is the room temperature (°C), and T_m is the melting temperature of the materials (°C).

In (3), the items $[A + B(\varepsilon^p)^n]$, $[1 + C\ln(\dot{\varepsilon}^p/\dot{\varepsilon}_1)]$, and $[1 - T^{*m}]$ are used to describe the work hardening effect, the strain rate effect, and the temperature effect, respectively. The parameter A is a material constant, B is a preexponential factor, n is a strain hardening coefficient, and C and m are strain rate effective factor and temperature effective factor, respectively.

4.2. Development of the Modified J-C Model. Based on the test results of the mechanical properties of corroded high strength rebars under dynamic loading, the nominal stress-strain relationship of the J-C model for corroded rebars under dynamic loading was proposed, as shown in Figure 12. In the proposed model, the average corrosion degree η_s was used, because it can be measured easily in practice and reflect overall corrosion of rebars. The nominal stress of the corroded rebar is the ratio of the applied load and the average corrosion degree, which can be calculated using the average corrosion degree and the original cross-sectional area before corrosion. When the corrosion degree is relatively small, the yield plateau does not disappear, and the J-C model is used. When the corrosion degree exceeds a critical value, the yield plateau $\eta_{s,cr}$ disappears, and the nonlinear model is used. Meanwhile, as the strain rate increases, the yield and ultimate strengths of corroded rebars increase, and the ultimate strain may be assumed to remain unchanged.

Based on the J-C model stress-strain relationship of uncorroded rebars, the stress-strain relationship of J-C model for corroded high strength rebars under dynamic loading can be expressed as

$$\sigma_{sc,d}$$
$$= \begin{cases} E\varepsilon_{sc,d} & 0 < \varepsilon_{sc,d} \le \varepsilon_{syc,d} \\ f_{syc,d} & \varepsilon_{syc,d} < \varepsilon_{sc,d} \le \varepsilon_{shc,d} \\ \left[A + B\left(\varepsilon^p\right)^n\right]\left[1 + C\ln\dfrac{\dot{\varepsilon}^p}{\dot{\varepsilon}_1}\right] & \varepsilon_{shc,d} < \varepsilon_{sc,d} \le \varepsilon_{suc,d}, \end{cases} \tag{4}$$

where $\sigma_{sc,d}$ is the Von Mises equivalent flow stress of the corroded rebars (MPa), A is the yield stress at a given reference strain rate (MPa), B is the coefficient of strain hardening, C is the coefficient of strain rate hardening, n is the strain hardening exponent, ε is the equivalent plastic strain, and $\dot{\varepsilon}^* = \dot{\varepsilon}/\dot{\varepsilon}_0$ is the dimension less strain rate with $\dot{\varepsilon}$ being the strain rate and $\dot{\varepsilon}_0$ the reference strain rate. Here, $\dot{\varepsilon}_0 = 10^{-3}\,\mathrm{s}^{-1}$. $E_{sc,d}$ is Young's modulus for corroded rebars under high strain rates; $\varepsilon_{syc,d}$ is the yielding strain; $\varepsilon_{shc,d}$ is the hardening strain; $\varepsilon_{suc,d}$ is the ultimate strain, which is assumed to be equal to the ultimate strain under static load, $\varepsilon_{suc,d} = \varepsilon_{suc,st}$; $f_{yc,d}$ and

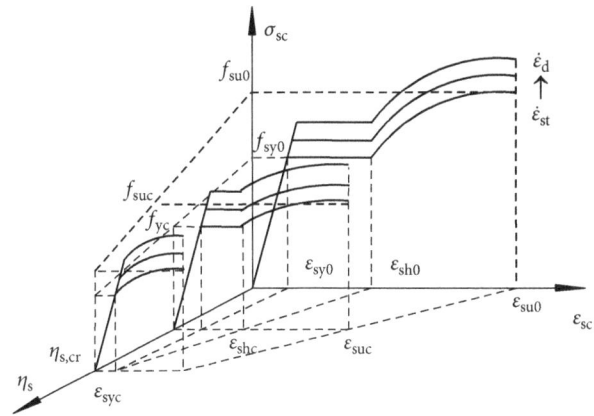

FIGURE 12: Johnson-Cook model of corroded high strength rebars under dynamic loading.

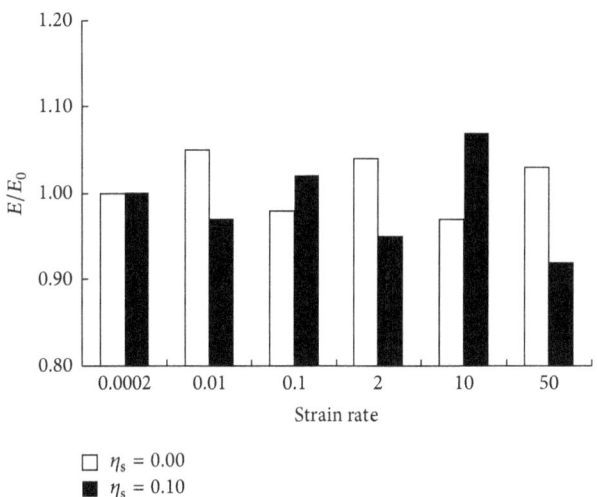

FIGURE 13: Influence of strain rate and corrosion degree on Young's modulus.

$f_{uc,d}$ are the yield and ultimate strengths of corroded rebars under dynamic loading (MPa), respectively.

4.3. Characteristic Parameters of Development of the Modified J-C Model. The relationships of the characteristic parameters, such as Young's modulus, the yield strength A, B, C, n, and the ultimate strain, with the degree of corrosion and the strain rate, can be obtained through appropriate analysis of the test results, which is discussed in the following section.

4.3.1. Young's Modulus. Figure 13 shows the dependence of the relative Young's modulus for rebars on the strain rate and the average corrosion degree. The relative Young's modulus is the ratio of Young's modulus of a corroded rebar to that of a corresponding uncorroded rebar. It can be seen from Figure 13 that Young's modulus was not sensitive to the strain rate or the corrosion degree. For simplicity, it is assumed that Young's modulus keeps constant for corroded rebars under static and dynamic loading. Ignoring the influence

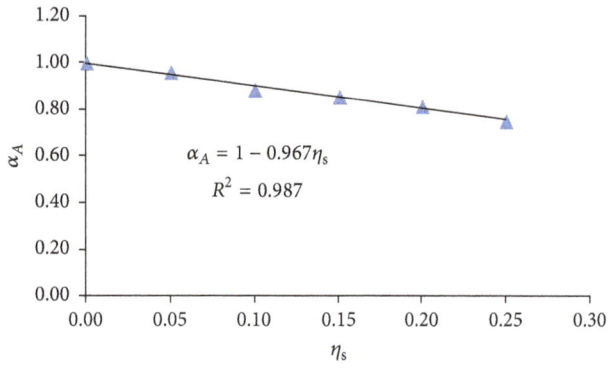

(a) Influence of corrosion degree on A for corroded rebars

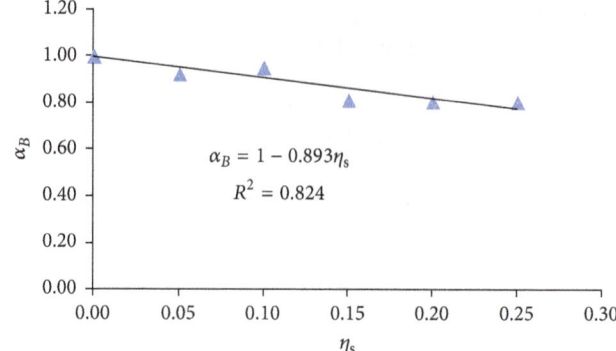

(b) Influence of corrosion degree on B for corroded rebars

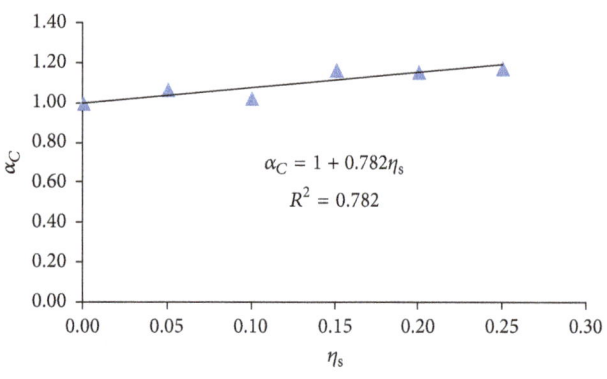

(c) Influence of corrosion degree on C for corroded rebars

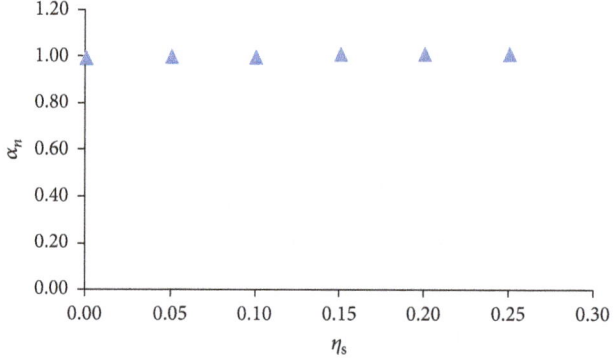

(d) Influence of corrosion degree on n for corroded rebars

FIGURE 14: Influence of corrosion degree on A, B, C, and n for corroded rebars.

TABLE 4: Parameters for J-C model.

Steel	η_s	A (MPa)	B (MPa)	C	n
	0.00	558.49	6.788	0.0113	5.455
	0.05	536.58	6.287	0.0121	5.493
HRB500	0.10	494.63	6.469	0.0116	5.479
	0.15	478.34	5.527	0.0132	5.551
	0.20	456.46	5.489	0.0131	5.554
	0.25	421.11	5.477	0.0133	5.555

of corrosion degree and strain rate on the elastic modulus, HRB500 steel bar takes $E = 2.0 \times 10^5$ MPa.

4.3.2. Plastic Stage Model Parameters. According to the nominal yield strength, ultimate strength, the strain of the corroded steel measured under different strain rates, and the same corrosion degree from the strain rate of $0.0002\,\text{s}^{-1}$, $0.1\,\text{s}^{-1}$, and $10\,\text{s}^{-1}$ the parameters of Johnson-Cook constitutive model of A, B, C, and n were got by using multivariate nonlinear least squares regression analysis, as shown in Table 4.

The impacts of corrosion degree on the relative value of the parameters of A, B, C, and n in the J-C model under dynamic loading are shown in Figure 14. It can be found that A and B decreased linearly with increases in the average corrosion degree as shown in (5a) and (5b); while the value

of B increased linearly with increases in the average corrosion degree and n was not obvious. It can be expressed as (5a), (5b), and (5c). It is concluded that the strain rate effect of corroded steel bars decreases with the increase of corrosion degree

$$\alpha_A = \frac{A}{A_0} = 1 - k_A \eta_s, \tag{5a}$$

$$\alpha_B = \frac{B}{B_0} = 1 - k_B \eta_s \tag{5b}$$

$$\alpha_C = \frac{C}{C_0} = 1 + k_C \eta_s, \tag{5c}$$

where A, B, and C are the basic parameters of corroded steel bars and A_0, B_0, and C_0 are parameters of the J-C model of the uncorroded steel bars; see Table 4. And the relative values of the parameters A, B, and C, respectively, k_A, k_B, and k_C, are the relative coefficients of corrosion degree of each parameter.

Equations (4), (5a), (5b), and (5c) can be used to determine the parameters A, B, C, and n, and the corresponding yield strength and ultimate strength can be obtained according to (4) according to the strain and ultimate strain of the corroded steel bar under dynamic loading.

4.3.3. Hardening Strain. With corrosion degree increased, the yield plateaus of corroded high strength rebars shortened. When the corrosion degree reached a critical value, $\eta_{s,cr}$, the yield plateaus of corroded high strength rebars disappeared. If the influence of the strain rate on the deformability of rebars was ignored and the yield plateau shortens linearly in length

$$
\varepsilon_{shc,d} = \begin{cases} \varepsilon_{syc,d} + \left(\varepsilon_{sh0,d} - \varepsilon_{sy0,d}\right) \cdot \left(1 - \dfrac{\eta_s}{\eta_{s,cr}}\right) = \dfrac{f_{yc,d}}{E_{sc,d}} + \left(\varepsilon_{sh0} - \dfrac{f_{y0,d}}{E_{s0,d}}\right) \cdot \left(1 - \dfrac{\eta_s}{\eta_{s,cr}}\right) & (\eta_s \leq \eta_{s,cr}) \\[3ex] \varepsilon_{syc,d} = \dfrac{f_{yc,d}}{E_{sc,d}} & (\eta_s > \eta_{s,cr}), \end{cases}
\tag{6}
$$

where $\varepsilon_{sy0,d}$ and $\varepsilon_{sh0,d}$ are the yielding strain and hardening strain for uncorroded rebars, respectively; $f_{y0,d}$ and $f_{yc,d}$ are the yielding strengths of the uncorroded and corroded rebar specimens, respectively; and, $E_{s0,d}$ is Young's modulus for uncorroded rebars.

4.3.4. Ductility. The ultimate strains for specimens under dynamic loading are presented in Figure 15(a). As shown in the figure, no significant dependence of the ultimate strain was observed on the strain rate. Ignoring the effect of the strain rate, the ultimate strain of corroded rebars decreased significantly with the increasing corrosion degree, as shown in Figure 15(b). It can be expressed as

$$
\varepsilon_{suc} = \alpha_{\delta c} \cdot \varepsilon_{su0},
\tag{7}
$$

where ε_{su0} is the ultimate strain of the uncorroded rebars, which can be chosen as 0.15 for HRB500 rebars based on the test results; ε_{suc} is the ultimate strain of the corroded rebars; $\alpha_{\delta c}$ is the ratio of the ultimate strain of corroded rebars to that of uncorroded rebars. Using regression analysis, it was determined that $\alpha_{\delta c} = e^{-2.586\eta_s}$ for corroded HRB500 rebars.

5. Comparison of J-C Model Results and Experimental Results

5.1. Comparison with Test Results. Table 5 shows the comparisons between the predicted values of the nominal yield strength f_{syc} and ultimate strength f_{suc} for the corroded high strength rebars of the modified J-C model with experiments. It can be seen that the predicted value and experimental value were in good agreement, which showed that the modified J-C model was effective. As can be seen from Table 5, most of the errors of high strength rebars under strain rates 2 to 50 s^{-1} were no more than 5%.

5.2. Comparison with the Existing Models. The predicted values of DIF (dynamic increase factor) based on the above modified J-C model were compared with the dynamic constitutive model of the yield strength and ultimate strength for uncorroded rebars with existing model in literature [23] and CEB model [24]. As shown in Figure 16, the DIF of yield strength and ultimate strength were slightly larger than the literature value. This may be caused by the difference of steel chemical composition and manufacturing process. The plastic strain rate is 0.0002 s^{-1} in the modified J-C model; the DIF of the static loading converges to 1; the J-C model can be

with the development of corrosion degree, the hardening strain of corroded rebars can be calculated by using

used for description of the mechanical behavior of HRB500 rebars with strain rates 0.0002–50 s^{-1}.

6. Conclusions

Corrosion has been found to be the most predominant cause for failures of RC structure. The tensile behaviors of corroded steel bars are important in the capacity evaluation of corroded reinforced concrete structures. High strength corroded steel bars were tested in this study to investigate their tensile behaviors under dynamic loading. The dynamic tensile testing system was used to study the mechanical behavior of the corroded high strength steel bars under dynamic loading, and the modified J-C Model was established based on Johnson-Cook equation. The following conclusions can be drawn.

(1) From the nominal stress-strain curves of corroded high strength steel bars, it was observed that both strength and deformation capacities decreased with increasing corrosion. However, this trend was less clear for the deformation capacity than for the strength capacity. As supported by 3D scan results, this is due to the fact that the strength capacity was affected mainly by the minimum cross-sectional area, while the deformation capacity was affected by more factors, including the minimum cross-sectional area and the distribution of cross-sectional areas along the bar.

(2) With increasing strain rates, the nominal yield and ultimate strengths increase, while its effects on the yield plateau and the ultimate deformation are not obvious. Based on the analysis of Johnson-Cook model parameters A, B, C, and n under different corrosion degrees, it is found that the strain rate effect of the yield strength and ultimate strength decreases with the increase of corrosion rate.

(3) The modified J-C model was used to describe the corroded high strength rebars under dynamic loading. The modified model fits well with experiments, and the max average relative error is 5%. Therefore, this modified J-C model provides a new modeling idea and approach for the corroded high steel under dynamic loading involving corrosion degree and strain rate.

Abbreviations

A_{Fe}: The atomic mass of iron ($A = 56$ g)
t_c: The corrosion duration (days)
Z: The valence of iron, which is 2 in this case (iron)
F: Faraday's constant (96500 A·s)

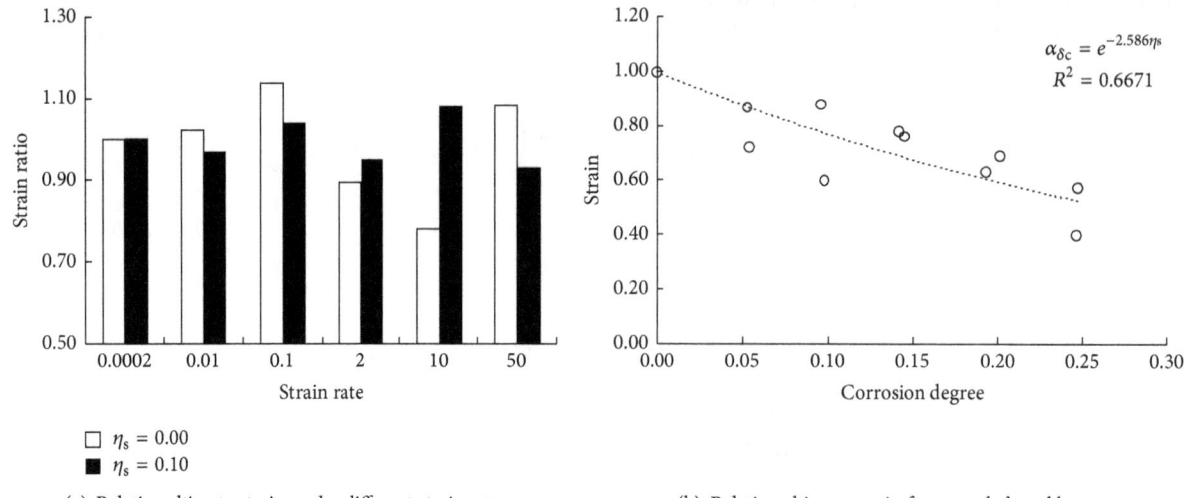

(a) Relative ultimate strain under different strain rates (b) Relative ultimate strain for corroded steel bars

FIGURE 15: Relative ultimate strain.

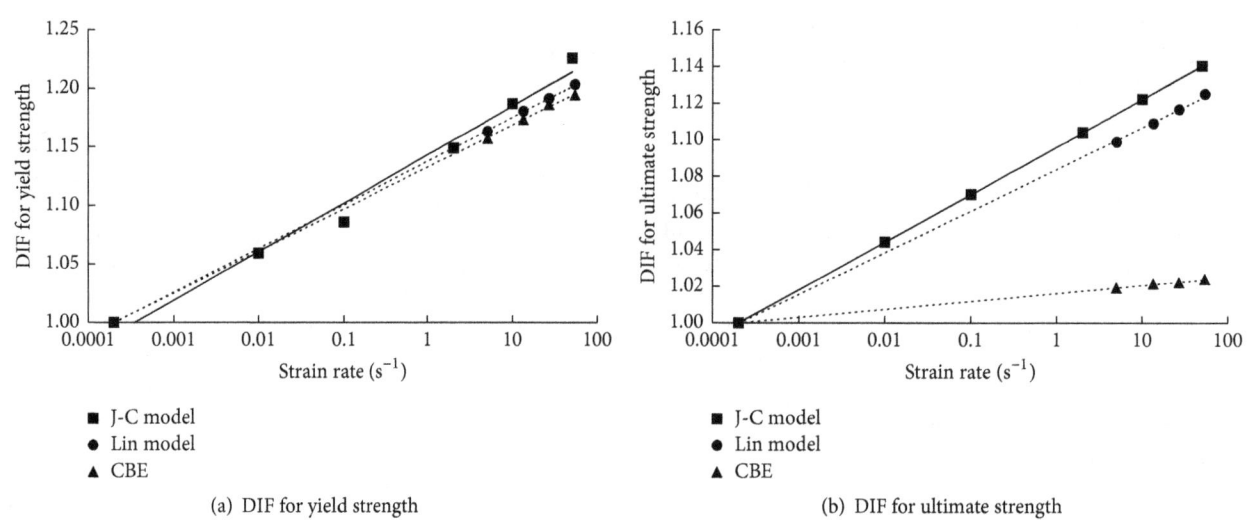

(a) DIF for yield strength (b) DIF for ultimate strength

FIGURE 16: Comparison of yield strength and ultimate strength between modified and existing models.

TABLE 5: Comparison of the results from J-C model and values from test.

Specimen	Test (MPa)	J-C (MPa)	Error (%)	Test (Mpa)	J-C (MPa)	Error %
HS00-2	573.47	599.62	4.6	696.21	687.68	−1.2
HS00-10	592.31	620.77	4.8	706.9	699.01	−1.1
HS00-50	611.66	636.93	4.1	727.78	710.34	−2.4
HS05-10	562.26	558.49	−0.7	664.63	622.86	−6.3
HS15-10	503.18	478.34	−4.9	580.09	591.58	2.0
HS20-10	479.25	456.46	−4.8	557.17	552.93	−0.8
HS25-10	444.1	456.46	2.8	519.99	521.10	0.2
HS10-2	496.56	517.48	4.3	611.66	612.01	0.1
HS10-10	522.54	546.05	4.5	625.92	622.33	−0.6
HS10-50	528.14	565.95	7.2	644.77	632.65	−1.9

Note. (1) Error = (Value from the model − value from tests)/value from tests.

i: The density of impressed current (A/cm^2)

ρ: The mass density of iron (g/cm^3)

r: The radius of to-be-corroded rebars (mm)

η_{sp}:   The targeted average corrosion degree

η_s: The average corrosion degree

$\eta_{s,max}$:   The maximum corrosion degree

$\eta_{s,cr}$:   The critical average degree of corrosion

E_{s0}:   Young's modulus for an uncorroded rebar under static loading (MPa)

$E_{s0,d}$:   Young's modulus for an uncorroded rebar under dynamic loading (MPa)

F_{y0}:   The yielding load of an uncorroded rebar or a corroded rebar (kN)

F_{u0}:   The ultimate load of an uncorroded rebar or a corroded rebar (kN)

f_{yc}:   The nominal yield strength of a corroded rebar (MPa)

f_{uc}:   The nominal ultimate strength of a corroded rebar (MPa)

$f_{yc,st}$:   The yielding strength of a corroded rebar under static load (MPa)

$f_{uc,st}$:   The ultimate strength of a corroded rebar under static load (MPa)

$f_{y0,st}$:   The yield strength of uncorroded rebar under static load (MPa)

σ: The corresponding average engineering stress (MPa)

$\tilde{\sigma}$:   The true stress (MPa)

σ_s: The Von Mises equivalent flow stress (MPa)

$\sigma_{sc,d}$:   The Von Mises equivalent flow stress of the corroded rebars (MPa)

A: The yield stress at a given reference temperature and a given reference strain rate (MPa)

B: The coefficient of strain hardening

C: The coefficient of strain rate hardening

n: The strain hardening exponent

A_0:   The parameter of the J-C model of the uncorroded steel bars (MPa)

B_0:   The parameter of the J-C model of the uncorroded steel bars

C_0:   The parameter of the J-C model of the uncorroded steel bars

k_A: The relative coefficient of corrosion degree of A

k_B: The relative coefficient of corrosion degree of B

k_C: The relative coefficient of corrosion degree of C

$\dot{\varepsilon}_{sc,st}$:   The strain rate of a rebar under static load

ε^p: The equivalent plastic strain

$\dot{\varepsilon}_0^*$:   The dimensionless strain rate with $\dot{\varepsilon}$ being the strain rate and $\dot{\varepsilon}_0$ the reference strain rate

ε: The corresponding average engineering strain

$\tilde{\varepsilon}$:   The true strain

ε_{su0}:   The ultimate strain of the uncorroded rebar

ε_{suc}:   The ultimate strain of the corroded rebar

$\alpha_{\delta c}$:   The ratio of the ultimate strain of corroded rebars to that of uncorroded rebars

T: The experimental temperature (°C)

T_r: The room temperature (°C)

T_m: The melting temperature of the materials (°C).

Conflicts of Interest

The authors declare that there are no conflicts of interest regarding the publication of this paper.

Acknowledgments

This study was financially supported by the National Natural Science Foundation of China (51608393).

References

[1] A. K. Azad, S. Ahmad, and B. H. A. Al-Gohi, "Flexural strength of corroded reinforced concrete beams," *Magazine of Concrete Research*, vol. 62, no. 6, pp. 405–414, 2010.

[2] C. A. Apostolopoulos, "The effect of ribs on the mechanical behavior of corroded reinforcing steel bars S500s under low-cycle fatigue," *Materials and Structures/Materiaux et Constructions*, vol. 41, no. 5, pp. 991–999, 2008.

[3] W. Zhang, X. Song, X. Gu, and S. Li, "Tensile and fatigue behavior of corroded rebars," *Construction and Building Materials*, vol. 34, pp. 409–417, 2012.

[4] J. A. González, C. Andrade, C. Alonso, and S. Feliu, "Comparison of rates of general corrosion and maximum pitting penetration on concrete embedded steel reinforcement," *Cement and Concrete Research*, vol. 25, no. 2, pp. 257–264, 1995.

[5] Y. G. Du, L. A. Clark, and A. H. C. Chan, "Residual capacity of corroded reinforcing bars," *Magazine of Concrete Research*, vol. 57, no. 3, pp. 135–147, 2005.

[6] J. Cairns, G. A. Plizzari, Y. Du, D. W. Law, and C. Franzoni, "Mechanical properties of corrosion-damaged reinforcement," *ACI Materials Journal*, vol. 102, no. 4, pp. 256–264, 2005.

[7] R. Palsson and M. S. Mirza, "Mechanical response of corroded steel reinforcement of abandoned concrete bridge," *ACI Structural Journal*, vol. 99, no. 2, pp. 157–162, 2002.

[8] Y.-C. Ou, H.-D. Fan, and N. D. Nguyen, "Long-term seismic performance of reinforced concrete bridges under steel reinforcement corrosion due to chloride attack," *Earthquake Engineering & Structural Dynamics*, vol. 42, no. 14, pp. 2113–2127, 2013.

[9] M. G. Stewart, "Mechanical behaviour of pitting corrosion of flexural and shear reinforcement and its effect on structural reliability of corroding RC beams," *Structural Safety*, vol. 31, no. 1, pp. 19–30, 2009.

[10] M. Beck, A. Burkert, J. Harnisch et al., "Deterioration model and input parameters for reinforcement corrosion," *Structural Concrete*, vol. 13, no. 3, pp. 145–155, 2012.

[11] I. Rohr, H. Nahme, and K. Thoma, "Material characterization and constitutive modelling of ductile high strength steel for a wide range of strain rates," *International Journal of Impact Engineering*, vol. 31, no. 4, pp. 401–433, 2005.

[12] A. Filiatrault and M. Holleran, "Stress-strain behavior of reinforcing steel and concrete under seismic strain rates and low temperatures," *Materials and Structures*, vol. 34, no. 238, pp. 235–239, 2001.

[13] P. Soroushian and K. B. Choi, "Steel mechanical properties at different strain rates," *Journal of Structural Engineering*, vol. 113, no. 4, pp. 663–672, 1987.

[14] F. Lin, X. L. Gu, X. X. Kuang, and X. J. Yin, "Constitutive models for reinforcing steel bars under high strain rates," *Journal of Building Materials*, vol. 11, no. 1, pp. 14–20, 2008.

[15] W.-P. Zhang, H. Chen, and X.-L. Gu, "Tensile behaviour of corroded steel bars under different strain rates," *Magazine of Concrete Research*, vol. 68, no. 3, pp. 127–140, 2016.

[16] G. R. Johnson and W. H. Cool, "A constitutive model and data for metals subjected to large strain, high strain rates and high temperatures," in *Proceedings of the 7th International Symposium on Ballistics*, The Hague, Netherlands, April 1983.

[17] ASTM, in *ASTM E8/E8M-09 Standard Test Methods for Tension Testing of Metallic Materials*, pp. 1–25, ASTM International, West Conshohocken, Pa, USA, 2009.

[18] A. A. Almusallam, "Effect of degree of corrosion on the properties of reinforcing steel bars," *Construction and Building Materials*, vol. 15, no. 8, pp. 361–368, 2001.

[19] Y.-C. Ou, Y. T. T. Susanto, and H. Roh, "Tensile behavior of naturally and artificially corroded steel bars," *Construction and Building Materials*, vol. 103, pp. 93–104, 2016.

[20] Y. S. Yuan, Y. S. Ji, and S. P. Shah, "Comparison of two accelerated corrosion techniques for concrete structures," *ACI Structural Journal*, vol. 104, no. 3, pp. 344–347, 2007.

[21] T. A. El Maaddawy and K. A. Soudki, "Effectiveness of impressed current technique to simulate corrosion of steel reinforcement in concrete," *Journal of Materials in Civil Engineering*, vol. 15, no. 1, pp. 41–47, 2003.

[22] ASTM, in *ASTM G1-03 Standard Practice for Preparing, Cleaning, and Evaluating Corrosion Test Specimens*, pp. 1–9, ASTM International, West Conshohocken, Pa, USA, 2003.

[23] F. Lin, Y. Dong, and X. Gu, "Dynamic constitutive models for high strength reinforcing steel bar HRB500," *Journal of Building Materials*, vol. 17, no. 4, pp. 592–597, 2014.

[24] J. Eibl, "Concrete structures under impact and impulsive loading: synthesis report," Tech. Rep., Comité Euro-International du Béton, Lausanne, Switzerland, 1988.

8

Regression Analysis of Bond Parameters between Corroded Rebar and Concrete Based on Reported Test Data

H. J. Zhou ⓘ, Y. F. Zhou, Y. N. Xu, Z. Y. Lin, F. Xing ⓘ, and L. X. Li ⓘ

Guangdong Provincial Key Laboratory of Durability for Marine Civil Engineering, Shenzhen University, Shenzhen 518060, China

Correspondence should be addressed to L. X. Li; lilixiao@szu.edu.cn

Academic Editor: Ramana M. Pidaparti

Reinforcement corrosion is a major cause of degradation in reinforced concrete structures. The fragile rust layer and cracking and spalling of the cover caused by splitting stress due to rust expansion can alter bond behaviors significantly. Despite extensive experimental tests, no stochastic model has yet incorporated randomness into the bond parameters model. This paper gathered published experimental data on the bond-slip parameters of pull-out specimens and beam-end specimens. Regression analysis was carried out to identify the best fit of bond strength and the corresponding slip value in the context of different corrosion levels from the recollected test results. An F-test confirmed the regression effect to be significant. Residual data were also analyzed and found to be well described by a normal distribution. Crack width data of the tested specimens were also collected. A regression analysis of the bond strength and maximum crack width was carried out given the comparative simplicity of measuring crack width versus rebar area loss. Results indicate that maximum crack width can also be used to predict bond strength degradation with similar variation magnitude.

1. Introduction

The civil engineering field generally accepts that reinforced concrete structures possess durability problems. One of the harshest environments for concrete structures is the marine environment, as it contains corrosive chloride ions [1]. Chloride ions can diffuse into concrete, accumulate at the reinforcement surface, and depassivate the protective layer. Corrosion of reinforcement occurs when chlorides surpass a threshold value; it is the main cause of performance degradation in aged concrete structures. Corrosion products of reinforcement expand in volume with a strength much lower than that of steel, reducing the effective reinforcement area. Expansion of reinforcement rust can also lead to cracking and spalling of the concrete cover when expansion stress surpasses the tensile strength of concrete [2].

The effects of reinforcing-bar corrosion on bond behaviors have been widely studied by many scholars [3–7] for over 30 years using different test setups. Abdullah et al. [8] studied the bond behavior of reinforced concrete members including ultimate bond strength, free-end slip, and failure modes in the precracking, cracking, and postfracture stages. Fang et al. [9] conducted a pull-out test to evaluate the effects of corrosion on bond and bond-slip behavior in specimens with and without stirrups that provided confinement. Wei et al. [10] designed beam-end specimens to study the bond between corroded steel and concrete. However, bond strength reduction due to the fragile corrosion products between concrete and steel has been often neglected in the field.

The research disparity persists between industry and academia in the absence of a unified and feasible model that considers the degradation of bond behaviors. Several factors influence bond performance and the complex nature of interface behaviors between concrete and reinforcement, especially when accounting for the effects of reinforcement corrosion [11]. The roughness of the reinforcement surface and confinement of concrete have been shown to exert significant effects on bond performance. Rebar corrosion can certainly change the surface between rebar and concrete; the expansion of corrosion products can also lead to cracking and spalling of the concrete cover, thus degrading confinement. Recent studies [12, 13] have revealed that stirrup corrosion can also degrade the confinement of rebar and concrete, potentially altering the bond failure type, strength, and ductility of confined concrete. In some research, these factors are considered collectively; as experiments often implement different test setups [9, 14], the results of each test are often unique. Additional studies on the effects of corrosion on bond parameters are necessary to further develop a widely applicable bond-slip model for assessment of corroded reinforcement concrete structures.

This preliminary study recollected the test data on bond strength and corresponding slip value and maximum crack width from published literatures. Regression analysis was applied to obtain the best fit for the above three parameters. An F-test [15] was carried out to verify the regression analysis results. Residuals were also analyzed and modeled as a normal distribution to consider variation in bond behaviors.

2. Pull-out Specimens

Two different rebars were used for pull-out specimen tests: plain rebar (Table 1) and deformed rebar (Table 2). These two types of rebar exhibit different bond behaviors; the rib of deformed rebar can hook to concrete and change the stress field around the rebar and concrete interface, whereas plain rebar bonds have no similar working mechanism. Each type will be discussed separately in the following sections.

2.1. Plain Rebar

2.1.1. Dimensionless Bond Strength. Table 1 lists the collected test parameters, where D is the rebar diameter, C is the concrete cover depth, and L_m is the bond length. "-" indicates the parameter was not reported in the literature. As setups involved test specimens with different levels of concrete strength, dimensionless bond strength was used in this paper. The dimensionless bond strength of a corroded rebar was defined as $\tau_{\max}^{\xi_s}/\tau_{\max}^0$, where $\tau_{\max}^{\xi_s}$ and $\tau_{\max}^0$ denote the tested bond strength of corroded steel and noncorroded steel, respectively. In tests with more than one noncorroded specimen, $\tau_{\max}^0$ represents the mean value of the bond strength of the tested noncorroded specimens.

Figure 1(a) shows the recollected test results of dimensionless bond strength and corresponding corrosion level (the mean rebar mass loss is identical to area loss). Nonlinear regression analysis of the test data included the following equation:

$$\frac{\tau_{\max}^{\xi_s}}{\tau_{\max}^0} = \frac{(5.57\xi_s + 5.48)}{\left((\xi_s + 0.240)^2 + 5.43\right)} \tag{1}$$

where ξ_s is the proportion (i.e., percentage) of the extent of steel corrosion. R^2 is 0.445, and $R^2 \subset [0, 1]$ is the coefficient of determination, interpreted as the proportionate reduction in total variation associated with the predictor variable; the closer it is to 1, the greater the degree of association between the predictor and response variables. As shown in Figure 1(a), the dimensionless bond strength first increased and then decreased with an increase in corrosion level.

An F-test was further applied to test the significance of the regression curve:

$$F = \frac{SS_R}{SS_E/(n-2)} \tag{2}$$

where $SS_R = \sum(\overline{Y}_i - \overline{Y})^2$ is the explained sum of squares, $SS_E = \sum(Y_i - \overline{Y}_i)^2$ is the residual sum of squares, $\overline{Y}_i$ is the predicted value corresponding to the abscissa, $\overline{Y}$ is the average of all predicted values, Y_i is the value of dimensionless bond strength, and n is the total number of data. When F is greater than $F_{1-\alpha} = (1, n-2)$, the regression effect is considered significant at α.

The F value of pull-out specimens with plain rebar was 21.0, greater than the corresponding $F_{1-\alpha} = (1, n-2) = 7.04$ at a significance level of $\alpha = 0.01$; hence the regression effect was significant. The bond between corroded steel bars and concrete is also affected by other factors, such as concrete strength, protective layer thickness, the existence of stirrups, the diameter of corroded steel bars, and test setup. The central limit theorem states that, under some conditions (including finite variance), the averages of sample observations of random variables drawn from independent distributions converge in a normal distribution; that is, they become normally distributed when the number of observations is sufficiently large. In the present study, dimensionless values have also been used to try to eliminate the effects of these factors; residuals were further assessed to verify the above assumption. Figure 1(b) plots the residuals after nonlinear regression analysis; the residuals are distributed randomly along the horizontal axis. Figure 1(c) presents a histogram of the distribution of residuals, which appear to fit a normal distribution with a mean value (μ) of -0.00738 and standard deviation (σ) of 0.355. The regression formula and normal distribution of residuals will be applied to model the bonds of corroded plain rebar and concrete later in this paper.

2.1.2. Dimensionless Slip Value Corresponding to Bond Strength. Given different setups of test specimens with different levels of concrete strength, the dimensionless slip value corresponding to bond strength was studied as follows. The dimensionless slip value corresponding to bond strength of corroded rebar was defined as δ^{ξ_s}/δ^0, where δ^{ξ_s} and δ^0 are the slip values corresponding to the bond strength of corroded steel and noncorroded steel, respectively. In cases with more than one noncorroded specimen, δ^0 denotes the mean value of the noncorroded specimens as that of bond strength.

Figure 2(a) shows the variation in dimensionless slip value corresponding to bond strength at certain corrosion levels. Nonlinear regression analysis of the test data employed the following equation:

$$\frac{\delta^{\xi_s}}{\delta^0} = \frac{1}{(0.354\xi_s + 1)} \tag{3}$$

R^2 is 0.540, indicating the existence of unexplained factors. Figure 2(b) shows that the residuals are randomly distributed along the horizontal axis on the negative and positive sides. Figure 2(c) indicates the residual data is well fitted by a

TABLE 1: Parameters of pull-out specimens with plain rebar.

References No.	Specimen size (mm)	Corrosion level (%)	f_c (MPa)	D (mm)	C (mm)	No. of specimens	L_m (mm)	Stirrup confinement	Reinforcement position	Test environment	Reinforcement type
[3]	100×100×100	0.27–5.01	22.13	12	44	14	80	No	Center	Accelerated corrosion	Plain bars
[4]	100×100×100	0–5.01	22.13	12	44	14	80	No	Center	Accelerated corrosion	Plain bars
[5]	100×100×100	0–8	13.98, 21.1, 23.63 39.8,	12	44	21	5d	No	Center	Accelerated corrosion	Plain bars
[6]	150×150×150	0–11.9	62.7, 73.4	12	69	52	60	No	Center	Current density of 0.25mA/cm^2	Plain bars
[9]	140×140×180	0–6.8	52.1	20	60	16	4d	Each half	Center	Direct current (0–2 A)	Plain bars
[16]	140×140×180	0–6.77	54.15	20	60	16	4d	Each half	Center	Current density of 1.214mA/cm^2	Plain bars
[17]	150×150×140	0–5.95	21.1	12	69	6	100	No	Center	Accelerated corrosion	Plain bars
[18]	150×150×150	0–10.41	24.81	18, 20, 22	30	17	100	No	Side	Accelerated corrosion	Plain bars
[19]	150×150×150	0–5.38	30	10	70	18	110	No	Center	Current density of 0.3mA/cm^2	Plain bars
[20]	100×100×150	0–12.2	33.5	16	42	5	50	No	Center	Accelerated corrosion	Plain bars
[21]	150×150×150	2.76–5.95	21.1	12	69	5	100	No	Center	Accelerated corrosion	Plain bars

TABLE 2: Parameters of pull-out specimens with deformed rebar.

References No.	Specimen size (mm)	Corrosion level (%)	f_c (MPa)	D (mm)	C (mm)	No. of specimens	L_m (mm)	Stirrup confinement	Reinforcement position	Test environment	Reinforcement type
[3]	100×100×100	0–9.95	22.13	12	44	14	80	No	Center	Direct current	Ribbed bars
[4]	100×100×100	0–9.72	22.13	12	44	14	80	No	Center	Accelerated corrosion	Ribbed bars
[5]	100×100×100	0–8	13.98, 21.1, 23.63	12	44	21	5d	No	Center	Electro-migration	Ribbed bars
[6]	150×150×150	0–4.66	62.7, 73.4	12	69	39	60	No	Center	Current density of 0.25mA/cm²	Ribbed bars
[7]	120×120×120	0–20	30.9	12	54	20	5d	Each half	Center	Current density of 200uA/cm²	Ribbed bars
[8]	152×254×280	0–79.74	30	12	63.5	No	102	Yes	Side	Direct current (0.4 A)	Ribbed bars
[9]	140×140×180	0–9	52.1	20	60	24	4d	Each half	Center	Direct current (0–2 A)	Ribbed bars
[14]	200×200×200	0–14.45	20.7, 44.4	18	91	40	80	Yes	Center	Current density of 150μA/cm²	Ribbed bars
[16]	140×140×180	0–9.02	54.15	20	60	24	4d	Each half	Center	Current density of 1.214mA/cm²	Ribbed bars
[17]	100×100×200	0–2.3709	38.3	16, 20	42, 40	20	5d	Each half	Center	Electro-migration (DC voltage of 60V)	Ribbed bars
[18]	150×150×150	0–9.06	24.81	18, 20, 22	30	17	50	No	Side	Accelerated corrosion	Ribbed bars
[20]	100×100×150	0–11.8	33.5	16	42	5	50	No	Center	Accelerated corrosion	Ribbed bars
[22]	140×140×180	0–6.1	52.1	20	60	24	4d	Each half	Center	Direct current (0–2 A)	Ribbed bars
[23]	150×150×150	0–5.95	41.9	18	66	12	100	Each half	Center	Electro-migration and wetting-drying cycle	Ribbed bars
[24]	140×140×180	0–27.9	50.8	20	60	31	4d	Each half	Center	Direct current (0–2 A)	Ribbed bars
[25]	150×150×150	0–2.3	21, 32	25	62.5	30	4d	No	Center	Accelerated corrosion	Ribbed bars
[26]	120×120×120	0–0.8	30.1	12	54	12	5d	No	Center	Current density of 0.25mA/cm²	Ribbed bars
	180×180×180	0–4.5	30.1	18	81	12	5d	No	Center	Current density of 0.25mA/cm²	Ribbed bars
[27]	140×140×180	0–15.27	31.4	20	60	20	3d	Yes	Center	Current density of 0.01–0.02mA/mm²	Ribbed bars
[28]	150×150×150	0–8.95	23, 51	14	15, 30, 45	90	50	No	Side	Direct current (0–5 A)	Ribbed bars
[29]	140×140×180	0–20	64	20	60	21	4d	Yes	Center	Maximum of 0.1 A per specimen	Ribbed bars
[30]	150×150×150	0–12	-	12	69	6	4d	No	Center	Direct current (17.5–59mA)	Ribbed bars

TABLE 2: Continued.

References No.	Specimen size (mm)	Corrosion level (%)	f_c (MPa)	D (mm)	C (mm)	No. of specimens	L_m (mm)	Stirrup confinement	Reinforcement position	Test environment	Reinforcement type
[31]	150×150×200	0–2.5	28.3	13	68.5	21	3d	No	Center	Accelerated corrosion	Ribbed bars
[32]	150×150×200	0–8.8	28.3	13	68	40	39	No	Center	Output of 24 V DC and 12 A	Ribbed bars
[33]	150×150×150	0–6.69	29	12	69	13	70	No	Center	Current density of $100\mu A/cm^2$	Ribbed bars
[34]	104×104×104	0–24	24.7, 33, 42.1	13	45.5	6	78	Each half	Side	Maximum of 1 A per specimen	Ribbed bars
[35]	10d×10d×10d	0–4.9	30.1	12,18	4.5d	24	5d	No	Center	Current density of $0.25mA/cm^2$	Ribbed bars
[36]	160×160×250	0–15.9	35.8	20	35, 45, 55	27	200	Yes	Side	Direct current (0–1 A)	Ribbed bars
[37]	150×150×150	0–9.8	24.83, 26.51, 25.04	20	65	45	100	No	Center	Direct current (0–2 A)	Ribbed bars
[38]	150×250×300	0–14.65	33	20	35	16	100	Yes	Side	Current density of $500\mu A/cm^2$	Ribbed bars
[39]	175×175×350	0–5.19	–	19	78	11	75	No	Center	Direct current (0.15 A)	Ribbed bars
[40]	150×150×150	0–9.06	24.81	18, 20, 22	30	34	100	No	Side	Direct current	Ribbed bars
[41]	150×250×300	0–10.19	50	20	35	43	100	Yes	Side	Current density of $400–600\mu A/cm^2$	Ribbed bars

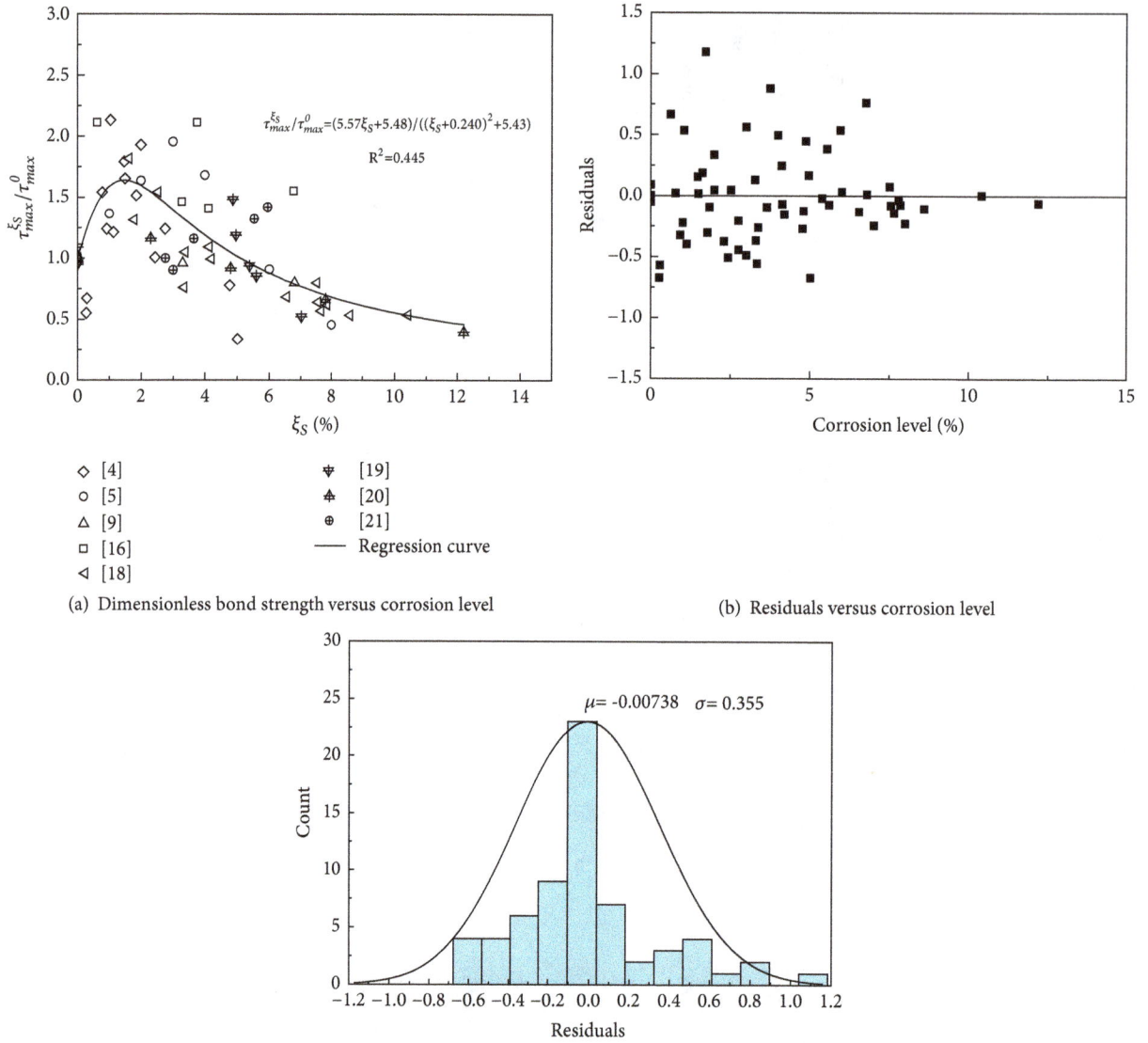

(a) Dimensionless bond strength versus corrosion level

(b) Residuals versus corrosion level

(c) Histogram of residuals of dimensionless bond strength

FIGURE 1: Dimensionless bond strength of pull-out specimens with plain rebar: (a) dimensionless bond strength versus corrosion level; (b) residuals versus corrosion level; (c) histogram of residuals of dimensionless bond strength.

normal distribution with an expectation (μ) of 0.0133, which is close to zero, and a standard deviation (σ) of 0.263.

2.2. Deformed Rebar. Table 2 shows collected references of pull-out tests with deformed rebars. Deformed rebar specimens can be divided into two subgroups: with and without stirrups. As confinement can change bond behaviors significantly, these subgroups will be discussed separately in subsequent sections. Stirrup density can also affect the bond properties of corroded steel bars. However, because stirrup density was randomly distributed in these tests, it was not considered in this preliminary study; a dimensionless value was used in relevant analyses instead.

2.2.1. Dimensionless Bond Strength without Stirrups. Figure 3(a) shows the variation in dimensionless bond strength

at different corrosion levels. Nonlinear regression analysis of the test data applied the following equation:

$$\frac{\tau_{max}^{\xi_s}}{\tau_{max}^0} = 0.278 + \frac{10.8}{\left(\xi_s^{1.8} + 13.9\right)} \tag{4}$$

R^2 is 0.384. The F value of pull-out specimens with deformed rebar (i.e., no stirrups) was 187, while the corresponding critical value was 6.72 at a significance level of $\alpha = 0.01$. Thus, the regression effect was significant, and the residuals could be analyzed.

Figure 3(b) illustrates that the residuals are distributed randomly around the horizontal axis. Figure 3(c) displays a histogram of residuals on a curve fitted with a normal distribution with a mean value (μ) of -0.000425 and standard

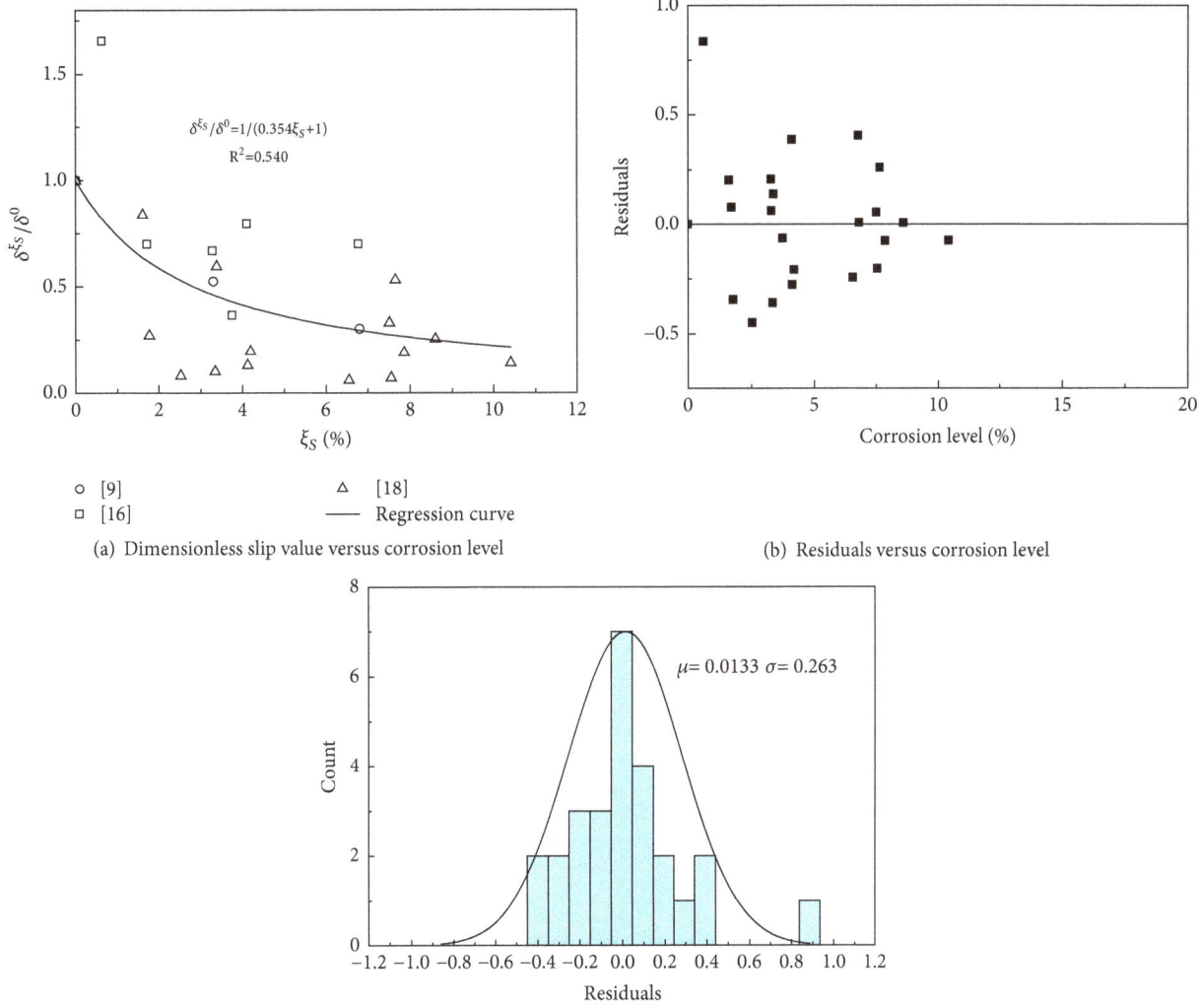

(a) Dimensionless slip value versus corrosion level

(b) Residuals versus corrosion level

(c) Histogram of residuals of dimensionless slip value

FIGURE 2: Dimensionless slip value corresponding to bond strength of pull-out specimens with plain rebar: (a) dimensionless slip value versus corrosion level; (b) residuals versus corrosion level; (c) histogram of residuals of dimensionless slip value.

deviation (σ) of 0.253, suggesting the residual data fit the proposed normal distribution well.

2.2.2. Dimensionless Slip Value Corresponding to Bond Strength without Stirrups.

Figure 4(a) shows the dimensionless slip value corresponding to bond strength compared to the corrosion level. The nonlinear regression analysis of the test data used the following equation:

$$\frac{\delta^{\xi_s}}{\delta^0} = \frac{1}{(0.123\xi_s + 1)} \quad (5)$$

The corresponding R^2 is 0.377. Figure 4(b) shows the residuals are nearly symmetrically distributed along the horizontal axis. Figure 4(c) presents the histogram of the residuals, which fit a normal distribution well with expectations (μ) of -0.00110, which is close to zero, and a standard deviation (σ) of 0.234.

2.2.3. Dimensionless Bond Strength with Stirrups.

Figure 5(a) shows the variation in dimensionless bond strength with the corrosion level. The nonlinear regression analysis of the test data employed the following equation:

$$\frac{\tau_{max}^{\xi_s}}{\tau_{max}^0} = 0.0760 + \frac{149}{\left(\xi_s^{1.8} + 161\right)} \quad (6)$$

R^2 is 0.505. The F value of pull-out specimens with a deformed rebar (stirrups) was 211, larger than the corresponding critical value of 6.76 at a significance level of $\alpha = 0.01$. Thus, the regression effect was significant, and the residuals could be analyzed.

Residuals are shown in Figure 5(b), randomly distributed well along the horizontal axis. Figure 5(c) plots the histogram of the residuals and indicates they fit with a normal distribution. The expected value (μ) is nearly zero (0.0000770), and the standard deviation (σ) is 0.190.

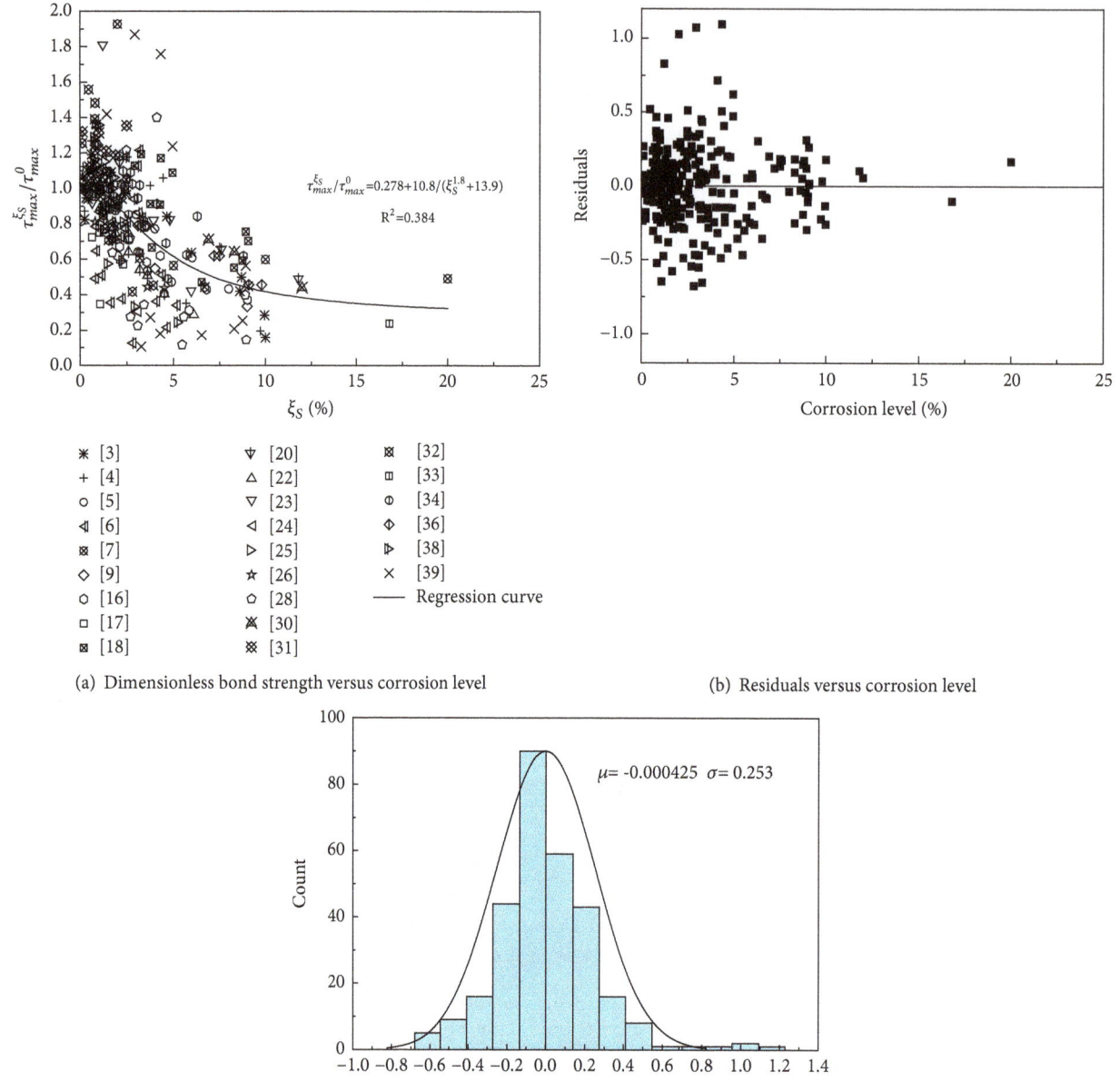

(a) Dimensionless bond strength versus corrosion level

(b) Residuals versus corrosion level

(c) Histogram of residuals of dimensionless bond strength

FIGURE 3: Dimensionless bond strength of pull-out specimens with deformed rebar (no stirrups): (a) dimensionless bond strength versus corrosion level; (b) residuals versus corrosion level; (c) histogram of residuals of dimensionless bond strength.

2.2.4. Dimensionless Slip Value Corresponding to Bond Strength with Stirrups. Figure 6(a) illustrates the variation in dimensionless slip value corresponding to bond strength with the corrosion level. Nonlinear regression analysis of the test data involved the following equation:

$$\frac{\delta^{\xi_s}}{\delta^0} = \frac{1}{(0.0870\xi_s + 1)} \qquad (7)$$

R^2 is 0.224, relatively smaller than the previously mentioned cases, presumably due to highly scattered data. Figure 6(b)

shows that the residuals are also nearly symmetrically distributed along the horizontal axis. Figure 6(c) presents the histogram of the residuals fitting to a normal distribution with a mean value (μ) of -0.0242 and standard deviation (σ) of 0.274.

3. Beam-End Specimens

3.1. Dimensionless Bond Strength. Table 3 lists the collected beam-end tests. Figure 7(a) shows the variation in the dimensionless bond strength of beam-end specimens by

TABLE 3: Parameters of beam-end specimens.

References No.	Specimen size (mm)	Corrosion level (%)	f_c (MPa)	D (mm)	C (mm)	No. of specimens	L_m (mm)	Stirrup confinement	Reinforcement position	Test environment	Reinforcement type
[3]	150×150×1140	0.47–6.05	22.13	12	-	17	360/100	Yes	Bottom	Direct current	Plain bars
[6]	150×240×560	0–15.88	62.7	16, 12	50	39	10d	Yes	Bottom	Current density of 250uA/cm²	Ribbed bars, plain bars
[10]	150×240×1260	0–4.63	-	20	50	6	200	Yes	Bottom	Wetting-drying cycle	Ribbed bars
[23]	200×120×980	0–10.04	30	18	30	10	140	Yes	Bottom	Wetting-drying cycle	Ribbed bars
[30]	125×160×1000	0.095–12.47	-	12	25	17	384/190	Yes	Bottom	Direct current	Ribbed bars
[37]	150×250×1200	0–20	33	20	40	16	150	Yes	Bottom	Wet-current density of 400uA/cm²	Ribbed bars
[41]	150×250×360	Crack width control	20	16	25	6	160	Yes	Bottom	Direct current	Ribbed bars
[42]	150×240×1260	0–4.6	38.3	20	50	6	200	Yes	Bottom	Wetting-drying cycle 100mA	Ribbed bars
[43]	150×240×1260	0–17.141	30, 40, 50	16, 20, 25	15, 20, 25	36	10d	Yes	Bottom	Accelerated corrosion	Ribbed bars
[44]	150×250×320	0–4	25	16	30	6	120	Yes	Bottom	Accelerated corrosion	Ribbed bars
[45]	200×300×380	0–12	-	12, 16	-	9	300	Each half	Bottom	Current density of 200uA/cm²	Ribbed bars
[46]	100×100×400	0–12.73	32.8	12	-	19	60	No	Bottom	Wetting-drying cycle current density	Ribbed bars
[47]	100×100×400	0–12.73	32.8	12	-	19	60	No	Bottom	Wetting-drying cycle current density	Ribbed bars
[48]	150×250×1200	0–20.86	30	20	40	18	150	Yes	Bottom	Current density of400uA/cm²	Ribbed bars
[49]	203×457×2440	0–10	33, 35	25	-	10	2400	Yes	Bottom	Direct current (2.5–10 A)	Ribbed bars
[50]	100×150×1100	0–5	45	10	20	7	100	No	Bottom	Current density of 800uA/cm²	Ribbed bars
[51]	150×240×600	0–10.28	36, 44.6, 52.7	20	30, 50	24	10d	Yes	Bottom	Wet-current density of 250uA/cm²	Ribbed bars
[52]	150×240×600	0–2.06	36, 44.6, 52.7	20	30, 50	8	10d	Yes	Bottom	Wet-current density of 250uA/cm²	Ribbed bars

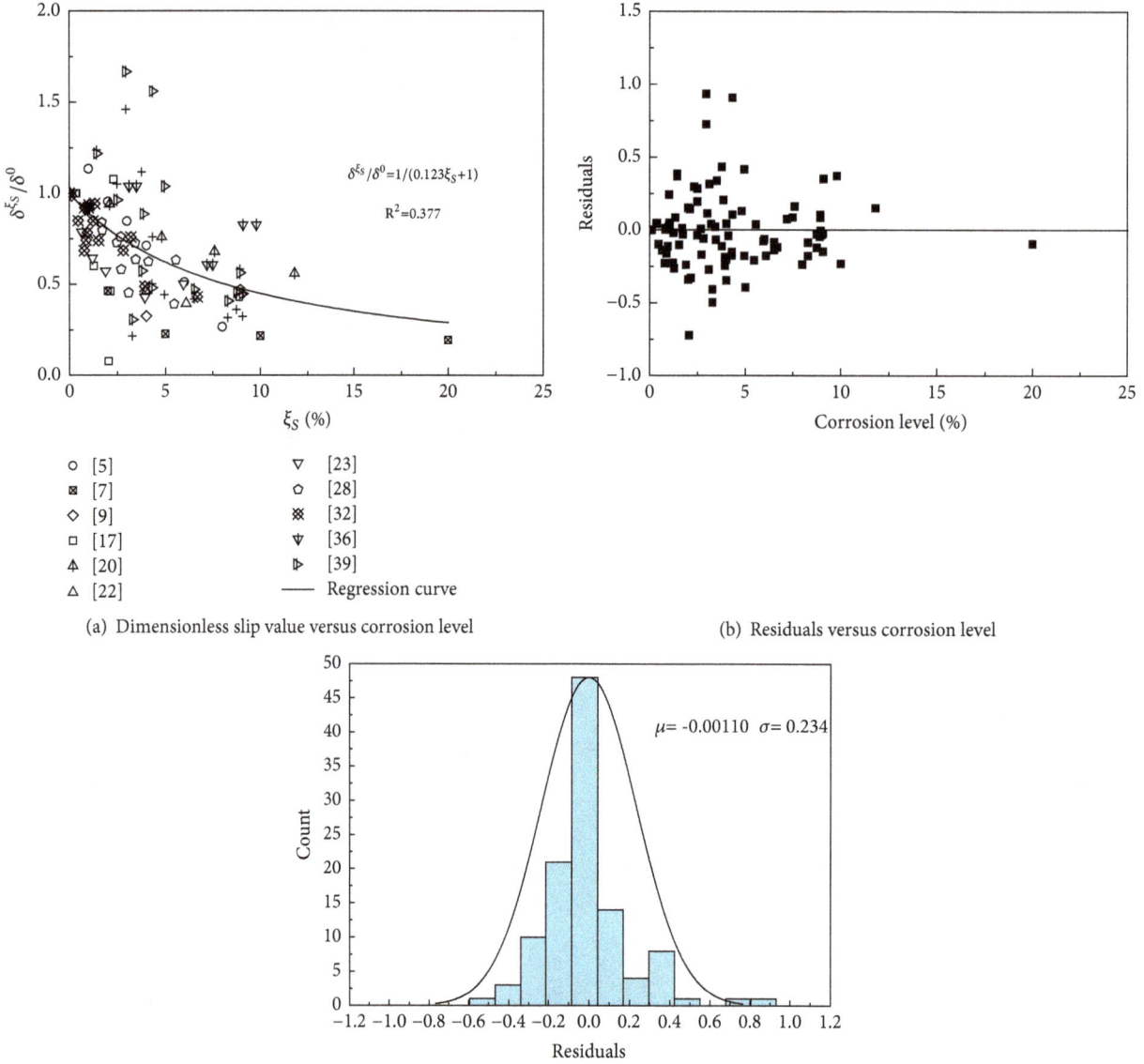

(a) Dimensionless slip value versus corrosion level

(b) Residuals versus corrosion level

(c) Histogram of residuals of dimensionless slip value

FIGURE 4: Dimensionless slip value corresponding to bond strength of pull-out specimens with deformed rebar (no stirrups): (a) dimensionless slip value versus corrosion level; (b) residuals versus corrosion level; (c) histogram of residuals of dimensionless slip value.

corrosion level. The nonlinear regression analysis of the test data employed the following equation:

$$\frac{\tau_{max}^{\xi_s}}{\tau_{max}^0} = 0.347 + \frac{41.1}{\left(\xi_s^{1.8} + 63.0\right)} \tag{8}$$

R^2 is 0.422. The F values of beam-end specimens confirmed the regression effect was significant, and the residuals could be analyzed. Figure 7(b) plots the residuals after nonlinear regression analysis, randomly distributed along the horizontal axis. Figure 7(c) is the histogram of the residuals, which fit a normal distribution well. The expected value (μ) is 0.000970, and the standard deviation (σ) is 0.170.

3.2. Dimensionless Slip Value Corresponding to Bond Strength. Figure 8(a) shows the variation in dimensionless slip value corresponding to bond strength with the corrosion level of the beam-end specimens. The nonlinear regression analysis of the test data shows the following equation:

$$\frac{\delta^{\xi_s}}{\delta^0} = \frac{1}{\left(0.138\xi_s + 1\right)} \tag{9}$$

The corresponding R^2 is 0.402. Residuals analysis suggested they are randomly distributed well along the horizontal axis as shown in Figure 8(b). Figure 8(c) provides the histogram of the residuals, which were well fitted to a normal distribution with a mean value (μ) of -0.00488, close to zero, and standard deviation (σ) of 0.215.

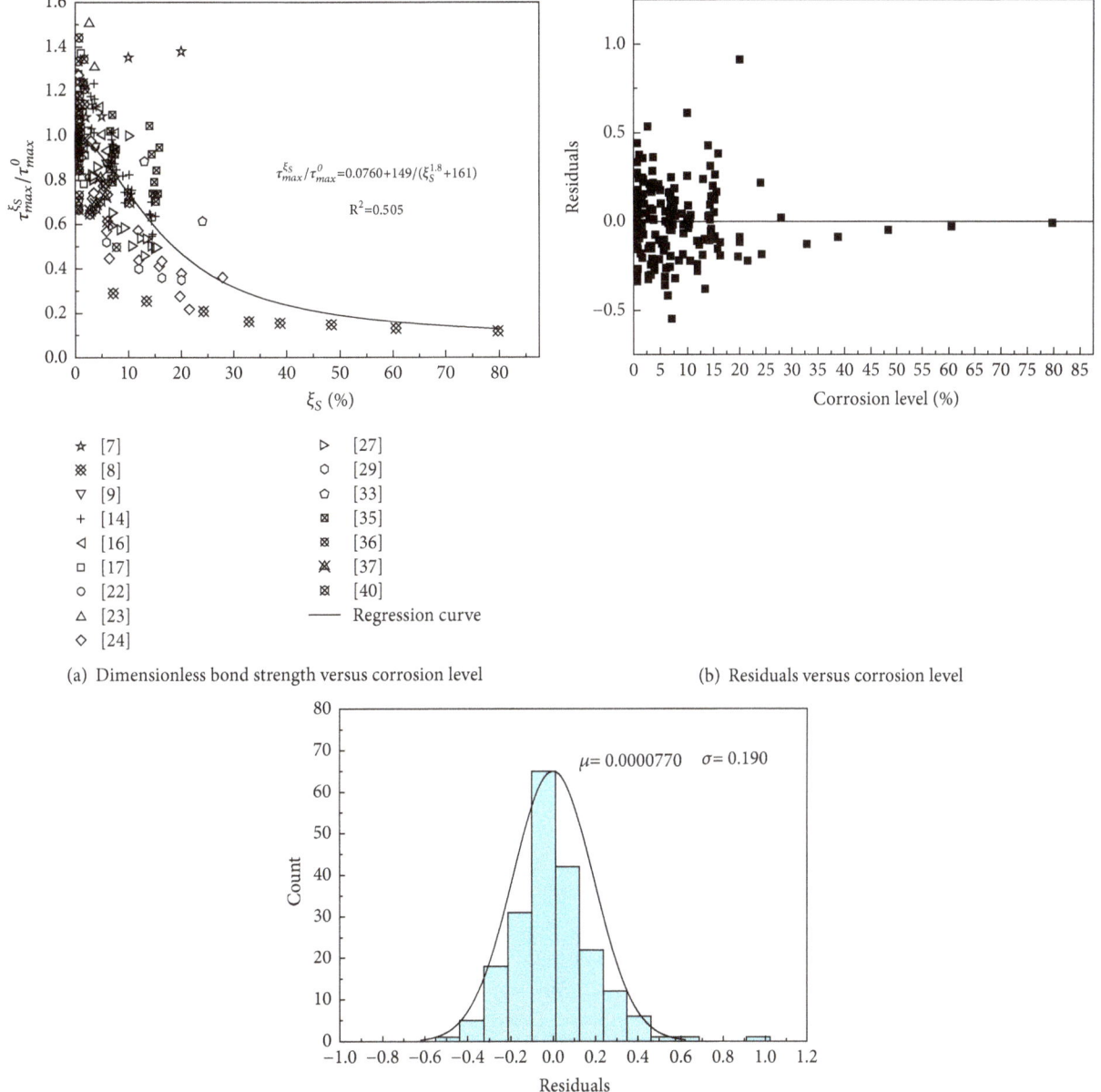

$$\tau_{max}^{\xi_S}/\tau_{max}^0 = 0.0760 + 149/(\xi_S^{1.8} + 161)$$

$$R^2 = 0.505$$

* [7]
⊠ [8]
∇ [9]
+ [14]
◁ [16]
□ [17]
○ [22]
△ [23]
◇ [24]

▷ [27]
○ [29]
⬠ [33]
⊠ [35]
⊠ [36]
✕ [37]
⊠ [40]
— Regression curve

(a) Dimensionless bond strength versus corrosion level

(b) Residuals versus corrosion level

$\mu = 0.0000770$ $\sigma = 0.190$

(c) Histogram of residuals of dimensionless bond strength

FIGURE 5: Dimensionless bond strength of pull-out specimens with deformed rebar (stirrups): (a) dimensionless bond strength versus corrosion level; (b) residuals versus corrosion level; (c) histogram of residuals of dimensionless bond strength.

4. Discussions

The above regression analysis outlines the degradation of bond parameters based on the measured rebar corrosion level. In engineering practice, it is nearly impossible to measure rebar area loss without structural damage to the site; however, crack width can be easily measured without resultant damage. Fortuitously, many recollected test results also recorded the maximum crack width involved in rebar corrosion. Figures 9(a) and 10(a) show dimensionless bond strength degradation with an increase in maximum crack width. The two different specimens, pull-out and beam-end, were discussed individually. The dimensionless bond strength clearly decreased as maximum crack width increased. The regression analysis involved the following formula for degradation:

$$\frac{\tau_{max}^{\omega}}{\tau_{max}^0} = \frac{1}{(0.968\omega^{1.35} + 1)} \tag{10}$$

$$\frac{\tau_{max}^{\omega}}{\tau_{max}^0} = \frac{1}{(0.461\omega^{1.23} + 1)} \tag{11}$$

where ω is the maximum crack width. The corresponding R^2 values are 0.231 and 0.397 for pull-out specimens and beam-end specimens, respectively. The mean value (μ) and

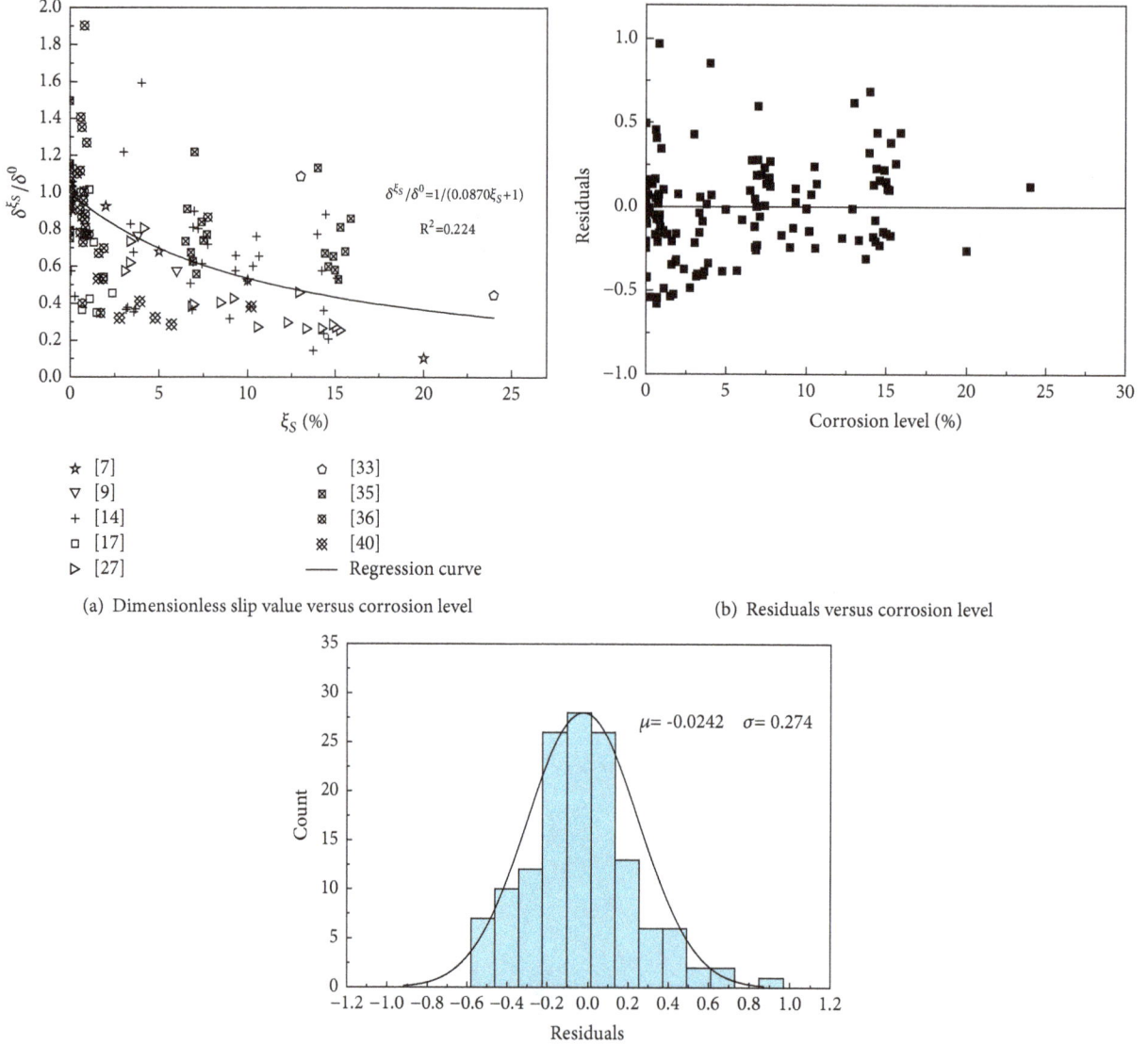

(a) Dimensionless slip value versus corrosion level

(b) Residuals versus corrosion level

(c) Histogram of residuals of dimensionless slip value

FIGURE 6: Dimensionless slip value corresponding to bond strength of pull-out specimens with deformed rebar (stirrups): (a) dimensionless slip value versus corrosion level; (b) residuals versus corrosion level; (c) histogram of residuals of dimensionless slip value.

standard deviation (σ) of the residuals are 0.00304 and 0.297 (see Figure 9(c)) for pull-out specimens. The mean value (μ) and standard deviation (σ) of the residuals are -0.0142 and 0.193 (see Figure 10(c)) for beam-end specimens. Results confirmed that the crack opening could serve as an index to stochastically evaluate bond strength degradation.

From the above studies, the degradation of dimensionless bond parameters can be modeled by the following stochastic process:

$$\frac{\tau_{max}^{\xi_s}}{\tau_{max}^0} = A + \frac{(B\xi_s + C)}{\left((\xi_s + D)^E + F\right)} + N\left(0, \sigma_{\tau_{max}^{\xi_s}}{}^2\right) \quad (12)$$

$$\frac{\delta^{\xi_s}}{\delta^0} = \frac{1}{(G\xi_s + 1)} + N\left(0, \sigma_{\delta^{\xi_s}}{}^2\right) \quad (13)$$

$$\frac{\tau_{max}^{\omega}}{\tau_{max}^0} = \frac{1}{(H\omega^I + 1)} + N\left(0, \sigma_{\tau_{max}^{\omega}}{}^2\right) \quad (14)$$

where parameters A, B, C, D, E, F, G, H, and I are the variables determined from regression analysis and $N(\mu, \sigma^2)$ is the normal distribution of residuals. Variables and standard deviations are listed in Table 4. The above analysis was based on recollected data indicating a general area loss of less than 20%; as such, these results are only applicable to corrosion levels below 20%. Additional data are still needed to refine the proposed formulas.

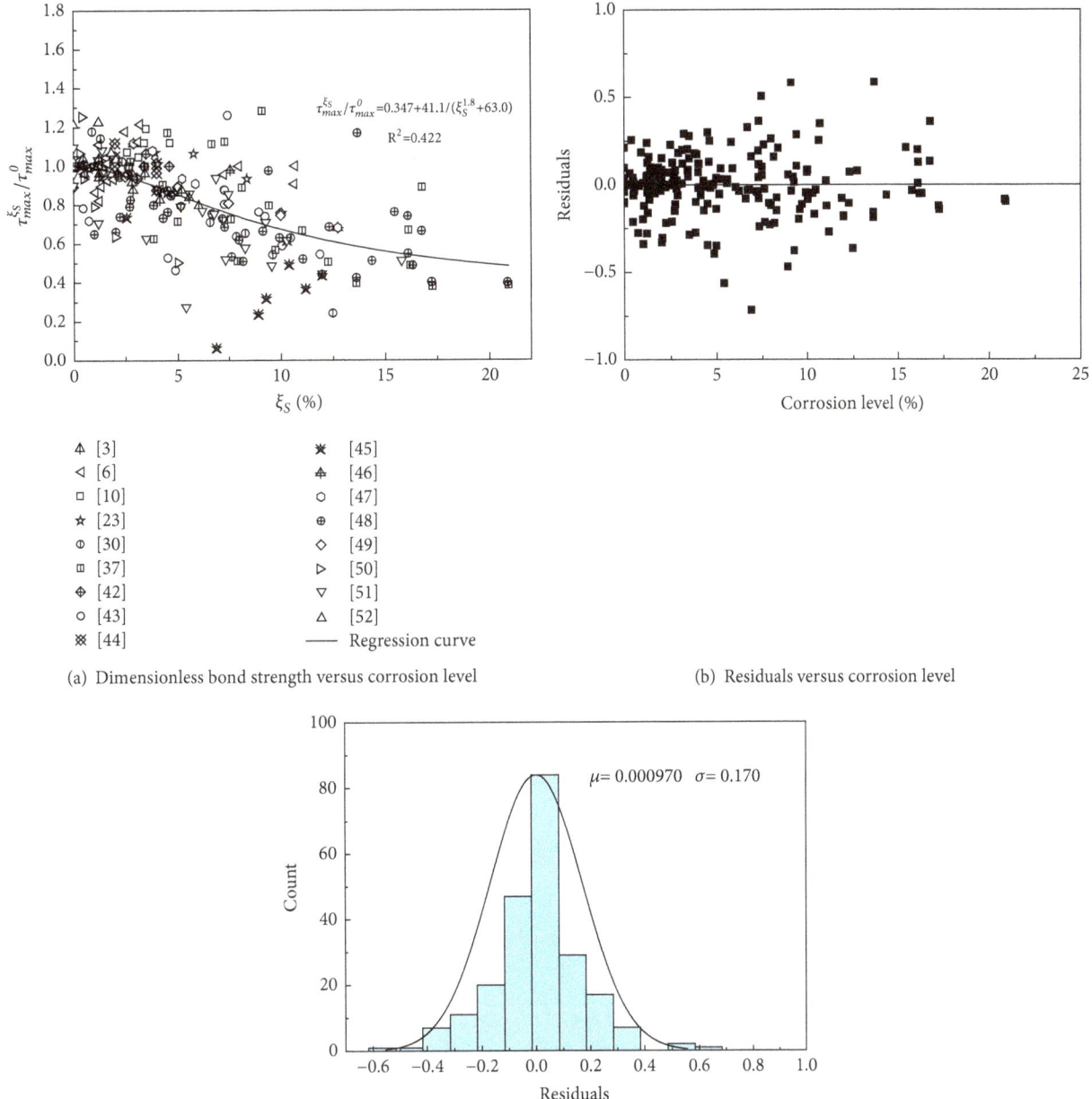

(a) Dimensionless bond strength versus corrosion level

(b) Residuals versus corrosion level

(c) Histogram of residuals of dimensionless bond strength

FIGURE 7: Dimensionless bond strength of beam-end specimens: (a) dimensionless bond strength versus corrosion level; (b) residuals versus corrosion level; (c) histogram of residuals of dimensionless bond strength.

5. Conclusions

This paper carried out nonlinear regression analysis of recollected test data on bond strength, corresponding slip value, and crack width following rebar corrosion. Four different test groups were discussed: pull-out tests of plain rebars, pull-out tests of deformed rebars with stirrup confinement, pull-out tests of deformed rebars without stirrups, and beam-end specimens. An F-test indicated the regression formulas were significant. The residuals were further analyzed and fitted by normal distributions. Unified formulas for a stochastic process describing dimensionless bond strength, corresponding slip value as corrosion level, and maximum crack width increases were proposed. Our findings suggest the following:

(1) Regarding dimensionless bond strength of plain bars, the regression formula indicates the bond strength first increased significantly and then decreased as the corrosion level increased. Maximum dimensionless bond strength was reached at an approximate corrosion level of 1.5%. Bond strength varied significantly,

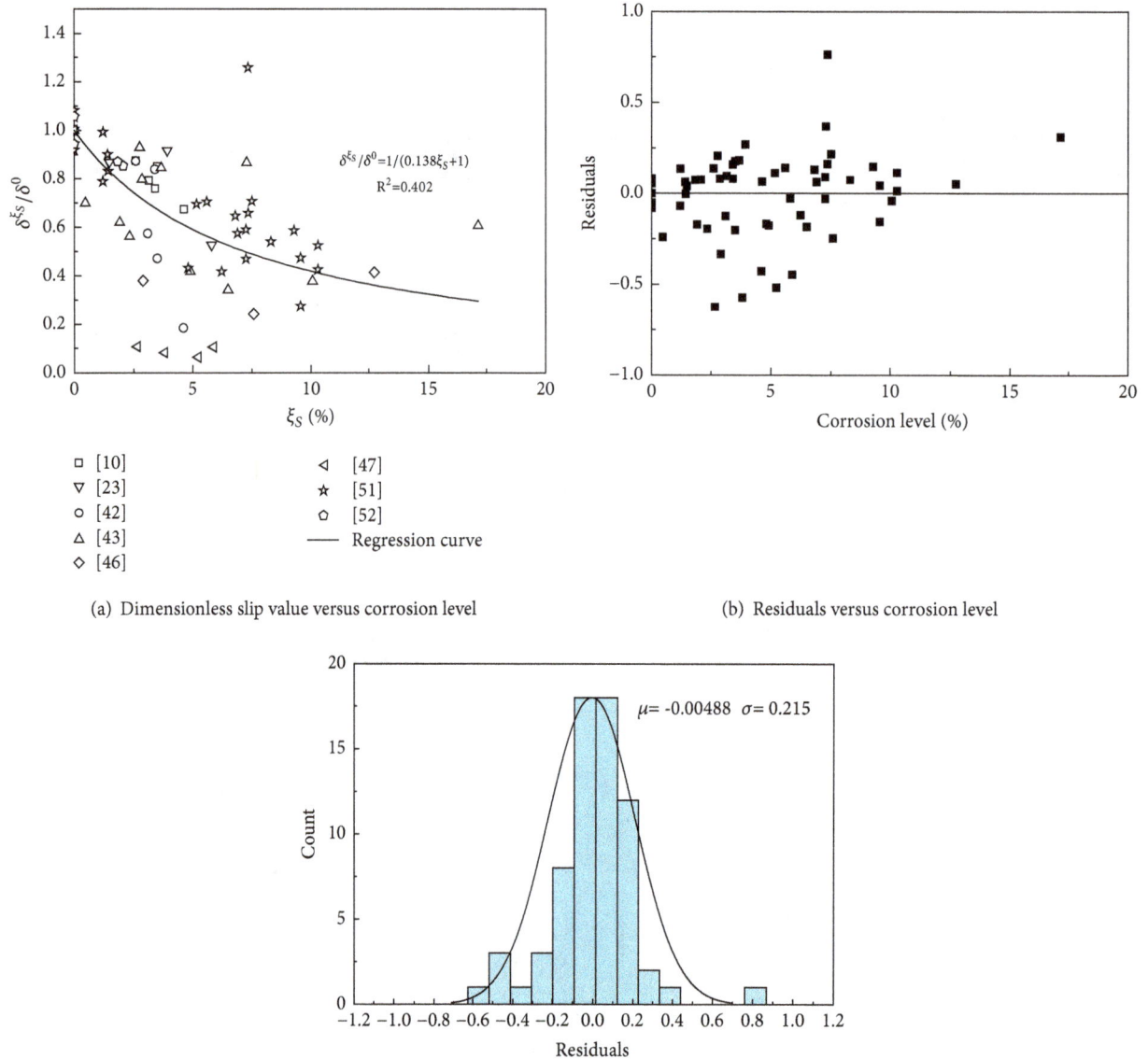

(a) Dimensionless slip value versus corrosion level

(b) Residuals versus corrosion level

(c) Histogram of residuals of dimensionless slip value

FIGURE 8: Dimensionless slip value corresponding to bond strength of beam-end specimens: (a) dimensionless slip value versus corrosion level; (b) residuals versus corrosion level; (c) histogram of residuals of dimensionless slip value.

TABLE 4: Variables and standard deviations of the proposed stochastic model.

	A	B	C	D	E	F	G	H	I	$\sigma_{\tau_{max}^{\xi_s}}$	$\sigma_{\delta^{\xi_s}}$	$\sigma_{\tau_{max}^{\omega}}$
Pull-out specimens with plain rebar	-	5.57	5.48	0.24	2.0	5.43	0.354	-	-	0.355	0.263	-
Pull-out specimens with deformed rebar (no stirrups)	0.278	-	10.8	-	1.8	13.9	0.123	-	-	0.253	0.234	-
Pull-out specimens with deformed rebar (with stirrups)	0.0760	-	149	-	1.8	161	0.087	0.968	1.35	0.190	0.274	0.297
Beam-end specimens with deformed rebar	0.347	-	41.1	-	1.8	623	0.138	0.461	1.23	0.170	0.215	0.193

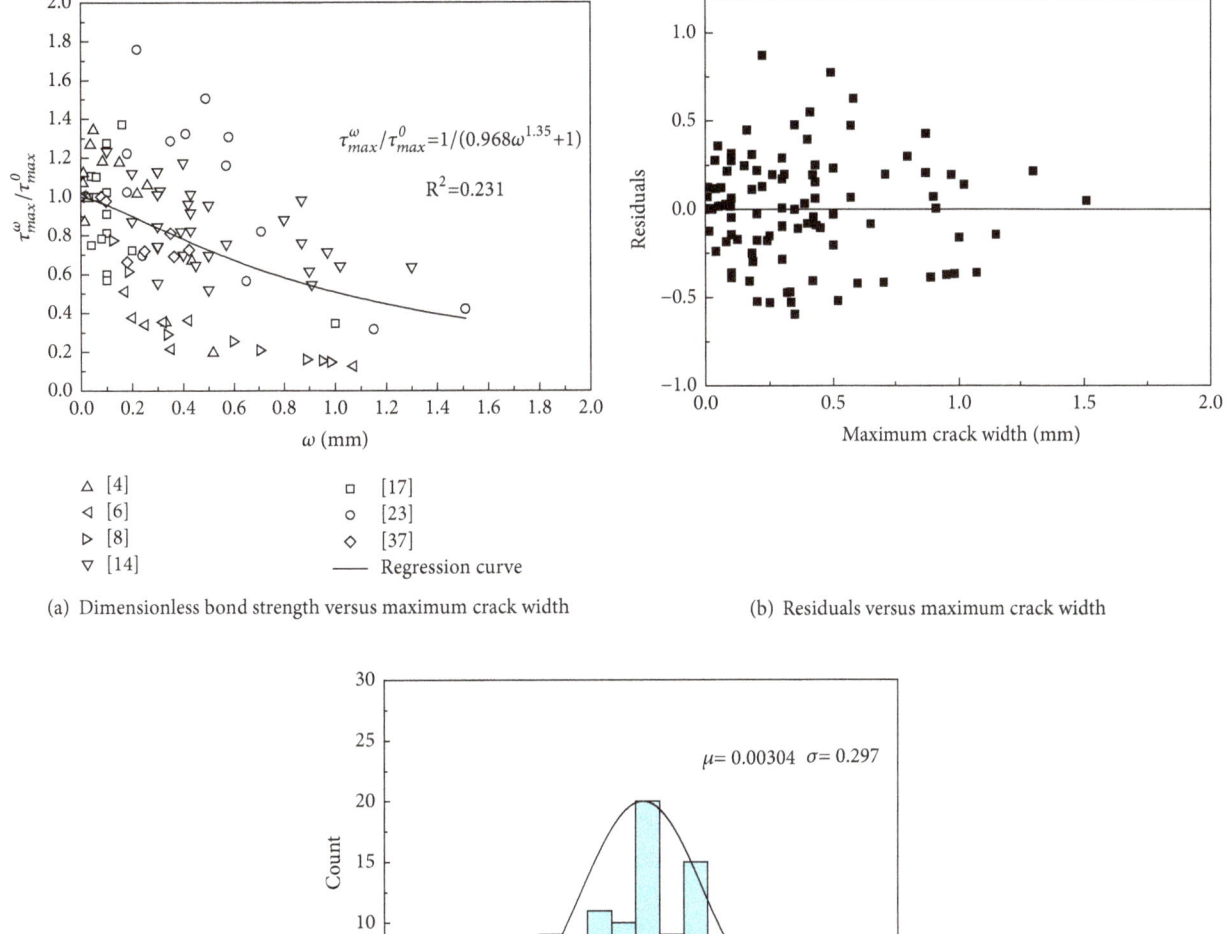

(a) Dimensionless bond strength versus maximum crack width

(b) Residuals versus maximum crack width

(c) Histogram of residuals of dimensionless bond strength

FIGURE 9: Dimensionless bond strength versus maximum crack width of pull-out specimens with deformed rebar: (a) dimensionless bond strength versus maximum crack width; (b) residuals versus maximum crack width; (c) histogram of residuals of dimensionless bond strength.

showing a standard deviation (σ) of 0.355. The dimensionless slip value decreased as the corrosion level increased.

(2) For deformed rebars of pull-out specimens and beam-end specimens, the regression analysis revealed the dimensionless bond strength initially degraded slightly as rebar corroded and then degraded significantly as the corrosion level continued to rise. Clear bond strength variations appeared between specimens with and without stirrups. As the corrosion level increased, the dimensionless slip value declined significantly.

(3) The dimensionless bond strength could also be predicted from the maximum crack width. Regression analysis showed the crack opening could serve as an index to stochastically evaluate bond strength degradation.

This paper represents a preliminary study based on recollected test specimens with a corrosion level below 20% (i.e., the proposed formula applies only to specimens with a corrosion level lower than 20%). The standard deviations reported in this paper were large in some cases, requiring additional experimental tests to further calibrate and improve the proposed formulas.

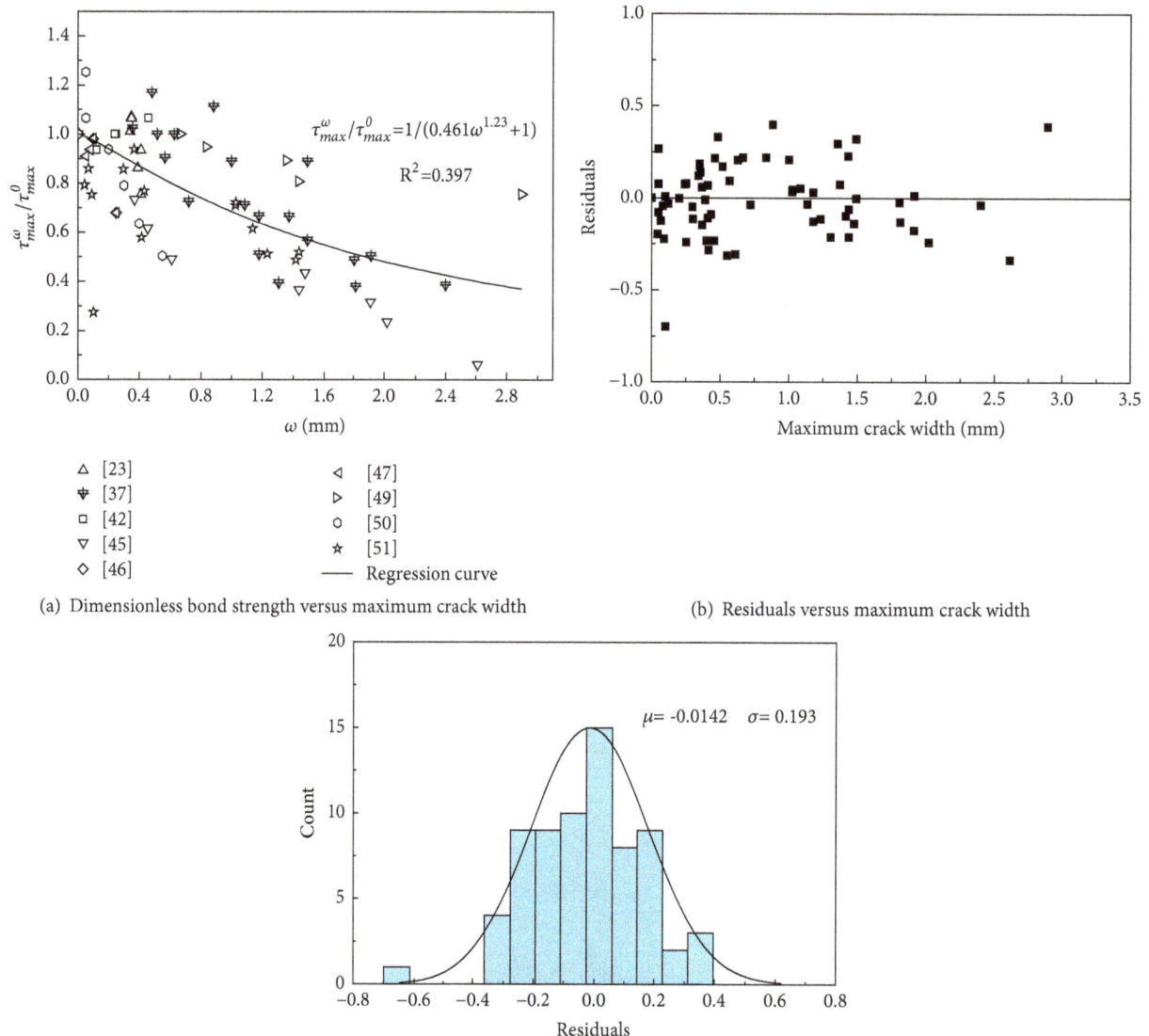

(a) Dimensionless bond strength versus maximum crack width

(b) Residuals versus maximum crack width

(c) Histogram of residuals of dimensionless bond strength

FIGURE 10: Dimensionless bond strength versus maximum crack width of beam-end specimens: (a) dimensionless bond strength versus maximum crack width; (b) residuals versus maximum crack width; (c) histogram of residuals of dimensionless bond strength.

Conflicts of Interest

The authors declare that there are no conflicts of interest regarding the publication of this paper.

Acknowledgments

The work described in this paper was financially supported by the National Natural Science Foundation of China (Grant no. 51378313) and the Ministry of Science and Technology for the 973-project (no. 2011CB013604). The first author gratefully acknowledges the support of the China Scholarship Council for a 1-year visit as a visiting research scientist in the Department of Civil Engineering and Engineering Mechanics, Columbia University.

References

[1] M. M. S. Cheung, K. K. L. So, and X. Q. Zhang, "Life cycle cost management of concrete structures relative to chloride-induced reinforcement corrosion," *Structure Infrastructure Engineering*, vol. 8, no. 12, pp. 1136–1150, 2012.

[2] I. Khan, R. François, and A. Castel, "Prediction of reinforcement corrosion using corrosion induced cracks width in corroded reinforced concrete beams," *Cement and Concrete Research*, vol. 56, pp. 84–96, 2014.

[3] W. Jin, "Effect of corrosion on bond behavior and bending strength of reinforced concrete beams," *Journal of Zhejiang University Science A*, vol. 2, no. 3, p. 298, 2001.

[4] Y. X. Zhao and W. L. Jin, "Test study on bond behavior of corroded steel bars and concrete," *Journal of Zhejiang University*

(Engineering Science), vol. 36, no. 4, pp. 352–356, 2002 (Chinese).

[5] Y. C. Xv, *Research on bond performance between the rusted bar and concrete by pullout tests*, Huazhong University of Science Technology, 2006.

[6] C. M. Li, *Experimental study on bond properties of non-uniformly corroded bar and concrete*, North China University of Water Resources and Electric Power, 2012.

[7] F. Tondolo, "Bond behaviour with reinforcement corrosion," *Construction and Building Materials*, vol. 93, pp. 926–932, 2015.

[8] A. A. Almusallam, A. S. Al-Gahtani, and A. R. Aziz, "Effect of reinforcement corrosion on bond strength," *Construction and Building Materials*, vol. 10, no. 2, pp. 123–129, 1996.

[9] C. Fang, K. Lundgren, L. Chen, and C. Zhu, "Corrosion influence on bond in reinforced concrete," *Cement and Concrete Research*, vol. 34, no. 11, pp. 2159–2167, 2004.

[10] J. Wei, H. Zhang, G. Xv et al., "Experimental Study on the bond behavior between concrete and corroded steel bars," *Journal of Railway Science and Engineering*, vol. 6, no. 4, pp. 28–31, 2009 (Chinese).

[11] H. Zhou, J. Lu, X. Xv, Y. Zhou, and F. Xing, "Experimental study of bond-slip performance of corroded reinforced concrete under cyclic loading," *Advances in Mechanical Engineering*, vol. 7, no. 3, pp. 1–10, 2015.

[12] H. J. Zhou, J. L. Lu, and X. Xv, "Effects of stirrup corrosion on bond-slip performance of reinforcing steel in concrete: An experimental study," *Construction Building Materials*, vol. 93, pp. 257–266, 2015.

[13] H. J. Zhou and X. B. Liang, "Variation and degradation of steel and concrete bond performance with corroded stirrups," *Construction Building Materials*, vol. 138, pp. 56–68, 2017.

[14] X. L. Zhang, *Experimental study on degradation of basic mechanical properties of corroded reinforced concrete*, Shenzhen University, 2016.

[15] J. B. Wang, Z. J. Qian, W. M. Qian et al., *Probability and Statistics: Engineering Mathematics*, Tongji University Press, 1994.

[16] L. G. Chen and C. Q. Fang, "Bond property of reinforced concrete with corroded reinforcement," *Industrial Construction*, vol. 34, no. 5, pp. 15–17, 2004.

[17] G. Xv, *Study on Bond Performance of Corroded Reinforced Concrete*, Huazhong University of Science Technology, 2007.

[18] L. Wang, Y. F. Ma, and J. R. Zhang, "Contrast experimental study on bond property of corroded reinforcement," *Joutnal of highway and transportation research and development*, vol. 27, no. 6, pp. 91–96, 2010.

[19] C. B. Bao, *Experimental study on the bond of corroded reinforced concrete*, Shandong University, 2015.

[20] J. P. Zhong, "Effect of rust corrosion on bond performance between steel bar and concrete," *Low Temperature Architecture Technology*, vol. 34, no. 11, p. 6, 2012.

[21] J. Wei, G. Xv, and Q. Wang, "Bond strength modeling for corroded reinforcing bar in concrete," *Journal of Building Structures*, vol. 29, no. 12, pp. 123–126, 2008.

[22] C. Fang, K. Lundgren, M. Plos, and K. Gylltoft, "Bond behaviour of corroded reinforcing steel bars in concrete," *Cement and Concrete Research*, vol. 36, no. 10, pp. 1931–1938, 2006.

[23] Y. Zhao, H. Lin, K. Wu, and W. Jin, "Bond behaviour of normal/recycled concrete and corroded steel bars," *Construction and Building Materials*, vol. 48, pp. 348–359, 2013.

[24] A. R. L. Kivell, *Effects of bond deterioration due to corrosion on seismic performance of reinforced concrete structures*, 2012.

[25] Y. S. Choi, S. T. Yi, and M. Y. Kim, "Effect of corrosion method of the reinforcing bar on bond characteristics in reinforced concrete specimens," *Construction Building Materials*, vol. 54, no. 3, pp. 180–189, 2014.

[26] D. Y. Tan, *Experimental study on bond behavior of corroded reinforced concrete*, Chongqing University, 2012.

[27] X. Q. Xiao, *Experimental study on bond behavior of corroded reinforced concrete*, Central South University, 2011.

[28] H. Yalciner, O. Eren, and S. Sensoy, "An experimental study on the bond strength between reinforcement bars and concrete as a function of concrete cover, strength and corrosion level," *Cement and Concrete Research*, vol. 42, no. 5, pp. 643–655, 2012.

[29] A. Kivell, A. Palermo, and A. Scott, "Complete model of corrosion-degraded cyclic bond performance in reinforced concrete," *Journal of Structural Engineering (United States)*, vol. 141, no. 9, Article ID 04014222, 2015.

[30] J. G. Cabrera, "Deterioration of concrete due to reinforcement steel corrosion," *Cement and Concrete Composites*, vol. 18, no. 1, pp. 47–59, 1996.

[31] L. Chung, J.-H. Jay Kim, and S.-T. Yi, "Bond strength prediction for reinforced concrete members with highly corroded reinforcing bars," *Cement and Concrete Composites*, vol. 30, no. 7, pp. 603–611, 2008.

[32] S. Coccia, S. Imperatore, and Z. Rinaldi, "Influence of corrosion on the bond strength of steel rebars in concrete," *Materials and Structures*, vol. 49, no. 1-2, pp. 537–551, 2016.

[33] H.-S. Lee, T. Noguchi, and F. Tomosawa, "Evaluation of the bond properties between concrete and reinforcement as a function of the degree of reinforcement corrosion," *Cement and Concrete Research*, vol. 32, no. 8, pp. 1313–1318, 2002.

[34] Y. Zeng, *Degradation of bond behavior of corroded reinforced concrete and its effect on bending stiffness of beams*, Chongqing University, 2014.

[35] Q. C. Mo, *Study on Bond and Slip Behavior of Corroded Reinforced Concrete and Ultrasonic Testing*, Harbin Institute of Technology, 2016.

[36] X. Li, *Experimental research on the effect of corroded deformed bars and concrete bonding properties*, Yanshan University, 2013.

[37] H. W. Lin, *Experimental study on bond behavior of corroded reinforced concrete under monotonic and repeated load*, Zhejiang University, 2017.

[38] Y. Auyeung, P. Balaguru, and L. Chung, "Bond behavior of corroded reinforcement bars," *ACI Structural Journal*, vol. 97, no. 2, pp. 214–221, 2000.

[39] Y. Ma, Z. Guo, L. Wang, and J. Zhang, "Experimental investigation of corrosion effect on bond behavior between reinforcing bar and concrete," *Construction and Building Materials*, vol. 152, pp. 240–249, 2017.

[40] H. Lin, Y. Zhao, J. Ožbolt, and R. Hans-Wolf, "The bond behavior between concrete and corroded steel bar under repeated loading," *Engineering Structures*, vol. 140, pp. 390–405, 2017.

[41] W. P. Zhang and Y. Zhang, "Experimental Study on the Degradation of Bonding Properties of Corroded Reinforced Bars and Concrete after Swelling," *Building Structure*, vol. 1, pp. 31–33, 2002.

[42] G. Xv, J. Wei, and Q. Wang, "Beam Test Study on Bond Behavior of Corroded Reinforcing Bar in Concrete," *Journal of Basic Science and Engineering*, vol. 17, no. 4, pp. 549–557, 2009.

[43] Y. L. Zhang, "Bond properties and Bearing capacity of corroded reinforced concrete member," *Xi'an University of Architecture Technology*, 2011.

[44] X.-J. Hong and M. Zhao, "Loading velocity effects on bond performance between corroded bar and concrete," *Tongji Daxue Xuebao/Journal of Tongji University*, vol. 30, no. 7, pp. 792–796, 2002.

[45] D. W. Law, D. Tang, T. K. C. Molyneaux, and R. Gravina, "Impact of crack width on bond: Confined and unconfined rebar," *Materials and Structures/Materiaux et Constructions*, vol. 44, no. 7, pp. 1287–1296, 2011.

[46] H. C. Wang, S. Q. He, and J. X. Hong, "Experimental studies on the bond character between corroded reinforcement and concrete subjected to freeze-thaw cycles," *Concrete*, vol. 8, p. 1, 2007.

[47] S. Q. He, *Experimental Study on Durability of Reinforced Concrete Members in Chloride Environment*, Dalian University of Technology, 2004.

[48] H. Lin and Y. Zhao, "Effects of confinements on the bond strength between concrete and corroded steel bars," *Construction and Building Materials*, vol. 118, pp. 127–138, 2016.

[49] A. Shetty, K. Venkataramana, and K. S. B. Narayan, "Effect of corrosion on flexural bond strength," *Journal of Electrochemical Science Engineering*, vol. 4, no. 3, 2014.

[50] P. S. Mangat and M. S. Elgarf, "Bond characteristics of corroding reinforcement in concrete beams," *Materials and Structures/Materiaux et Constructions*, vol. 32, no. 216, pp. 89–97, 1999.

[51] M. M. Yang, *Experimental study on the influence of the concrete cover thickness and steel bar position on the bond behaviors between corrosion steel bars and concrete*, Dalian University of Technology, 2016.

[52] H. N. He, M. M. Yang, and J. X. Hong, "Influence of cover thickness on bond behaviors between concrete and slight corroded bars," *Journal of Water Resources and Architectural Engineering*, vol. 14, no. 4, pp. 25–30, 2016 (Chinese).

9

Tribocorrosion of Passive Materials: A Review on Test Procedures and Standards

A. López-Ortega ⓘ,[1] **J. L. Arana,**[2] **and R. Bayón**[1]

[1]*IK4-TEKNIKER, Eibar, Spain*
[2]*Department of Metallurgical and Materials Engineering, University of the Basque Country, Spain*

Correspondence should be addressed to A. López-Ortega; ainara.lopez@tekniker.es

Academic Editor: Ramesh Chinnakurli

This paper reviews the most recent available literature relating to the electrochemical techniques and test procedures employed to assess tribocorrosion behaviour of passive materials. Over the last few decades, interest in tribocorrosion studies has notably increased, and several electrochemical techniques have been adapted to be applied on tribocorrosion research. Until 2016, the only existing standard to study tribocorrosion and to determine the synergism between wear and corrosion was the ASTM G119. In 2016, the UNE 112086 standard was developed, based on a test protocol suggested by several authors to address the drawbacks of the ASTM G119 standard. Current knowledge on tribocorrosion has been acquired by combining different electrochemical techniques. This work compiles different test procedures and a combination of electrochemical techniques used by noteworthy researchers to assess tribocorrosion behaviour of passive materials. A brief insight is also provided into the electrochemical techniques and studies made by tribocorrosion researchers.

1. Introduction

In accordance with the ASTM G 40 Standard [1], tribocorrosion can be defined as a synergetic process involving the simultaneous action of contact between surfaces in relative motion with the chemical reactions in the environment, where each process is affected by the action of the other and, in many cases, accelerated.

The first steps in the field of tribocorrosion date back to 1875, when Edison observed alterations in the coefficient of friction with different applied potentials [2]. The effect of surface chemistry on the mechanical response of materials has been investigated since the beginning of the twentieth century [3]. Between the late 1970s and early 1980s, the effect of wear on corrosion was studied by several researchers in different industrial application systems, i.e., abrasion-corrosion, erosion-corrosion, or sliding-corrosion [4]. But it was not until the 1990s that tribocorrosion mechanisms in sliding contact were proposed by Mischler et al. (1993) [5] and Madsen (1994) in a standard form [6].

Tribocorrosion encompasses several industrial sectors, e.g., material processing, energy conversion, transportation, oil and gas exploration, medical and dental implants, surgical devices, among others [2, 7, 8]. Due to its impact on daily life and potential economic benefits, interest in the study of tribocorrosion phenomenon has increased over the last few decades [3, 9, 10]. As a consequence, several electrochemical techniques have been adapted to be applied to tribocorrosion research. Thus, crucial improvements in the study of tribocorrosion have been achieved through a better interpretation of triboelectrochemical results [7, 11].

Despite growing interest and the enhancement of electrochemical techniques, standardized testing methodology for tribocorrosion evaluation has only been made available recently. Prior to 2016, the only existing standard was the ASTM G119 [6], which describes a method to determine the synergism between wear and corrosion. However, this standard has drawn criticism from several authors who developed different approaches to study the tribocorrosion behaviour of passive materials [3, 12, 13]. In 2016, a new standard (UNE 112086 [14]) was published, with a different test protocol to address the drawbacks of the older standard. The ASTM standard involves wear testing under anodic and cathodic polarization, whereas the new standard approach involves wear tests at open circuit potential.

Up to now, knowledge acquired on tribocorrosion has been achieved by performing and combining different electrochemical techniques before, during, and after the wear process, to evaluate the influence of wear on corrosion, and vice versa. This work has compiled the combination of electrochemical techniques used by noteworthy researchers to assess tribocorrosion behaviour of passive materials and coatings in sliding contacts and divided them into different test procedures.

2. Tribocorrosion: Description and Background

2.1. Tribocorrosion. Tribocorrosion can be defined as the irreversible transformation of materials resulting from the simultaneous action of mechanical loading (e.g., friction, erosion, abrasion) and chemical/electrochemical interactions with the surrounding environment, i.e., corrosion attack. In other words, it is the interaction of chemical, electrochemical, and tribological factors in materials in mechanical contact with each other, under relative motion in a corrosive environment. It combines two major scientific areas, i.e., tribology and corrosion. The former comprises the study of friction, wear, and lubrication, whereas the latter is related to the chemical aspects of material degradation [2–5, 7, 8, 10–13, 15, 16].

Tribocorrosion involves a synergism between wear and corrosion, since the degradation caused by the combined action of mechanical and electrochemical processes is larger than the sum of each of them acting separately [2–8, 10–13, 15, 16]. This synergism can be either beneficial or detrimental, depending on the surface reactions that take place in the tribological contact [2, 7]. The reaction products formed on the surface can protect the surface by forming self-lubricating layers, or accelerating the material degradation by third-body effect, for instance [2, 7–9, 17–20].

Degradation of material due to tribocorrosion may occur under a variety of wear mechanisms (erosion, abrasion, microabrasion, fatigue, fretting, sliding wear, etc.) interacting with corrosion. Furthermore, contacts between surfaces in tribocorrosion can be both two-body or three-body contacts, and there are different contact modes such as sliding, fretting, rolling, impact, etc. The relative motion between surfaces can be either unidirectional or reciprocating [2, 3, 7, 8, 10, 16].

2.2. Tribocorrosion Test Apparatus. For tribocorrosion, the test equipment shall allow monitoring and control of both mechanical and electrochemical parameters. The apparatus used to measure tribological properties is named tribometer. A tribometer creates relative motion, either unidirectional (rotatory) or bidirectional (reciprocating), rubbing two surfaces against each other. On the other hand, electrochemical cells are used to record and control the electrochemical

parameters. These cells are usually composed of three electrodes: a reference electrode, a counterelectrode, and the working electrode. The reference electrode has a stable, well-defined potential, and it is used to register the potential of the working electrode, i.e., the test sample. Typical reference electrodes are Saturated Camomel Electrodes (SCE) and Silver/Silver Chloride electrodes (Ag/AgCl). The counterelectrode is used to measure or control the current and is usually made of inert materials such as platinum, gold, or graphite. The electrodes are connected to a potentiostat to register the potential between the reference electrode and the working electrode, or the current between the counterelectrode and the working electrode. A typical tribocorrosion test setup is schematically shown in Figure 1, for a unidirectional tribometer under a ball-on-disc configuration.

There is no standardized test apparatus for tribocorrosion tests, which makes the interlaboratory comparability difficult for results [2]. Generally, existing tribometers are modified to incorporate the electrochemical cell. The electrochemical cells and all parts of the tribometer in contact with the sample or the test solution, e.g., the counter-body holder, are usually made of insulating materials such as nylon or Teflon, to provide adequate insulation for current leakages. Furthermore, the electrochemical cell must be designed so as to avoid any electrolyte leakage.

Finally, the test conditions should be selected to be as close to real conditions as possible. The contact geometry determines the contact area, and some factors such as the normal load and the sliding velocity are critical aspects, since they determine the depassivation rate of the system [7].

2.3. Historical Background. The earliest studies in the passivation phenomenon were performed using acidic solutions (e.g., sulphuric acid), and they were focused on understanding the formation of the passive film and its behaviour under corrosion-wear solicitations [16, 21–23]. The electrolyte and materials used in tribocorrosion tests were reasonably well-adapted to the technological interests, so saline solutions began to be used more in order to reproduce industrial applications [24–28] and thus predict the useful life of components more accurately. Remarkable developments in the marine industry have also promoted tribocorrosion studies in synthetic seawater in later years [29–34]. On the other hand, developing coatings with enhanced wear and corrosion properties has also aroused interest, and a great number of tribocorrosion studies have taken place over the last decade to evaluate the performance of coatings produced by different depositing techniques (HVOF, PVD, etc.) [27, 35–37]. Finally, the increasing impact and growing interest of biomedicine over the last decade have led to a great number of studies involving biomedical alloys used in orthopaedic and dental implants [2, 38–40], which are subjected to both corrosion and wear as a result of human activity. Biomedical materials such as the CoCrMo [41–43] alloy or titanium alloys [44–46] have been widely studied in simulated body fluids (SBF), e.g., artificial saliva, NaCl solutions, phosphate buffer saline (PBS) solutions, or foetal bovine serum (FBS) solutions.

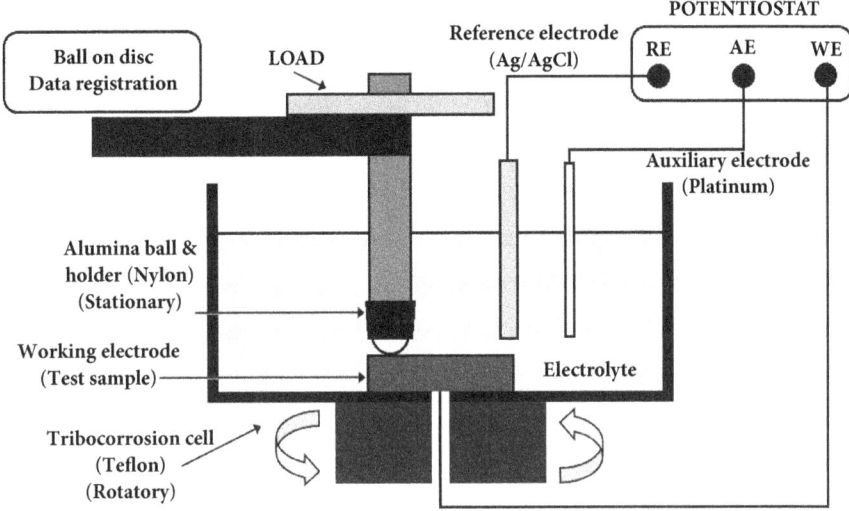

FIGURE 1: Schematic of a unidirectional tribocorrosion ball-on-disc experimental setup.

2.4. Tribocorrosion Test Procedures. The test procedures employed to assess tribocorrosion behaviour of passive materials involve performing and combining different electrochemical techniques. In general, tribocorrosion tests consist of analysing the electrochemical behaviour of surfaces before and during the tribological process, in order to evaluate how wear influences the electrochemical response of materials exposed to specific corrosive media, under determined mechanical conditions. The most widely used electrochemical techniques in tribocorrosion evaluation are potentiodynamic and potentiostatic tests, electrochemical impedance spectroscopy (EIS), open circuit potential registration (OCP), and electrochemical noise (EN) analysis.

The following sections explain how the electrochemical techniques used in tribocorrosion studies are combined. The sections have been divided into tests under controlled potential and tests under open circuit conditions. The different electrochemical techniques and their usefulness are described, and studies performed in tribocorrosion fields and relevant findings are included at the end of each section. The following section collects the current standardized tribocorrosion test procedures. The last sections provide a brief insight into the different approaches and models for tribocorrosion phenomenon proposed over the last few decades plus future perspectives.

3. Tribocorrosion Tests under Controlled Potential

3.1. Potentiodynamic Polarization Tests. The potentiodynamic polarization technique is one of the most widely used test methods to evaluate corrosion. It consists of putting a potential between the reference and working electrodes, in a potential difference range from the cathodic to the anodic domain at a constant sweep rate, while registering the current

being produced. In these tests, the current represents the rate of the anodic or cathodic reactions that are taking place on the working electrode surface, i.e., the studied metal exposed to the corrosive electrolyte. The registered current is typically expressed in terms of current per unit area of the working electrode, that is, the current density [47]. This technique provides information about the different corrosion processes taking place on the surface of a metal, such as pitting occurrence susceptibility, passive layer formation, and the cathodic behaviour of an electrochemical system. It is the most useful method to evaluate the active/passive behaviour of the materials at different potentials and to determine the kinetics of ongoing reactions, i.e., the corrosion rate [10, 20, 48].

Performing potentiodynamic polarization tests during wear experiments can be used to evaluate the effect of wear on the electrochemical reactions on the surface of the working electrode, as a function of the potential applied, while being rubbed against an insulating body. Therefore, it is a quick and useful tool to detect the possible effects of friction on the electrochemical kinetics of the system, and vice versa [11, 49]. As a first approach, the current registered during potentiodynamic polarization in a material subjected to a sliding process is the sum of two components, namely, the currents of the worn and unworn areas. Furthermore, the coefficient of friction usually varies with applied potential during a potentiodynamic scan, due to the electrochemical changes taking place in the surface state of the material in the tribological contact [3, 50].

Figure 2(a) shows an example of a polarization curve obtained for a Monel K500 alloy in artificial seawater under corrosion-only and tribocorrosion (wear and corrosion) conditions. It can be clearly seen that, under tribocorrosion conditions, the corrosion potential is shifted to more cathodic potentials, and the corrosion current density is increased by two orders of magnitude [29]. Furthermore, there is a

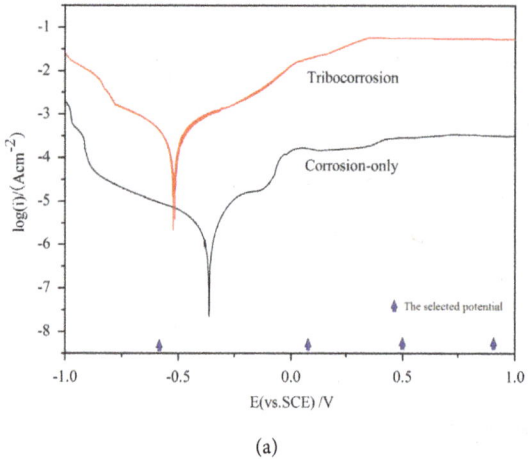

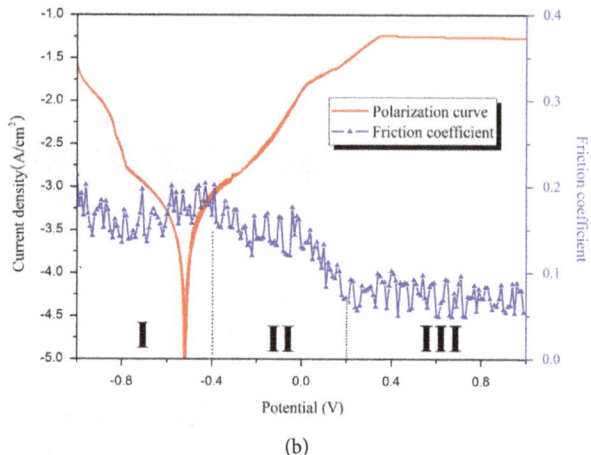

(a)   (b)

FIGURE 2: (a) Polarization curves of a Monel K500 alloy in artificial seawater obtained in corrosion-only and tribocorrosion conditions. (b) Polarization curve of Monel K500 alloy in artificial seawater and evolution of coefficient of friction obtained during measuring. The corrosion potential under tribocorrosion conditions is shifted to more cathodic potentials, and the corrosion current density increases by two orders of magnitude. The coefficient of friction decreases with higher anodic potentials as a consequence of the presence of oxide film on the surface [29].

correlation between the coefficient of friction and the current density arising from the polarization imposed (Figure 2(b)). In this case, the coefficient of friction decreases at higher anodic potentials, as a consequence of oxide films forming on the surface [29]. Thus, the potentiodynamic polarization technique performed with and in the absence of sliding provides interesting information on both the effect of wear on corrosion and the influence of corrosion on wear behaviour.

3.2. Potentiostatic Polarization Tests. Electrochemical polarization is used to simulate the oxidizing action of a corrosive environment. This technique consists of imposing a fixed potential between the reference and the working electrodes. The current is measured as a function of time, to evaluate the evolution of electrochemical kinetics of reactions occurring on the electrode surface [11]. The potential value being applied determines the dominant electrochemical reactions taking place. The polarization curve of passive materials can be divided into three regions:

(i) Active region, where the metal dissolves directly in contact with the solution

(ii) Passive region, where a protective passive film of few nanometres is grown on the surface, protecting the bare material from dissolution

(iii) Transpassive region, at high anodic potentials, where the current increases sharply with potential, as a consequence of the nonstable state of the oxide layer and the breakdown of this film.

These three regions are clearly observed in the polarization curve represented in Figure 3, with i_{corr} the corrosion current density, E_{corr} the corrosion potential, i_p the current density in the passive state, and E_{pit} the potential at which pitting processes begin in the passive layer.

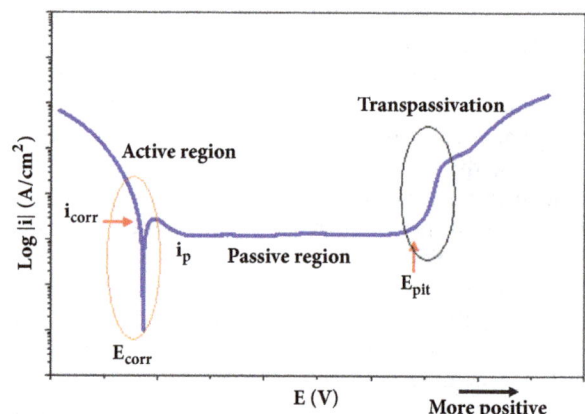

FIGURE 3: Typical Polarization curve of a passive material showing the active, passive, and transpassive regions.

Imposing a fixed potential on rubbing surfaces gives information on the potential effect on the wear behaviour [7]. The current registered during rubbing mainly flows through the wear track area, which constitutes a very small area compared to the complete metal surface exposed to the electrolyte [11]. The measured current value corresponds to the sum of the anodic and cathodic currents, due to all electrochemical reactions occurring on the exposed surface. It is thus possible to simulate different corrosion conditions by imposing appropriate potentials [7, 11, 49]. At cathodic potentials, the corrosion is inhibited, and the material loss after the tribocorrosion tests is attributed to pure mechanical wear. On the contrary, at anodic or passive potentials, the metal is covered by an oxide film whose thickness is controlled by the applied potential. This layer reduces the dissolution rate of the metal to negligible values and affects the friction response of the material.

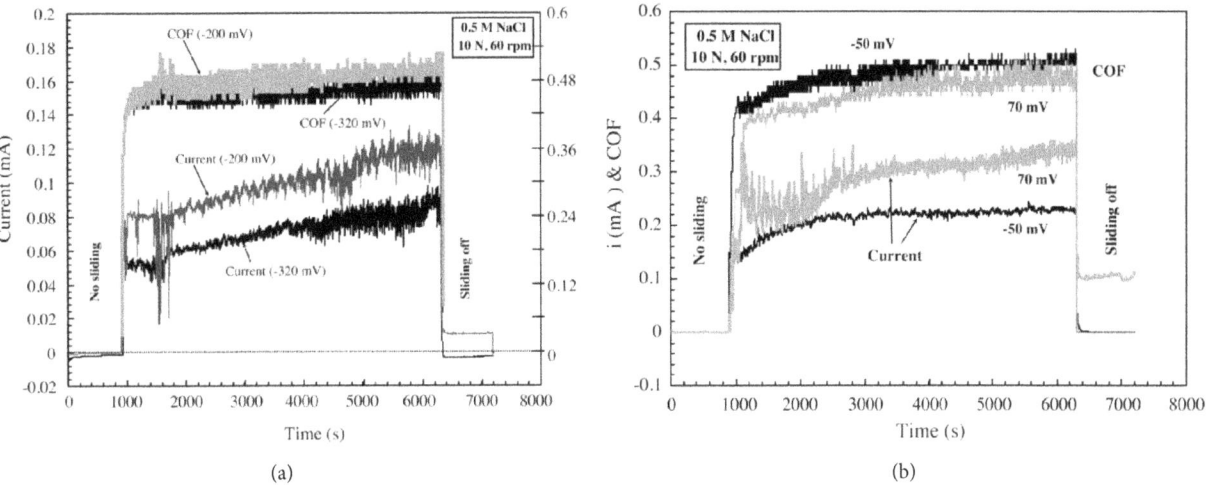

FIGURE 4: Current transients and coefficients of friction (COF) recorded before, during, and after sliding wear test at fixed potentials for an AISI 304 stainless steel in 0.5M NaCl solution: (a) -320 mV and -200 mV; (b) -50 mV and 70 mV. Removing the passive layer generates an increase in the current transient, with higher shifts at higher applied potentials. The coefficient of friction also varies with applied potential [24].

During sliding, the registered anodic current increases as a consequence of the local damage or removal of the passive layer exposing the base material surface to the electrolyte and, thus, to an active dissolution [3].

The evolution of current transient and coefficient of friction during sliding wear tests at different fixed potentials is shown in Figure 4, for a AISI 304 stainless steel in 0.5M NaCl solution [24]. Sliding results in a shift of current transient towards more positive values as a consequence of removing the passive layer, and this shift becomes higher as greater potential is applied. Moreover, the coefficient of friction value is also variable depending on the potential.

Additionally, this technique can also quantify the metal dissolution rate, i.e., the corrosion rate, from the corrosion current density (i_{corr}) registered using Faraday's law [4, 7, 11, 13, 21, 22, 48]:

$$v_{corr} = \frac{i_{corr}.M}{n.F} \qquad (1)$$

where v_{corr} is the corrosion rate, M the atomic mass, n the number of electrons taking part in the process, and F the Faraday constant (96500 coulomb per electron mol). More details on calculating corrosion rates from electrochemical measurements can be found in the ASTM G102 [51] Standard.

3.3. Combination of Potentiodynamic and Potentiostatic Polarization Techniques to Assess Tribocorrosion.

A wide number of researchers have combined polarization techniques for tribocorrosion evaluation of different passive materials in a wide variety of solutions. The polarization curves obtained from the potentiodynamic scan provide information on the electrochemical state of the material in a specific electrolyte at the different potential ranges. From the curves obtained in these tests, different potentials can be selected, namely, cathodic or passive potentials, to perform the potentiostatic tests under wear conditions. Some authors have also

performed sliding wear tests at open circuit potential in combination with these two techniques, in order to acquire information on the potential evolution with sliding and evolution of the friction coefficient under free conditions. This technique is explained in more detail in the following section.

Former studies in the tribocorrosion field consisted of analysing the passivation phenomenon. As a first approach, understanding the formation of the passive film and its behaviour under corrosion-wear solicitations was the main interest. In the following studies, the influence of different chemical and tribological parameters on the tribocorrosion response was assessed, i.e., the effect of potential, velocity and normal load, or the test solution. Some studies performed by combining these two techniques over the last few decades and their major findings are detailed below.

3.3.1. Study of Passivation Phenomenon in Tribocorrosion.

Understanding passivation drew attention from a great number of researchers in the tribocorrosion field from the outset. P. Jeremy et al. (2000) [23] studied the influence of passivation on tribocorrosion for an iron-chromium (AISI 430) alloy in sulphuric acid by measuring the current during rubbing under applied cathodic potential. They modelled the shape of the current transient considering the ohmic effects and the film growth kinetics and demonstrated that the repassivation kinetics determined the chemical metal removal rate under tribocorrosion conditions. A similar deduction was made by Mischler et al. (2001) [22] for a 34CrNiMo6 carbon steel in different aqueous solutions (NaOH and borate buffer solutions). They observed that passivity determined chemical degradation of both the metal and the detached metal wear particles and also played a crucial role in the mechanical properties of the metal surface. In 2001, Mischler and Ponthiaux [16] confirmed the reproducibility and

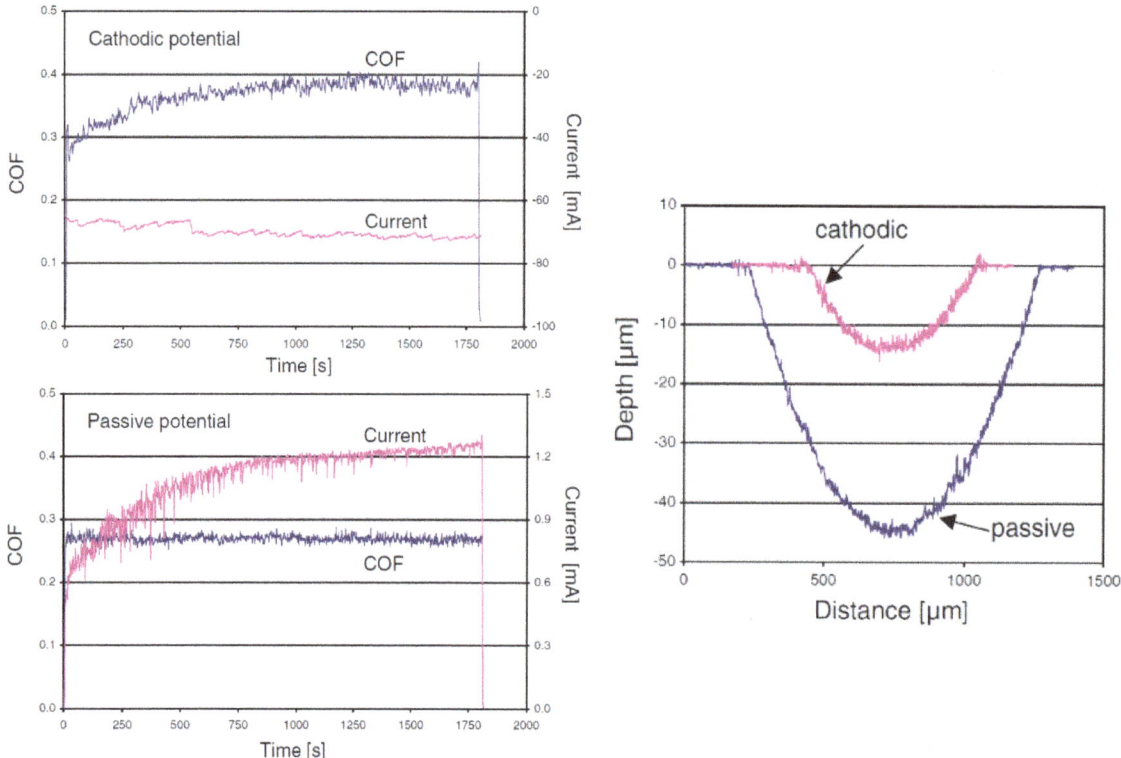

FIGURE 5: Effect of potential on the coefficient of friction and current during sliding (left) and on the wear track volume for a 316L steel in H$_2$SO$_4$ at cathodic (-1.5V) and passive (0V) potentials. The current values recorded are higher and the coefficient of friction is lower for passive potentials. The material loss at passive potentials is considerably higher due to wear-accelerated corrosion [55].

compatibility of tribocorrosion experiments analysing the electrochemical and tribological response of the AISI 316 alloy in sulphuric acid in seven different laboratories. They observed identical trends in the tribocorrosion behaviour of the alloy, an increase in passive current during rubbing, which they found to be proportional to the extent of the wear. For the same material/electrolyte system, García et al. (2001) [52] also observed sliding to cause both a potential shift to more anodic values in the polarization curves and an increase in the current values in the potentiostatic tests. They proposed the concept of active wear track to describe the depassivation-repassivation processes taking place in the contact and demonstrated that there was no need to calculate the real wear contact area. They asserted that it was possible to determine this active area by electrochemical means, using the anodic current registered during sliding corrosion-wear tests under applied potential.

3.3.2. Influence of the Electrochemical Potential, Solution Chemistry, and Sliding Velocity and Normal Load on Wear and Wear-Accelerated Corrosion. Independently of the material and solution studied, the applied potential has been found to have an important effect on wear and wear-accelerated corrosion, as a consequence of the electrochemical state of the surface, i.e., free of oxides at cathodic potentials or covered by an oxide film in the passive domain. Stemp et al. (2003) [21] investigated the effect of contact configuration

and observed that the metal loss rate was governed by the rate of mechanical depassivation. The repassivation rate, in turn, was found to be critically affected by sliding velocity and by the behaviour of the material in the wear track, which depended on the applied potential [53]. Barril et al. (2005) [54] and Favero et al. (2006) [55] observed that the extent of wear was profoundly affected by the applied potential, with higher material losses due to wear-accelerated corrosion at potentials above the corrosion potential (Figure 5). Similar observations were made for different material/electrolyte systems, by other researchers [29, 56]. Bidiville et al. (2007) [57] continued the study of Favero [55], and they further observed that the presence of a passive film affected the plastic deformation and the wear track behaviour. The metal under cathodic potential, free of oxides, was rapidly work-hardened under the applied load, and the plastic deformation resulted in limited wear. Furthermore, the wear rate was greater under passive potential due to subsurface cracking and limited plastic flow as a consequence of the passive film. Therefore, the wear rate and friction of materials can be controlled with the use of the right electrochemical potential, which could be a useful tool in industrial applications to reduce friction in certain components, as confirmed by Ismail et al. (2009) [58].

On the other hand, several studies have shown that the tribocorrosion rate also depends on the chemistry of the test solution. Espallargas and Mischler (2010) [59] investigated the passivation of an overlay-welded Ni-Cr 625 alloy in

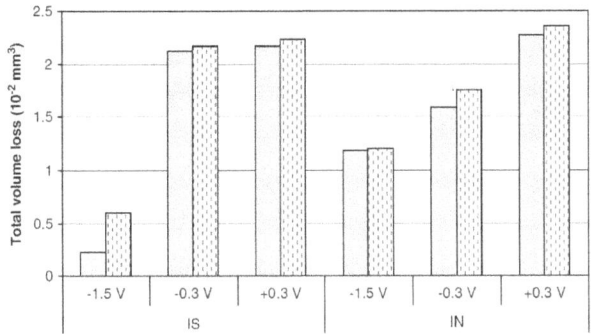

Figure 6: Total volume loss for an overlay-welded Ni-Cr 625 alloy under 15N at different applied potentials in sulphuric acid (IS) and nitric acid (IN) solutions. The material loss in IS at cathodic potentials is considerably smaller than at passive potentials, whereas this effect is less pronounced in IN, where the wear rate increases steadily with applied potential [59].

sulphuric and nitric acids, at different potentials (Figure 6). The tribocorrosion damage in sulphuric acid was observed to be less under cathodic conditions, and no effect of potential in the passive region was found. In nitric acid, however, similar wear was observed in passive and cathodic potentials. This was attributed to the oxidizing action of nitrate ions that helped form an oxide film layer even under cathodic polarization, which was found to cause wear. Muñoz and Julián (2010) [48] studied the corrosion resistance of the CoCrMo alloy in NaCl and a bovine serum (BS) solution under potentiostatic control at different applied potentials. They observed that the tribocorrosion rate depended on both the applied potential and solution chemistry. Wear was almost negligible at cathodic potentials and increased critically with increasing potentials, and the properties of the passive film were different depending on the solution chemistry, thus modifying the tribocorrosion behaviour of the alloy. Chen and Yan (2012) [31] performed tribocorrosion tests on Ti-6Al-4V and Monel K500 alloys against a 316 stainless steel in artificial seawater and distilled water. They found that the coefficient of friction was lower in seawater, as a consequence of the increase in surface roughness due to higher corrosion, which led to a reduction in the contact area. Golvano et al. (2015) [60] evaluated the effect of fluoride content and solution pH on the corrosion and tribocorrosion behaviour of titanium alloys in artificial saliva. The increase in fluoride content and the acidification of the saliva were found to have a negative effect on corrosion resistance of the alloys.

In addition to electrochemical potential and solution chemistry, tribological conditions such as sliding velocity or applied normal load also affect the tribocorrosion behaviour of materials. Radice and Mischler (2006) [61] found that the wear rate increased with increasing load and electrochemically applied potential. Sun and Rana (2011) [24] reported that the cathodic shift in the potential due to the sliding was greater with higher sliding speeds and contact loads. Dahm (2007) [62] observed that the corrosion current did

not depend linearly on the wear scar area, and the electrochemical contribution increased with increasing sliding distance. Therefore, the wear-corrosion synergies were not a fixed proportion of the total wear. Yue et al. (2014) [30] analysed the influence of microstructure evolution on the tribocorrosion behaviour of a stainless steel, and the strain-induced microstructure changes occurring during sliding were found to affect the synergistic degree of wear and corrosion.

3.3.3. Bio-Tribocorrosion. Interest in studying tribocorrosion in the biomedical field has increased over the last few decades. The major bio-tribocorrosion areas are orthopaedic science and dentistry. In both cases, the main goal is to develop or select a biocompatible material that can resist both wear and corrosion conditions for the human body, assuring certain durability with the lowest metal ion release. However, the study of bio-tribocorrosion is quite complicated, since it involves the composition of human fluids containing bacteria, proteins, and so on, leading to a more complex system with several influencing parameters.

As mentioned in the previous section, the electrode potential can affect the wear rate of materials in tribocorrosion systems. This effect has also been observed in bio-tribocorrosion studies, where the metal ion release under wear-corrosion conditions of the human body critically depends on the electrode potential, as confirmed by Espallargas et al. (2015) [63]. On the other hand, the proteins present in human body fluids have been found to have a beneficial effect on the tribocorrosion response for biomedical materials. These proteins have been found to form a biofilm on the alloy surface, composed of inorganic graphitic carbon, which acts as a boundary lubricant [64]. However, high applied potentials and normal loads can inhibit the lubricating ability of the proteins, as observed in several studies such as Liau et al. [64], Hiromoto and Mischler (2006) [65], Mathew et al. (2011) [66], or Sadiq et al. (2015) [67]. As reported by Runa et al. (2013) [68], the effect of proteins can be either favourable or undesired, depending on the characteristics of the passive film formed at different applied potentials under mechanical exposure.

4. Tribocorrosion Tests at Open Circuit Potential

The open circuit potential (OCP) is the spontaneous potential established between the working and the reference electrodes, at which anodic and cathodic reactions take place in an electrochemical system. This parameter does not give information on electrochemical reaction kinetics but indicates whether a material is noble or active in a determined electrolyte. Tribocorrosion tests at open circuit potential consist of performing sliding wear on a material immersed in a specific electrolyte, without imposing any anodic or cathodic potential [3, 11, 20]. There are three main techniques used in a wear tests at open circuit potential: OCP monitoring with and without wear, electrochemical impedance spectroscopy (EIS) measurements, and electrochemical noise (EN) technique.

4.1. Open Circuit Potential Monitoring. The open circuit potential monitoring is the simplest electrochemical method for corrosion evaluation, although it does not provide quantitative information on the interaction between wear and corrosion. To study the effect of wear, the potential is monitored before, during and after the sliding. Before the sliding test, the samples are immersed in the electrolyte until a steady state is achieved. Several authors have performed experiments with an hour of stabilisation [3, 10, 11, 20]. However, the stabilisation period depends on the material and electrolyte used and should be selected considering the system being studied. The potential evolution registered during the stabilisation period provides information on the electrochemical reactivity of the material in the test solution. An increase in the potential with immersion time reaching stable values after several minutes of immersion indicates the formation of a passive oxide layer on the surface. This layer of a few nanometres protects the material underlying from corrosion. This phenomenon is known as passivation [3, 10, 11, 20]. On the other hand, a decrease of potential with immersion time suggests that general corrosion has occurred. Finally, short-term potential fluctuations are attributed to localized corrosion processes such as pitting. These fluctuations are a consequence of the successive breakage of the passive layer and subsequent growth of the film in the affected zone [10, 12, 38]. The reaction time can be calculated using the potential evolution curve meaning the time needed upon immersion to reach a stable potential for the tested material in the selected media [3, 10, 11, 20].

Measuring potential during sliding can also provide information on the evolution of surface state when the material is subjected to wear-corrosion conditions. Once sliding begins, the potential has been observed to shift towards more negative values, indicating the initiation of electrochemical activity. Change in potential is a consequence of removing the passive layer in the tribological contact, thus exposing the bare material to corrosion. During wear, galvanic coupling between the active worn area and passive unworn area is generated [3, 10, 11, 20]. According to Ponthiaux et al. [20], there are four parameters that affect the corrosion potential during rubbing:

(i) The intrinsic corrosion potentials of the unworn and worn surfaces where the electrochemical state of the worn surface is disturbed by the removal of the passive layer and the mechanical strain caused by sliding

(ii) The ratio between worn and unworn surface areas

(iii) The relative position between worn and unworn areas

(iv) The mechanisms and kinetics of the anodic and cathodic reactions involved in the worn and unworn areas.

After sliding, the potential usually tends to increase again reaching values close to presliding figures, as a consequence of the reformation of the passive layer in the depassivated area. The potential restoration process gives information on the material's ability to recover after sliding, which is known as repassivation.

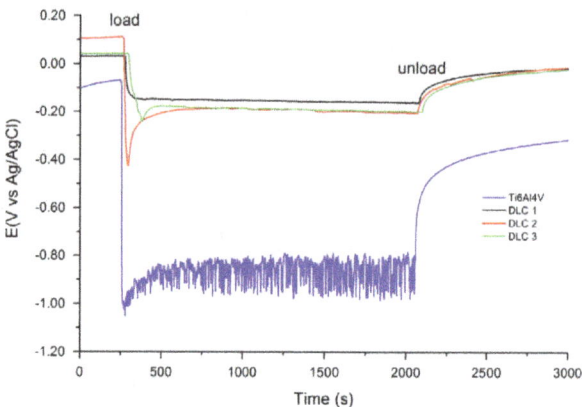

FIGURE 7: Evolution of the open circuit potential before, during, and after sliding wear test for a Ti6Al4V alloy with different DLC coatings in PBS solution. Sliding removes the passive layer on the surface, leading to a decrease in the potential value. At the end of sliding, the material is repassivated, and the potential increases to reach values close to those before sliding [69].

The shift in potential due to the action of wear can be clearly observed in Figure 7, for a Ti6Al4V alloy with several DLC (Diamond Like Carbon) coatings in Phosphate Buffered Solution (PBS) [69]. As soon as the countermaterial begins to slide against the test materials, a depassivated area is created in the surface leading to a decrease of potential due to the galvanic couple generated between the worn and unworn surfaces. Once sliding ends, the potential increases again to reach potential values close to those before wear process, as a consequence of the repassivation of the surface.

4.2. Electrochemical Impedance Spectroscopy. The electrochemical impedance spectroscopy (EIS) technique consists of exciting the system using an AC potential or small amplitude current sinusoidal signal over a wide range of frequencies and measuring the current or potential response obtained. This technique is usually performed at free potential, with the requirement that the material should be electrochemically stable, in terms of the open circuit potential. Therefore, a small amplitude, usually 5-10 mV, is employed to maintain the system in a (quasi)equilibrium state. As a consequence, the system can be evaluated without imposing significant perturbations. The impedance spectrums are obtained by plotting the electrochemical impedance, i.e., the potential and current ratio, over the frequency range being studied. The spectrums are then modelled using equivalent electric circuits that combine passive electric elements such as resistors, inductors, and capacitors. Combining the elements reproduces the electrochemical behaviour of the electrode surface. Information on the elementary steps occurring in the electrochemical reactions and their kinetics can also be obtained from the impedance diagrams [3, 10, 20, 70–72].

One of the diagrams used to represent impedance data is the so-called Nyquist plot, where the imaginary impedance, indicative of capacitive response, is represented versus the real impedance, which indicates a resistive response. Figure 8

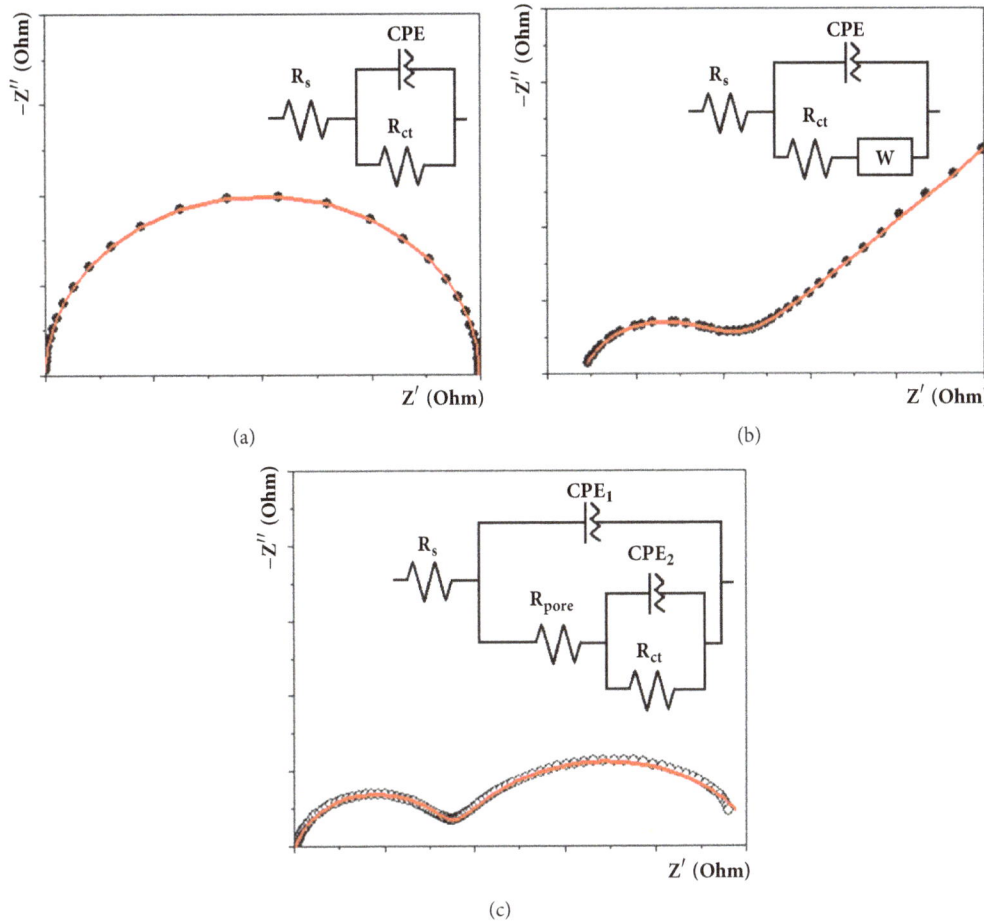

FIGURE 8: Nyquist plots for impedance data and equivalent circuits used to fill each curve for a metal or alloy (a) without any coating, (b) without a coating showing diffusion of reactants, and (c) with a passive oxide layer or a porous coating on the surface.

shows typical Nyquist diagrams, together with the electro-chemical equivalent circuit used in each curve to fit the impedance data. Figure 8(a) represents a bare metal without any coating, (b) shows the response of a metal or alloy with mixed activation and diffusion control, and (c) is the case of a metal with a protective oxide layer or a porous coating on the surface. In the equivalent circuits, R_s represents the solution resistance, CPE is the constant phase element representing the capacitive properties of the electrolyte/metal interface, and R_{ct} is the charge transfer resistance in the electrolyte/metal interface, which determines the kinetics of the reaction. The Warburg impedance (W) in Figure 8(b) indicates the occurrence of reactant diffusion to the corroding surface, as a consequence of reactant concentration gradients in the solution. In the last circuit (Figure 8(c)), CPE_1 corresponds to the capacitance of the coating, whereas CPE_2 is the capacitive properties of the electrolyte/metal interface. Finally, R_{pore} is the resistance of the paths generated in the coating, i.e., the resistance in the pores.

This technique can be used to study the role of intermediate species adsorbed in the surface, the properties of the passive films generated, and also the changes in the metal/electrolyte interface. The high frequency loop is associated with the charge transfer taking place in the interface, whereas the low frequency loop is attributed to diffusion of dissolved oxygen from the electrolyte to the metal/electrolyte interface [20, 48].

By performing impedance measurements during sliding, it is possible to study the influence of wear on the elementary processes involved in the corrosion mechanism. Analysing the changes in the impedance diagrams with sliding parameters, i.e., normal force or sliding speed, it is possible to develop a model taking into account the effects of sliding in the corrosion mechanism [10, 50]. On the other hand, the EIS measurements after sliding provide information on the recovery of the material in post-wear damage. The approximate value of corrosion current (I_{corr}) or passivation current (I_{pass}) of a metal or alloy can be determined from the polarization resistance (R_p). This value is obtained from the impedance diagrams. EIS is the most accurate technique among the different methods available to measure the polarization resistance [10].

The effect of wear on corrosion can be observed in the Nyquist and Bode plots in Figure 9. The results correspond to EIS measurements performed before and during sliding of a TaN-coated and uncoated titanium alloy in PBS solution [73]. The electrochemical data was fitted with the so-called Randles equivalent circuit.

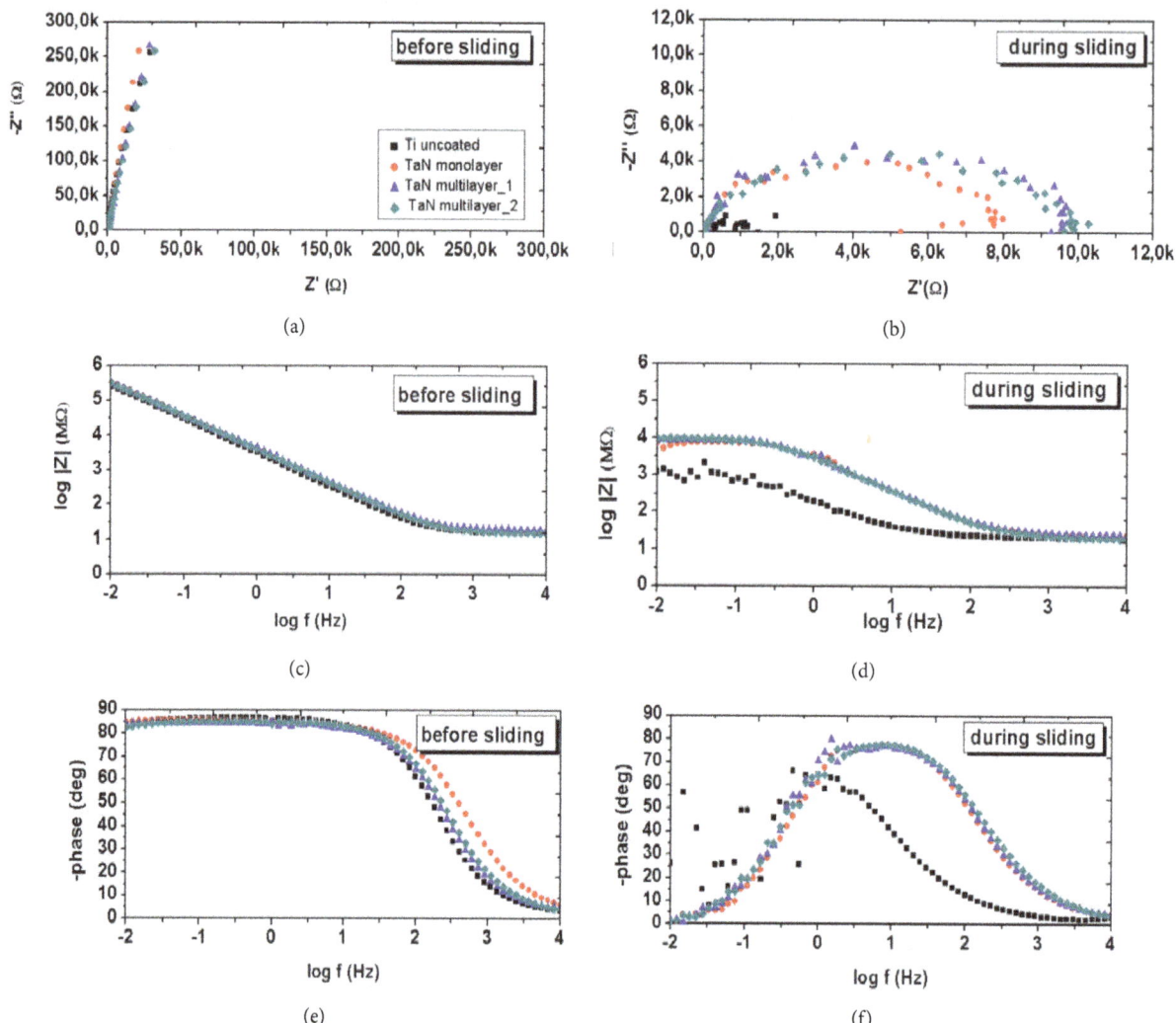

FIGURE 9: Nyquist and Bode plots of the EIS measurements performed before (a,c,d) and during (b,d,f) sliding process, for a titanium alloy and a TaN-coated titanium alloy in PBS solution. A decrease in the resistance is observed due to wear processes in both titanium and TaN coatings, but the resistance of the coatings was an order of magnitude higher than for the uncoated alloy [73].

4.3. Combination of Open Circuit Potential Monitoring and Impedance Measurements. Due to its simplicity, the combination of OCP monitoring and EIS measurements has been widely used in tribocorrosion studies. These tests usually consist of monitoring the potential until a steady state is reached, and registering the potential response during the wear test, and several minutes after the end of wear to study the depassivation and repassivation phenomena. These measurements are combined with EIS measurements before, during, or after sliding, to evaluate the effect of wear on the corrosion response of the system.

By combining OCP and EIS measurements, M. Azzi and J.A. Szpunar (2007) [39] observed that the nitriding process applied on a titanium alloy enhanced the corrosion resistance of the alloy. R. Bayón et al. (2009) [74] compared the tribocorrosion response of different Cr/CrN multilayer coatings in NaCl solution and observed the influence of the layers' thickness on the wear-corrosion behaviour. R. Bayón

et al. (2015) [69] studied the enhancement on corrosion and tribocorrosion performance of different Ti-DLC coatings deposited on Ti6Al4V biomedical alloy in simulated body fluids. M.J. Runa et al. (2015) [75] evaluated the role of organic materials (osteoblast cells) on the bio-tribocorrosion performance of a Ti6Al4V alloy in simulated body fluids and observed that the layer formed by the osteoblastic cells improved the wear and corrosion performance of the alloy. V. Saenz de Viteri et al. (2016) [76] analysed the tribocorrosion behaviour of TiO_2 coatings containing osseointegration enhancing (calcium and phosphorous) and antibacterial (iodine) elements generated by plasma electrooxidation (PEO) technique and observed an improved wear resistance for one of the developed coatings (Figure 10).

Several researchers have also combined potentiodynamic polarization techniques with open circuit potential and electrochemical impedance spectroscopy measurements. N. Papageorgiou et al. (2014) [77] analysed the tribocorrosion

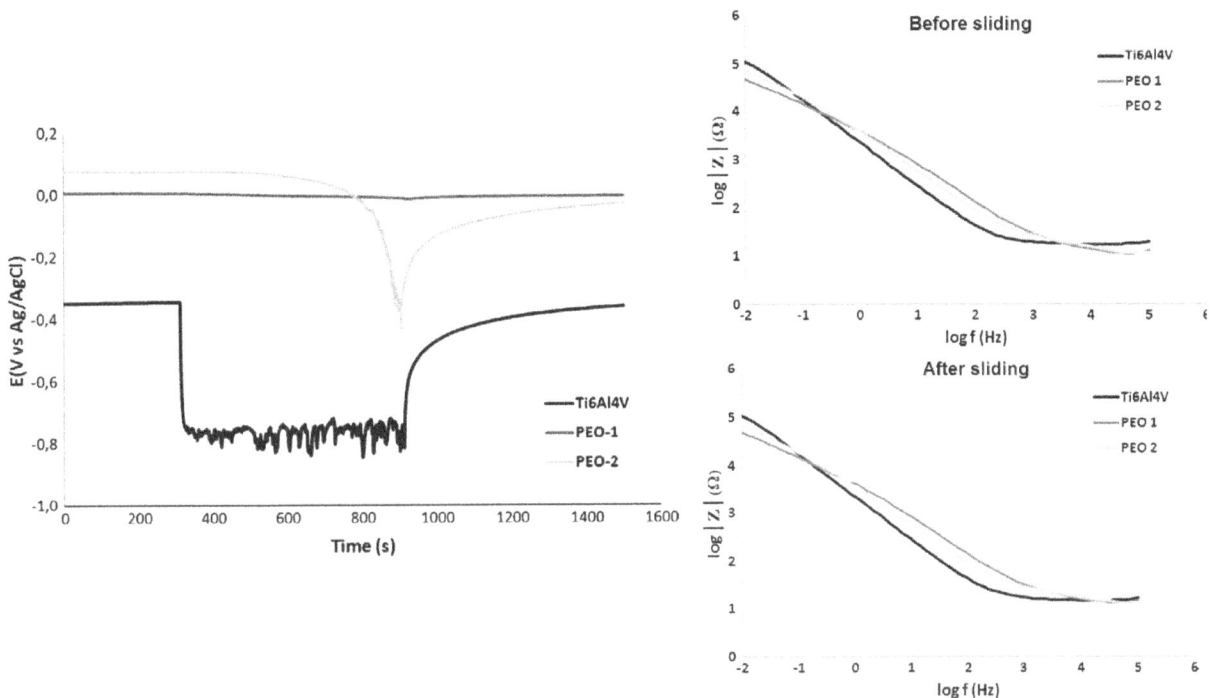

FIGURE 10: Potential shift towards more negative values due to a sliding process for a Ti6Al4V alloy with and without TiO_2 coatings generated by PEO technique in PBS solution (left) and impedance spectroscopy measurements obtained before and after sliding (right). The electrochemical state of the TiO_2 coating surface was not highly affected by sliding, and no influence of sliding on the corrosion resistance of either sample was observed [76].

performance of NiCrMo625 alloy in NaCl aqueous solution and proposed a triboelectrochemical model that proved the presence of debris particles. R. Alemón et al. (2015) [78] investigated the improvement of ion release of a CoCrMo alloy by a $TiAlVCN/CN_x$ multilayer coating. M. Fazel et al. (2015) [79] observed an enhancement in the corrosion and tribocorrosion behaviour of titanium alloys treated by micro-arc oxidation process.

Finally, the literature also includes tribocorrosion assessment studies by just performing open circuit potential measurements before, during, and after sliding. S. Rossi et al. (1999) [80] compared the wear-corrosion behaviour of (Ti,Cr)N and Ti/TiN PVD coatings in sodium chloride solution. S.A. Alves et al. (2015) [81] investigated the response of titanium oxide films with different compositions to enhance the tribocorrosion behaviour of a titanium biomedical alloy. M.Buciumeanu et al. (2016) [82] studied the tribocorrosion behaviour of hot pressed CoCrMo biomedical alloy in artificial saliva and found the processing conditions to play a relevant role on the tribocorrosion response of the alloy.

4.4. Electrochemical Noise Analysis. Electrochemical noise (EN) can be defined as the spontaneous potential and current fluctuations taking place in a metal exposed to an aggressive medium. Most corrosion processes in metals are electrochemical and, thus, are likely to generate electrochemical noise. The use of the electrochemical noise technique to assess corrosion as an *in situ* nondestructive technique dates back

to the late 70s and early 80s [40, 83, 84]. Electrochemical impedance spectroscopy and electrochemical noise technique are commonly used when the information obtained from direct current (DC) techniques is not enough to characterise the system. In other words, when the corrosion process takes place in a series of stages, other techniques such as potentiodynamic polarization just provide information on the corrosion rate of the controlling reaction, which is not enough to understand the mechanism of the reaction taking place [83].

The previously explained electrochemical techniques, i.e., potentiodynamic polarization, potentiostatic tests and electrochemical impedance spectroscopy, require an external potential source to perturb the system, by either accelerating or inhibiting the corrosion kinetics. The main advantage of the electrochemical noise technique is that the corrosive system is not disturbed externally by the measurement process, so the system is kept in the natural corrosion potential. Thus, it allows potential and current fluctuations to be registered simultaneously, so it is possible to obtain information on the thermodynamics from the potential noise and kinetics information from the current noise. EN measurements allow the early stages of localized corrosion to be detected and studied and can isolate the individual events related to film breakdown in the sliding zone [10, 40, 85, 86].

The electrochemical noise is a low frequency (<10 Hz) potential or current fluctuation of small amplitude, originating from the variation of electrochemical reaction rates in a

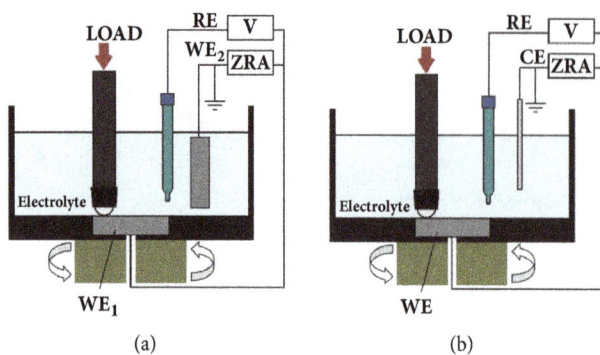

FIGURE 11: Schematic setup configuration for potential and current noise registration using: (a) two nominally identical electrodes and (b) asymmetrical electrodes.

corrosion process. There are several noise sources in corrosion [40, 85, 86]: atoms exchange kinetics in the electrode surface, the formation and release of bubbles on the surface, mechanical effects, formation of pits, transport fluctuations, temperature variations, variations on the solution resistance due to concentration gradients derived from the corrosion process, or the appearance of events controlled by nucleation, among others.

There are three main experimental configurations for simultaneous registration of potential and current signals. The configuration that has been most widely used in tribocorrosion assessment is schematically shown in Figure 11(a). It consists of using two working electrodes made of the same material (WE_1, WE_2). CE and RE are the counterelectrode and reference electrode, respectively. It is necessary to connect the two working electrodes through a Zero Resistance Ammeter (ZRA), which can measure the current while maintaining both working electrodes at a negligible potential difference. Two electrodes coupled through a ZRA will basically behave as a single electrode. The registered current signal corresponds to the current flowing between the two working electrodes, whereas the potential measured is the potential difference between both working electrodes and the reference electrode [84–87].

The local potential variations of each working electrode generate small variations in the mixed potential of the system. Therefore, the working electrodes must be nominally identical with the same surface preparation, since any difference would generate a galvanic couple, accelerating the oxidation of the more active electrode [84, 85]. In a tribocorrosion test, however, the electrodes are identical right until sliding wear process is initiated. Once wear is generated, the removal of the passive layer makes the worn electrode more active compared to the undamaged one. Therefore, the mechanical activation leads to galvanic coupling between the worn (anode) and unworn (cathode) electrodes. The latter acts as the cathode of the reaction, accelerating the corrosion-wear process on the worn sample [3, 40, 86, 88].

Sometimes, the different electrochemical characteristics presented in nominally identical electrodes generate asymmetry between them. In order to assess this drawback, several researchers have used asymmetric electrodes intentionally,

measuring the electrochemical noise and studying the electrochemical processes taking place just in one of the working electrodes. For this purpose, platinum microcathodes have been widely used [40, 84, 86]. The configuration is schematically represented in Figure 11(b). The platinum microelectrode area must be small enough to avoid polarization of the working electrode, ensuring that the tribocorrosion process is not accelerated, and thus, current variations registered are the result of the mechanically influenced electrochemical process. Therefore, the electrochemical noise registered mainly belongs to the worn working electrode.

Figure 12 shows the response of current and potential before, during, and after sliding measured by electrochemical noise technique for a bare and duplex treated (plasma nitriding and deposition of CrN coatings) AISI 304 stainless steel in a Hank's Balanced Salt Solution (HBSS) [89]. In this study, the working electrode was coupled with a platinum microcathode through a ZRA.

4.4.1. Tribocorrosion Assessment by Electrochemical Noise Technique. Some investigators have used the abovementioned configurations to evaluate the tribocorrosion behaviour of passive materials using the electrochemical noise technique. In these tests, the shifts in current and potential signals due to the sliding process are simultaneously registered. Several authors have also performed potentiodynamic tests in addition to EN measurements.

Galliano et al. (2001) [88] connected two identical working electrodes to a ZRA and performed the wear test on one of them. A reference electrode was used to measure the open circuit potential of the worn sample. The two originally identical samples became significantly different once a wear process started, generating galvanic corrosion between the two samples which, in turn, accelerated the tribocorrosion process. In one of the test configurations employed by Salasi et al. (2015) [90], they used the ZRA setup, coupling the test sample galvanically to a nominally identical electrode, and recording the current difference between the two samples. They observed an increase in the current and a cathodic shift of the free potential with the onset of abrasive particles to the sample. Thus, it was confirmed that the amount of current

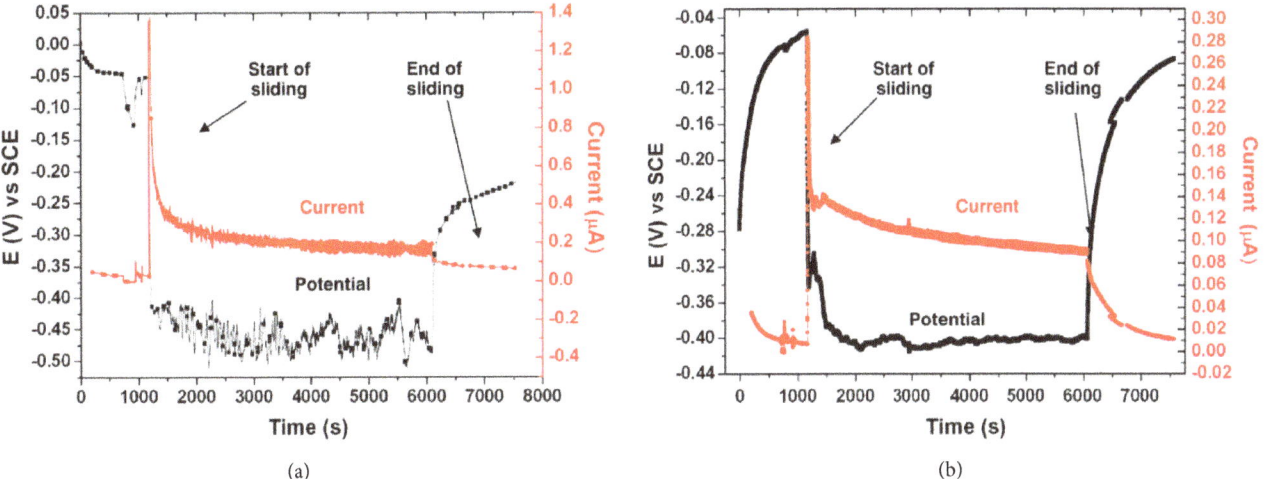

FIGURE 12: Potential and current response measured by electrochemical noise technique on (a) AISI 304 stainless steel and (b) duplex treated 304 against corundum in HBSS. In both cases, the potential decreases and the current increases during sliding, as a consequence of removing the passive layer and exposing the bare material to electrolyte [89].

discharge for this contact configuration was proportional to the number of particles that could successfully disrupt the passive film. Stachowiak et al. (2015) [9] studied the three-body abrasion-corrosion using several electrochemical techniques, discussing the benefits and limitations of each technique. They used two different methods based on measuring the EN: monitoring the galvanic current and potential noise of a freely corroding metal connecting two nominally identical electrodes and monitoring the current noise at an imposed anodic potential.

In order to avoid the asymmetry generated in the identical electrodes by the action of wear, Wu and Celis (2004) [86] coupled the working electrode to a platinum microelectrode. The microelectrode was used to record current variations during fretting experiments resulting from modifying the working electrode induced by sliding. This configuration allowed both potential and current responses to be measured during a fretting corrosion test, without accelerating the corrosion-wear of the tested materials. Therefore, the electrochemical noise technique was found to be useful to identify and/or unravel materials modification processes taking place during corrosion-wear sliding tests on passivating materials.

Several investigations have therefore been made by connecting the working electrode to a platinum microelectrode through the ZRA [28, 91–93]. Quan et al. (2006) [94] evaluated the corrosion-wear of TiN coated AISI 316 stainless steel in sulphuric acid solution and found that the corrosion-wear mechanism of the coatings depended on their substrate properties. They affirmed that the EN technique was a promising on-line monitoring tool that allowed delamination of a coating to be detected, since the interface between the coating and the substrate could be detected sensitively. Berradja et al. (2006) [95] used in situ electrochemical noise measurements to investigate the dependence of tribocorrosion on applied normal force and sliding velocity. The increase in the normal force and sliding velocity was found

to induce an increase in the current and a decrease in the potential, accelerating the depassivation rate of the tested materials. From the electrochemical noise registered during tribocorrosion tests, A. de Frutos et al. (2010) [89] observed the activation of the worn surface when sliding started, and they attributed local current and potential variations during the steady-state phase to the oxide removal and repassivation process taking place in the wear track (Figure 12).

In an attempt to overcome a major drawback of ZRA measurements, i.e., the loop back effect of current lines that do not go through the ZRA apparatus but remain confined to the working electrode, Silva et al. (2011) [96] suggested a new test configuration. This setup consisted of exposing only the wear track to the electrolyte, by covering the working electrode with an insulating polymer film. Besides, they used a cylinder of the same material concentrically placed just above the working electrode. With this configuration, most of the cathodic current was forced to flow through the ZRA, since the WE acted as a continuously depassivated anode. With this setup, the load increase was found to increase the anodic current between the working and counterelectrode, which remained passive and reduced the potential.

Espallargas et al. (2013) [97] proposed a new experimental technique based on galvanic current and potential measurements through a ZRA, for quantifying the electrode potential and the anodic current inside the wear track during rubbing at OCP. The proposed experimental setup allowed them, for the first time, to determine the electrochemical conditions inside the wear track, i.e., electrode potential and anodic current. This was achieved by physically separating the cathode from the anode (wear track), by using two working electrodes: a coated steel sample where just the wear track was exposed to the electrolyte and an uncoated steel sample acting as the unworn area (Figure 13). The experimental setup was found to be suitable as a quantitative tool to determine both galvanic coupling parameters during

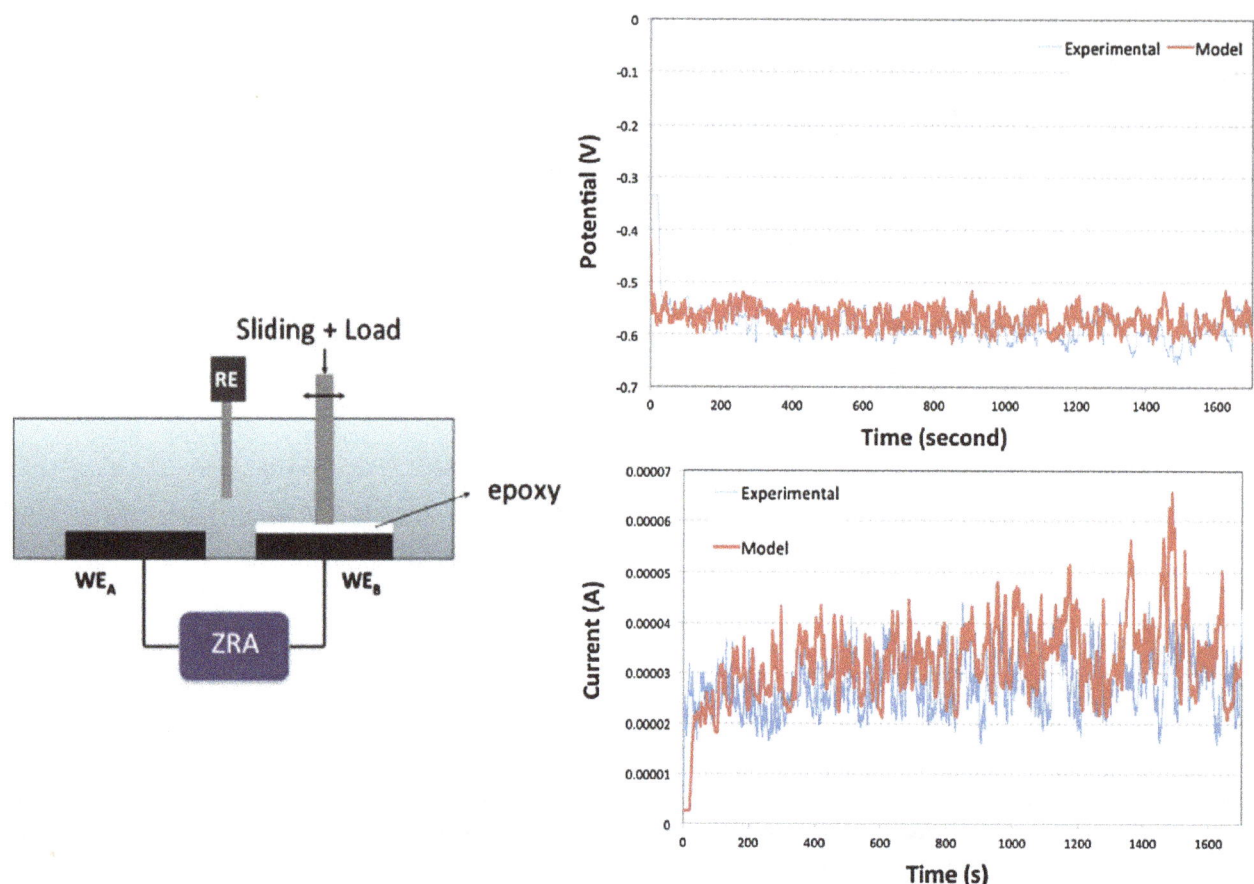

FIGURE 13: Left: schematic of the electrochemical setup used by Espallargas et al. [97] connecting the uncoated sample (unworn surface) and coated sample (wear track) through a ZRA. Right: potential and current evolution for a Ti6Al4V rubbed against an Al_2O_3 ball at 540 MPa in artificial saliva, obtained by Licausi et al. [98] using this setup. The experimental results and the simulated values.

tribocorrosion at OCP and the wear-accelerated corrosion. Based on the electrochemical technique and configuration proposed by Espallargas and coworkers [97], Licausi et al. (2015) [98] studied the tribocorrosion of Ti6Al4V biomedical alloy in artificial saliva at open circuit potential, measuring the galvanic potential and current between the wear track (anode) and the passive material (cathode). They could accurately simulate the current and potential during wear with the galvanic coupling model for tribocorrosion, which is based on the cathodic reaction kinetics (Figure 13).

Finally, a totally different configuration was used by Bryant et al. (2014) [99], who connected two working electrodes made of two different materials to investigate the galvanic coupling. They connected both electrodes through a ZRA, to measure the current flowing between the two materials. Galvanic coupling was found to significantly increase both pure wear and the wear-enhanced corrosion.

5. Standardized Tribocorrosion Test Produces

Although abrasion-corrosion and erosion-corrosion mechanisms of metals were described during the 1980s, no tribocorrosion mechanism in sliding contacts was proposed until the 1990s [3]. The tribocorrosion mechanisms have been evaluated through different approaches: synergistic, mechanistic, third-body, and nanochemical wear approaches [3]. In 1994, Madsen developed a standard guide to determine the synergism between wear and corrosion [6]. This is known as the synergistic approach, where the material loss is the sum of the material loss due to corrosion, the material loss due to wear, and a synergetic component. In order to assess some drawbacks of the synergistic approach collected in the ASTM G119 standard reported by some authors [3, 12, 13, 38], a test procedure based on the mechanistic approach was developed. According to this approach, the material loss due tribocorrosion is considered to be composed of two main contributions: anodic dissolution and mechanical removal of material. This test procedure was standardized in 2016 as a UNE 112086 standard [14]. The third-body approach in tribocorrosion was proposed by Mischler, et al. in 2001 [22], based on the third-body concept defined by Godet [100] in 1984 for dry sliding contacts. In this approach, the metal volume loss due to tribocorrosion is the sum of three contributions, namely, the mechanically detached material (abrasion, adhesion, or delamination), the chemically removed material (metal ions dissolved in the electrolyte), and the metal oxidized to form the passive film. In turn, the metal particles in the track can be ejected from the contact, oxidized, or smeared and transferred to the metal surface. Finally, the most recent nanochemical wear approach considers subsurface deformation of the material in the contact under the mechanical solicitations. In dry sliding contacts, the plastic deformation leads to formation and movement of dislocations, leading to a recrystallization of the material and the modification of the mechanical behaviour in the contact [3, 101, 102]. In this section, the synergistic and mechanistic approaches collected in the ASMT G119 and UNE 112086 standards are explained in detail.

5.1. Wear and Corrosion Synergism Evaluation Procedure (ASTM G119).

From approximately 1980 onwards, many researchers studied the synergism between wear and corrosion [4]. In 1981, Kim et al. (1981) [103] studied the synergism by examining the effect of load on the open circuit potential. It was found that wear increased the corrosion rate of passive materials, in environments and conditions in which the material loss due to pure corrosion was negligible in the absence of wear. The potential shifted to more active values and the current increased during sliding, indicating the increase in the kinetics of corrosion processes. Batchelor et al. (1988) [104] saw that the synergism of abrasion-corrosion was four times the corrosion rate even with negligible abrasion. Barker and Ball (1989) [105] observed that abrasion accelerated material degradation by removing the corrosion products or protective layer formed on top of the material and exposing the bare material to the corrosive environment. Kotlyar et al. (1988) [106] found that the magnitude of the synergistic effect between abrasion and corrosion was considerable, even when the influence of corrosion was small. Furthermore, they observed that the total abrasive-corrosive wear increased with increasing applied load, abrasive hardness, and decreasing solution pH. After several previous studies, Madsen [4, 107, 108] developed a standard guide in 1994 [6], to quantify the synergism between wear and corrosion. The ASTM standard consisted of four tests where the mechanical and electrochemical conditions were modified: corrosion only, wear only, and a combination of corrosion and wear.

According to the standard, the total material loss (T) corresponds to the total degradation due to wear and corrosion, i.e., tribocorrosion. The term includes the contributions from mechanical wear, corrosion dissolution, and the interaction between them and can be defined as follows:

$$T = W_0 + C_0 + S \tag{2}$$

with W_0 being the rate of material loss in the absence of corrosion, C_0 the electrochemical corrosion rate when no mechanical wear is applied, and S the synergetic component.

Since wear affects corrosion, and corrosion affects wear, the synergetic component can be further divided into two components, namely, the increase in the mechanical wear due to corrosion (ΔW_c) and the increase in corrosion due to mechanical wear (ΔC_w):

$$S = \Delta W_c + \Delta C_w, \tag{3}$$

where ΔW_c and ΔC_w can be calculated as follows:

$$\Delta W_c = W_c - W_0 \tag{4}$$

$$\Delta C_w = C_w - C_0 \tag{5}$$

with W_c being the total wear component of T and C_w the electrochemical corrosion during corrosive wear.

The above-mentioned parameters can be calculated, as specified in the ASTM G119 standard, as follows:

(i) T is material loss after a corrosion-wear test at open circuit potential.

(ii) W_0 can be obtained from the material loss after a wear test in the absence of corrosion. For this aim, the sample must be cathodically polarized one volt with respect to the free corrosion potential, so that corrosion is inhibited.

(iii) C_w is calculated from the corrosion current obtained in a potentiodynamic polarization test under mechanical wear, by using Faraday's law (Eq. (1)).

(iv) C_0 is calculated similarly to C_w from the corrosion current obtained from a potentiodynamic polarization test without mechanical wear.

Finally, the standard defines three dimensionless factors to describe the degree of wear-corrosion synergism:

(i) The total synergism factor: $\dfrac{T}{(T-S)}$ (6)

(ii) The corrosion augmentation factor: $\dfrac{(C_0 + \Delta C_w)}{C_0}$ (7)

(iii) The wear augmentation factor: $\dfrac{(W_0 + \Delta W_c)}{W_0}$ (8)

Figure 14 shows a graphic representation of the synergetic contributions of wear and corrosion on an AISI 304 stainless steel in seawater under different applied loads [30], obtained according to the ASTM G119 standard.

5.1.1. Tribocorrosion Studies Performed with the ASTM G119 Procedure.

The ASTM G119 standard has been widely used to quantify the synergism of a large number of materials in different environments [33, 34, 109–117]. Neville and Hu (2001) [118] performed erosion-corrosion tests and observed that the resistance to material loss did not increase with the increase in hardness of the studied alloy, since corrosion was found to have an important synergistic effect on erosion. Following the Standard procedure, Gant et al. (2004) [119] obtained a synergistic component that was larger than the material loss due to pure corrosion and pure wear only, by an order of magnitude. Thakare et al. (2007) [120] performed abrasion-corrosion tests and obtained negative synergism. They attributed this effect to the surface chemistry modification effects such as passive film formation or depletion of most active surface compositions in contact with the electrolyte, which lowered abrasion due to corrosion. Lu et al. (2011) [121] investigated

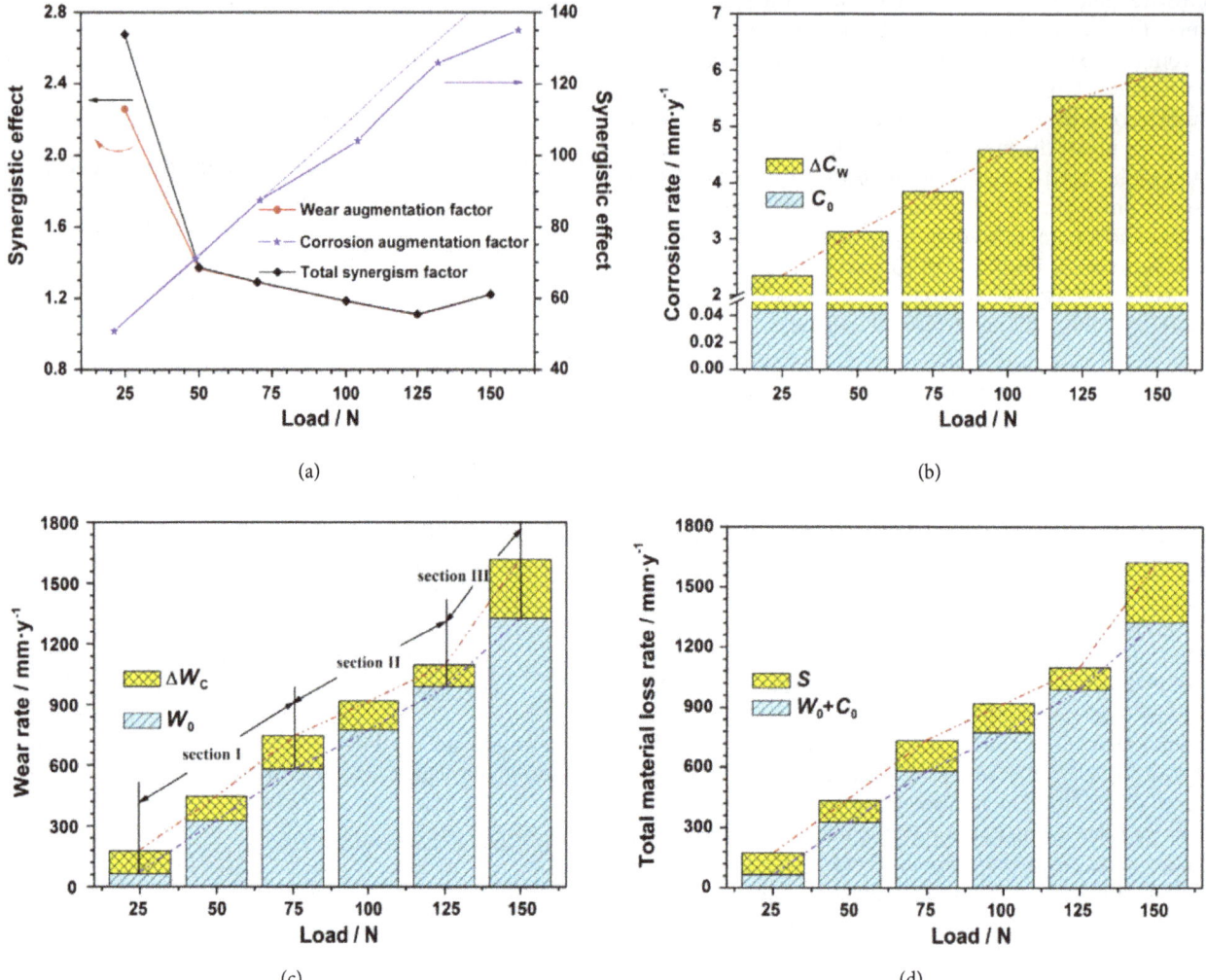

FIGURE 14: Synergetic contributions of mechanical wear and corrosion to each other and total material loss for a AISI 304 stainless steel at various applied loads in artificial seawater [30].

the synergism of mechanical and electrochemical factors in erosion-corrosion and observed that the synergism resulted mainly from corrosion-enhanced erosion. Yue Zhang et al. (2014) [30] investigated the influence of microstructure evolution on tribocorrosion of a 304 stainless steel, and the wear-corrosion synergistic degree was found to be dependent on the characteristics of microstructural evolution. Hardening of the steel due to plastic deformation, by both twinning and friction-induced martensite formation, led to enhanced wear resistance but, in turn, the galvanic corrosion of the thus-formed martensite led to a decrease in the surface hardness (corrosion-accelerated wear). Abedini and Ghasemi (2014) [122] performed erosion-corrosion tests and evaluated the synergism at different impingement angles. They observed that the higher rates corresponded to an angle of 40∘, and the total synergistic effect was mainly an effect of corrosion on erosion.

Several authors have used this protocol to build wear-corrosion maps, which helped to evaluate the influence of different variables. Bello et al. (2007) [123] investigated the

synergistic effects of abrasion and corrosion and developed a mechanistic abrasion-corrosion map to establish the influence between abrasion and corrosion processes. Stack et al. (2011) [124] performed microabrasion corrosion tests to evaluate the effect of load and solution pH and constructed synergism maps showing transitions between microabrasion and corrosion regimes, as a function of load and solution pH.

5.2. Tribocorrosion Test Procedure for Passive Materials (UNE 112086:2016). In spite of growing interest and enhancement of electrochemical techniques, there was no standardized testing methodology for tribocorrosion evaluation until 2016. The only existing standard was the ASTM G119 [6], which describes a method to determine the synergism between wear and corrosion. However, different approaches were developed to evaluate the tribocorrosion behaviour of passive materials, in order to assess some drawbacks of the ASTM standard, reported by some authors [3, 12, 13, 38]. Based on the test procedures suggested by several researchers, a new standard was developed in 2016. The ASTM [6] standard involves wear testing under anodic and cathodic polarization, whereas the new UNE 112086 [14] standard comprises wear tests at open circuit potential.

As reported by several authors [3, 12, 13, 38], the main shortcoming of the ASTM G119 standard is that the methodology employed does not separate measurements of the different contributions to the total material loss. According to the ASTM standard, material loss due to wear is obtained by performing sliding tests at cathodic potentials, where corrosion is negligible. However, the material loss measured does not necessarily represent the real conditions correctly since, in the absence of corrosion, the contribution of corrosion products to the mechanical response of the surface cannot be considered, as explained by Diomidis and coworkers [10, 12]. Furthermore, Akonko et al. (2005) [125] demonstrated that the wear volume depended on the cathodic potential selected and could vary by an order of magnitude due to the sensitivity to hydrogen embrittlement of the metal. On the other hand, the standard does not consider the galvanic couple generated between the worn and unworn areas during sliding either. Another considerable drawback, according to Diomidis et al. [10, 12], is that the standard does not provide information on the different processes that might take place in the different parts of the wear track, since it does not consider the different processes occurring within the wear track.

The protocol suggested in the UNE standard to assess the shortcomings of the ASTM standard can be used to identify and quantify the different mechanisms leading to material loss, i.e., mechanical and electrochemical mechanisms. Furthermore, it can also be used to identify the origin of material loss from different parts of the wear track, that is, from the active areas and partially or fully repassivated areas inside the track [3, 12, 13, 38].

The new approach consists of a series of successive steps during which essential data on tribocorrosion behaviour of materials is acquired:

(1) Open circuit potential registration until reaching a steady state

(2) Electrochemical impedance spectroscopy: after a stable potential is reached, an electrochemical impedance spectroscopy measurement is performed in OCP conditions. The polarization resistance (R_p) value obtained from this measurement is then used to calculate the corrosion current density (i_{pass}) of the passivated material as follows:

$$r_{pass} = R_p . A_0 \qquad (9)$$

$$i_{pass} = \frac{B}{r_{pass}} \qquad (10)$$

where r_{pass} is the specific polarization resistance of the passive material, A_0 the exposed surface area, and B a constant with typical values between 13 and 15 mV for metallic materials.

(3) First sliding wear test: the next step consists of determining the corrosion rate of the depassivated material

inside the wear track. For this purpose, a sliding wear test should be performed in order to remove the passive layer and keep the material in the wear track in a continuous active state. The time between two successive contact events during sliding (t_{rot}) needs to be low enough to avoid passivation of the material in the wear track. The potential is registered during the sliding process, and the stable potential during sliding (E_{oc}^s) is calculated. This potential is a mixed potential resulting from the galvanic coupling of the active and passive areas.

(4) Second sliding wear test and second EIS measurement: the effect of sliding on the corrosion resistance can be evaluated by performing an EIS measurement during sliding, once the sliding steady state has been achieved. In order to assure the stable state during sliding, some researchers have registered the EIS data by imposing a fixed potential on the system, which corresponds to the average potential value during sliding (E_{oc}^s) calculated in the previous step. The polarization resistance during sliding (R_{ps}) obtained is a combination of two polarization resistances, namely, the resistances corresponding to the active (R_{act}) and passive areas (R_{pass}):

$$\frac{1}{R_{ps}} = \frac{1}{R_{act}} + \frac{1}{R_{pass}} \qquad (11)$$

$$R_{act} = \frac{r_{act}}{A_{act}} \qquad (12)$$

$$R_{pass} = \frac{r_{pass}}{A_0 - A_{act}}. \qquad (13)$$

Since r_{pass} is known from (9), r_{act} can be calculated as follows:

$$r_{act} = \frac{A_{tr} . R_{ps} . r_{pass}}{r_{pass} - R_{ps} . (A_0 - A_{act})}. \qquad (14)$$

The corrosion current density of the active material (i_{act}) can be obtained from the next equation:

$$i_{act} = \frac{B}{r_{act}}. \qquad (15)$$

This test procedure can be used to quantify the reduction in the corrosion resistance of the studied material during sliding, by comparing the polarization resistances from the EIS data obtained before and during sliding.

The material loss due to corrosion in the wear track (W_{act}^c) can be calculated using Faraday's law:

$$W_{act}^c = i_{act} . A_{act} . \frac{M}{n.F.\rho} N.t_{lat}, \qquad (16)$$

with M being the molecular weight, n the number of electrons involved in the anodic process, ρ the density, and F the Faraday constant (96500 C). N corresponds to the number

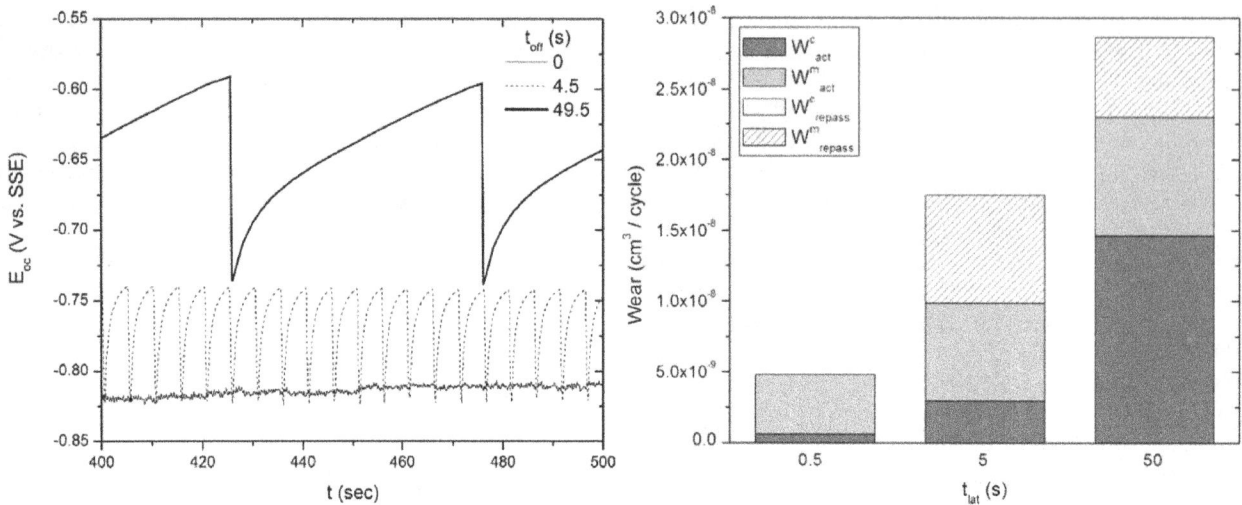

FIGURE 15: Left: evolution of continuous (t_{lat}=0) and intermittent (t_{lat}=4.5s and t_{lat}=49.5s) potential for an AISI 316 stainless steel in 0.5M H_2SO_4 at a rotation period (t_{rot}) of 0.5s. Right: contribution of the tribocorrosion components to the total volumetric material loss at the different latency times [13].

of cycles and t_{lat} is the latency time of the sliding test. The material loss due to wear and corrosion (W_{tr}) can be obtained from the volume of the wear track generated after the tribocorrosion test. Finally, the material loss due to mechanical wear in the wear track (W_{act}^m) can be calculated from the following equation:

$$W_{tr} = W_{act}^c + W_{act}^m. \tag{17}$$

In the case of intermittent sliding, W_{tr} is expressed as the sum of components related to both passive and active areas present in the wear track:

$$W_{tr} = W_{act}^c + W_{act}^m + W_{repass}^c + W_{repass}^m \tag{18}$$

with W_{repass}^c and W_{repass}^m being the material loss of the repassivated material in the wear track due to corrosion and mechanical wear, respectively. W_{repass}^c is calculated similarly to W_{act}^c in (16), but for the repassivated area

$$W_{repass}^c = i_{pass} . A_{repass} . \frac{M}{nF\rho} . N . t_{lat}. \tag{19}$$

Finally, W_{repass}^m is obtained from (18), once the other values are known.

5.2.1. Tribocorrosion Studies Performed with the UNE 112086 Procedure. In 2009, Diomidis et al. proposed a testing protocol [12] to assess the drawbacks of the preexisting ASTM standard. Several studies have been performed following this procedure, before it was standardized in 2016 (UNE 112086). In 2010, they investigated the tribocorrosion of an AISI 316 stainless steel in a diluted sulphuric acid solution [13]. They applied the protocol in order to determine the contributions of the different mechanical and corrosive components involved in material loss (Figure 15). They also quantified the synergism between wear and corrosion by calculating the

corrosion current density from the polarization resistance values obtained in the impedance measurements. V. Saenz de Viteri et al. (2015) [126] studied the tribocorrosion behaviour of a Ti6Al4V alloy with several Ti-C-N coatings in simulated body fluids and found that the coating provided the system with improved performance against tribocorrosion and fretting. Lucia Mendizabal et al. (2015) [73] studied the tribocorrosion response of a multilayer TaN film deposited on titanium alloy in simulated body fluids. The corrosion resistance of the coatings was found to decrease during sliding, and the wear rate was observed to improve compared with the uncoated titanium alloy. Fatma Ben Saada et al. (2015) [127] investigated the tribocorrosion behaviour of an AISI 304L stainless steel in a mixture of olive pomace and tap water filtrate, by performing both continuous and intermittent sliding tests. They observed that the damage caused by tribocorrosion was more noticeable under intermittent sliding, due to the partially active state of the wear track, since it was constantly being passivated and repassivated. They [128] also studied the improvement of the tribocorrosion properties for the alloy by using severe shot peening.

6. Tribocorrosion Modelling

Development of theoretical models to predict the material loss due to the synergistic effect of wear and corrosion has been of great interest over the last two decades. However, the implication of different influencing parameters in tribocorrosion systems hinders the postulation of models. In 1998, Mischler et al. [129] developed a model to describe the effect of mechanical and material parameters on the wear-assisted corrosion rate of passive metals in two-body sliding contacts. They considered the effect of normal load and the passivation rate of the materials. Mischler et al. [129] and Landolt et al. [8] correlated the passivation charge density and the depassivation rate. Several models have been proposed thenceforth, and some of the most recent ones are compiled in this section.

Jiang et al. (2002) [130] developed a model to describe interactions between wear and corrosion in sliding contacts. They incorporated different factors such as experimental and environmental conditions and material properties. In 2006, Jiang and Stack (2006) [131] proposed mechanisms for wear debris generation and validated the mathematical models which were developed, based on the proposed mechanisms. In 2010, von der Ohe et al. (2010) [132] developed a model to describe the multidegradation mechanism of tribocorrosion and static and cyclic fatigue taking place in a piston rod. In 2012, Papageorgiou and Mischler (2012) [133] suggested a quantitative galvanic coupling model that predicted potential evolution during tribocorrosion. Cao and coworkers developed [134] a predictive model considering the mechanical wear, wear-accelerated corrosion, and hydrodynamic lubrication applied to metal-on-metal implants. The model predictions were found to be consistent with tribocorrosion tests [135] and simulator results [136]. Recently, A. Dalmau et al. (2018) [137] developed a numerical contact model based on a Boundary Element Method (BEM) to describe the contact pressure distribution and quantification of the worn material as a function of time, taking into account the plastic behaviour of the material during the first cycles. They validated the model for a AISI 316L steel in a NaCl solution.

7. Conclusions and Future Perspective

A great number of studies have been carried out to assess the tribocorrosion behaviour of passive materials in order to get a closer insight into the mechanisms of this complex phenomenon. For this purpose, electrochemical techniques have been combined in different ways leading to several test procedures. This work compiles different test procedures and a combination of electrochemical techniques used by noteworthy researchers to assess tribocorrosion behaviour of passive materials. A brief explanation of the electrochemical techniques and studies made by tribocorrosion researchers has also been provided.

The early tribocorrosion studies were performed by combining potentiodynamic and potentiostatic polarization techniques, and they were focused on understanding the formation of the passive film and its behaviour under corrosion-wear solicitations. Other widely employed procedures combine open circuit potential and electrochemical impedance spectroscopy measurements. Electrochemical noise technique has been employed in more recent researches. Unlike polarization techniques, open circuit potential, and electrochemical impedance spectroscopy, no standard procedure has been developed by considering this technique so far. Electrochemical noise has been proven to be effective when evaluating tribocorrosion without externally disturbing the system. This technique provides information on both the corrosion potential and current variations taking place as a consequence of sliding. Besides, the configuration proposed by some researches [96–98] allows the worn and unworn areas to be physically separated. Hence, it could be interesting

to run further analysis on the effectiveness of this technique with such a novel configuration and even standardize the procedure.

In the same vein, the standardization of an acceptable test procedure is a major concern for tribocorrosionists. Current standards can be used to characterise and evaluate tribocorrosion of passive materials, but there are still several drawbacks that have drawn criticism and disagreement among experts in the field. Until 2016, the only existing standard for tribocorrosion tests was the ASTM G119, which can be used to appraise the synergism between wear and corrosion. The synergistic approach focuses on the dependence of wear and corrosion but does not provide information on the nature of the interactions. A mechanistic approach was developed to overcome these shortcomings, and the test procedure was standardized in 2016. The UNE 112086 standard can be used to evaluate the combined action of wear and corrosion, taking into consideration the galvanic couple generated between the worn and unworn areas. This approach quantifies the mechanical and electrochemical contributions to material loss. However, there is no direct way to determine the mechanical contribution to material loss in the track, and it has commonly been calculated by subtracting the chemical contribution to the total material loss, as specified in (17). In turn, the selection of the oxidation state (n number) in (16) to calculate the chemical contribution has been found to affect the results considerably [8, 18, 21, 89]. In cases where the valence of oxidation products varies with the applied potential or for alloys, the value of n is uncertain, and detailed surface analysis such as X-ray photoelectron spectroscopy should be employed to determine it [3]. On the other hand, wear and corrosion are not independent processes, although they interact with each other, and calculating these two contributions alone is not enough to describe the tribocorrosion process in sliding contacts. Generation of the third-body particles and their effect during rubbing must be considered, since they play an important role. This debris can aggravate the extent of wear by acting as abrasives, or they can form a lubricating layer reducing the friction [8, 9, 17–19]. In this regard, the third-body approach considers the behaviour of these particles in the contact, but no understanding on the formation, transformation, and ejection is provided [3]. Finally, the microstructural transformations taking place in the contacts under the applied mechanical solicitations should also been considered, as proposed in the nanochemical wear approach. Therefore, one of the most significant challenges in the field is developing a more accurate tribocorrosion mechanism that considers all the aspects mentioned above that could allow the phenomenon to be predicted. Furthermore, the need for standardized test equipment is also a drawback in the advance of the knowledge generation process, since the lack of a reliable apparatus leads to discrepancies in the results and how they are interpreted.

Another challenge in the study of tribocorrosion to be addressed is adapting test procedures for passive materials to assessment of tribocorrosion in active materials. Unlike passive materials, active materials do not generate a protective oxide layer on the surface when in contact with a corrosive media. On the contrary, the oxide layer in the surface is rather porous with low adherence. Therefore, the material loss is a consequence not only of wear and corrosion processes taking place in the wear track, but also of pure corrosion in the unworn surface. Tribocorrosion of passive materials has been widely studied, whereas the response of active materials is less documented. Recently, A. López et al. (2015) [138] studied the tribocorrosion behaviour of High-Strength Low-Alloyed (HSLA) steels in synthetic seawater by means of potentiodynamic and potentiostatic polarization techniques and evaluated the wear-corrosion synergism according to the ASTM G119 standard. In a following study, A. López-Ortega et al. (2018) [139] evaluated the effect of temperature on the tribocorrosion of HSLA steels in synthetic seawater, using a test protocol based on the UNE 112086 standard. The results showed that, unlike the observation for passive materials, sliding resulted in a potential shift towards more positive values and a current shift to more negative values. The material in the wear track was found to be more cathodic than the unworn surface, and thus, the galvanic couple generated between worn and unworn areas promoted the corrosion of the unworn surface. This opposing behaviour of active materials compared to passive alloys demonstrates the need to develop a test protocol to evaluate the tribocorrosion behaviour of active materials. Furthermore, since the unworn area is actively corroding during immersion tests, the methods currently used to quantify mechanical and electrochemical process contributions to the total material loss in passive materials are not appropriate. A closer understanding on the tribocorrosion degradation mechanism of active metals and alloys could provide better awareness of the use of these materials, with or without protective coatings, in order to enlarge their useful life in applications where passive materials are not suitable for use, i.e., higher costs and lower mechanical properties.

Conflicts of Interest

The authors declare that they have no conflicts of interest.

Acknowledgments

The authors would like to acknowledge the FRONTIERS III project (ELKARTEK 2017, KK-2017/00096) financed by the Basque Country. The authors would also like to acknowledge the Education, Linguistic Politics and Culture Department of the Basque Government for its support through the "*Programa Predoctoral de Formación de Personal Investigador No Doctor (PRE_2017_2_0088)*" grant awarded to the primary author.

References

[1] ASTM G40-17, *Standard Terminology Relating to Wear and Erosion*, ASTM International, West Conshohocken, PA, USA, 2017.

[2] M. T. Mathew, P. Srinivasa Pai, R. Pourzal, A. Fischer, and M. A. Wimmer, "Significance of Tribocorrosion in Biomedical Applications: Overview and Current Status," *Advances in Tribology*, vol. 2009, Article ID 250986, pp. 1–12, 2009.

[3] D. Landolt and S. Mischler, *Tribocorrosion of Passive Metals and Coatings*, ISBN: 978-1-84569-966-6, Woodhead Publishing, Cambridge, UK, 2011.

[4] S. W. Watson, F. J. Friedersdorf, B. W. Madsen, and S. D. Cramer, "Methods of measuring wear-corrosion synergism," *Wear*, vol. 181–183, no. 2, pp. 476–484, 1995.

[5] S. Mischler, E. A. Rosset, and D. Landolt, "Effect of Corrosion on the Wear Behavior of Passivating Metals in Aqueous Solutions," *Tribology and Interface Engineering Series*, vol. 25, no. C, pp. 245–253, 1993.

[6] G119-09 ASTM, *Standard Guide for Determining Synergism Between Wear and Corrosio*, ASTM International, West Conshohocken, PA, USA, 2016.

[7] D. Landolt, "Electrochemical and materials aspects of tribocorrosion systems," *Journal of Physics D: Applied Physics*, vol. 39, no. 15, article no. S01, pp. 3121–3127, 2006.

[8] D. Landolt, S. Mischler, and M. Stemp, "Electrochemical methods in tribocorrosion: A critical appraisal," *Electrochimica Acta*, vol. 46, no. 24-25, pp. 3913–3929, 2001.

[9] G. Stachowiak, M. Salasi, and G. Stachowiak, "Three-Body Abrasion Corrosion Studies of High-Cr Cast Irons: Benefits and Limitations of Tribo-electrochemical Methods," *Journal of Bio- and Tribo-Corrosion*, vol. 1, no. 1, article no. 6, 2015.

[10] J.-P. Celis and P-. Ponthiaux, *Testing Tribocorrosion of Passivating Materials Supporting Research and Industrial Innovation*, ISBN: 978-1-907975-20-2, Handbook. Maney Publishing, UK, 2012.

[11] S. Mischler, "Triboelectrochemical techniques and interpretation methods in tribocorrosion: A comparative evaluation," *Tribology International*, vol. 41, no. 7, pp. 573–583, 2008.

[12] N. Diomidis, J.-P. Celis, P. Ponthiaux, and F. Wenger, "A methodology for the assessment of the tribocorrosion of passivating metallic materials," *Lubrication Science*, vol. 21, no. 2, pp. 53–67, 2009.

[13] N. Diomidis, J.-P. Celis, P. Ponthiaux, and F. Wenger, "Tribocorrosion of stainless steel in sulfuric acid: Identification of corrosion-wear components and effect of contact area," *Wear*, vol. 269, no. 1-2, pp. 93–103, 2010.

[14] UNE 112086:2016, Ensayos de tribocorrosión en materiales pasivos, 07 September 2016.

[15] H. Shih, *Corrosion Resistance*, ISBN 978-953-51-0467-4, InTech, 2012.

[16] S. Mischler and P. Ponthiaux, "A round robin on combined electrochemical and friction tests on alumina/stainless steel contacts in sulphuric acid," *Wear*, vol. 248, no. 1-2, pp. 211–225, 2001.

[17] R. I. Trezona, D. N. Allsopp, and I. M. Hutchings, "Transitions between two-body and three-body abrasive wear: Influence of test conditions in the microscale abrasive wear test," *Wear*, vol. 225-229, no. I, pp. 205–214, 1999.

[18] D. Landolt, S. Mischler, M. Stemp, and S. Barril, "Third body effects and material fluxes in tribocorrosion systems involving a sliding contact," *Wear*, vol. 256, no. 5, pp. 517–524, 2004.

[19] N. Diomidis and S. Mischler, "Third body effects on friction and wear during fretting of steel contacts," *Tribology International*, vol. 44, no. 11, pp. 1452–1460, 2011.

[20] P. Ponthiaux, F. Wenger, D. Drees, and J. P. Celis, "Electrochemical techniques for studying tribocorrosion processes," *Wear*, vol. 256, no. 5, pp. 459–468, 2004.

[21] M. Stemps, S. Mischler, and D. Landolt, "The effect of contact configuration on the tribocorrosion of stainless steel in reciprocating sliding under potentiostatic control," *Corrosion Science*, vol. 45, no. 3, pp. 625–640, 2003.

[22] S. Mischler, A. Spiegel, M. Stemp, and D. Landolt, "Influence of passivity on the tribocorrosion of carbon steel in aqueous solutions," *Wear*, vol. 250-251, no. 2, pp. 1295–1307, 2001.

[23] P. Jemmely, S. Mischler, and D. Landolt, "Electrochemical modeling of passivation phenomena in tribocorrosion," *Wear*, vol. 237, no. 1, pp. 63–76, 2000.

[24] Y. Sun and V. Rana, "Tribocorrosion behaviour of AISI 304 stainless steel in 0.5 M NaCl solution," *Materials Chemistry and Physics*, vol. 129, no. 1-2, pp. 138–147, 2011.

[25] X. Y. Wang and D. Y. Li, "Application of an electrochemical scratch technique to evaluate contributions of mechanical and electrochemical attacks to corrosive wear of materials," *Wear*, vol. 259, no. 7-12, pp. 1490–1496, 2005.

[26] Y. Sun and R. Bailey, "Improvement in tribocorrosion behavior of 304 stainless steel by surface mechanical attrition treatment," *Surface and Coatings Technology*, vol. 253, pp. 284–291, 2014.

[27] R. Pileggi, M. Tului, D. Stocchi, and S. Lionetti, "Tribocorrosion behaviour of chromium carbide based coatings deposited by HVOF," *Surface and Coatings Technology*, vol. 268, pp. 247–251, 2015.

[28] A. C. Vieira, A. R. Ribeiro, L. A. Rocha, and J. P. Celis, "Influence of pH and corrosion inhibitors on the tribocorrosion of titanium in artificial saliva," *Wear*, vol. 261, no. 9, pp. 994–1001, 2006.

[29] J. Chen, J. Wang, F. Yan, Q. Zhang, and Q.-A. Li, "Effect of applied potential on the tribocorrosion behaviors of Monel K500 alloy in artificial seawater," *Tribology International*, vol. 81, pp. 1–8, 2015.

[30] Y. Zhang, X. Yin, J. Wang, and F. Yan, "Influence of microstructure evolution on tribocorrosion of 304SS in artificial seawater," *Corrosion Science*, vol. 88, pp. 423–433, 2014.

[31] J. Chen and F.-Y. Yan, "Tribocorrosion behaviors of Ti-6Al-4V and Monel K500 alloys sliding against 316 stainless steel in artificial seawater," *Transactions of Nonferrous Metals Society of China*, vol. 22, no. 6, pp. 1356–1365, 2012.

[32] J. Chen, Q. Zhang, Q. Li, S. Fu, and J. Wang, "Corrosion and tribocorrosion behaviors of AISI 316 stainless steel and Ti6Al4V alloys in artificial seawater," *Transactions of Nonferrous Metals Society of China*, vol. 24, no. 4, pp. 1022–1031, 2014.

[33] Y. Zhang, X. Yin, and F. Yan, "Effect of halide concentration on tribocorrosion behaviour of 304SS in artificial seawater," *Corrosion Science*, vol. 99, pp. 272–280, 2015.

[34] Y. Zhang, X.-Y. Yin, and F.-Y. Yan, "Tribocorrosion behaviour of type S31254 steel in seawater: Identification of corrosion-wear components and effect of potential," *Materials Chemistry and Physics*, vol. 179, pp. 273–281, 2016.

[35] Y. Sun and E. Haruman, "Tribocorrosion behaviour of low temperature plasma carburised 316L stainless steel in 0.5M NaCl solution," *Corrosion Science*, vol. 53, no. 12, pp. 4131–4140, 2011.

[36] B. A. Obadele, M. L. Lepule, A. Andrews, and P. A. Olubambi, "Tribocorrosion characteristics of laser deposited Ti-Ni-ZrO2 composite coatings on AISI 316 stainless steel," *Tribology International*, vol. 78, pp. 160–167, 2014.

[37] A. C. Vieira, L. A. Rocha, N. Papageorgiou, and S. Mischler, "Mechanical and electrochemical deterioration mechanisms in the tribocorrosion of Al alloys in NaCl and in NaNO3 solutions," *Corrosion Science*, vol. 54, no. 1, pp. 26–35, 2012.

[38] Y. Yu, *Bio-Tribocorrosion in Biomaterials and Medical Implants*, ISBN: 978-0-85709-540-4, Woodhead Publishing, 2013.

[39] M. Azzi and J. A. Szpunar, "Tribo-electrochemical technique for studying tribocorrosion behavior of biomaterials," *Biomolecular Engineering*, vol. 24, no. 5, pp. 443–446, 2007.

[40] A. de Frutos Rozas, *Tribocorrosión de biomateriales metálicos modificados superficialmente mediante técnicas de vacío [Ph.D. thesis]*, Universidad Autónoma de Madrid, Facultad de Ciencias, June 2010.

[41] X. Luo, X. Li, Y. Sun, and H. Dong, "Tribocorrosion behavior of S-phase surface engineered medical grade Co-Cr alloy," *Wear*, vol. 302, no. 1-2, pp. 1615–1623, 2013.

[42] Z. Guo, X. Pang, Y. Yan, K. Gao, A. A. Volinsky, and T.-Y. Zhang, "CoCrMo alloy for orthopedic implant application enhanced corrosion and tribocorrosion properties by nitrogen ion implantation," *Applied Surface Science*, vol. 347, Article ID 30147, pp. 23–34, 2015.

[43] A. M. Ribeiro, A. C. Alves, L. A. Rocha, F. S. Silva, and F. Toptan, "Synergism between corrosion and wear on CoCrMo-Al2O3 biocomposites in a physiological solution," *Tribology International*, vol. 91, 2017.

[44] Z. Doni, A. C. Alves, F. Toptan et al., "Dry sliding and tribocorrosion behaviour of hot pressed CoCrMo biomedical alloy as compared with the cast CoCrMo and Ti6Al4V alloys," *Materials and Corrosion*, vol. 52, pp. 47–57, 2013.

[45] V. G. Pina, A. Dalmau, F. Devesa, V. Amigó, and A. I. Muñoz, "Tribocorrosion behavior of beta titanium biomedical alloys in phosphate buffer saline solution," *Journal of the Mechanical Behavior of Biomedical Materials*, vol. 46, pp. 59–68, 2015.

[46] I. Hacisalihoglu, A. Samancioglu, F. Yildiz, G. Purcek, and A. Alsaran, "Tribocorrosion properties of different type titanium alloys in simulated body fluid," *Wear*, vol. 332-333, 2016.

[47] R. Baboian, Ed., *ASTM corrosion tests and standards*, ISBN: 0-8031-2098-2, ASTM International, 2013.

[48] A. Igual Muñoz and L. Casabán Julián, "Influence of electrochemical potential on the tribocorrosion behaviour of high carbon CoCrMo biomedical alloy in simulated body fluids by electrochemical impedance spectroscopy," *Electrochimica Acta*, vol. 55, no. 19, pp. 5428–5439, 2010.

[49] S. Mischler, A. Spiegel, and D. Landolt, "The role of passive oxide films on the degradation of steel in tribocorrosion systems," *Wear*, vol. 225-229, no. II, pp. 1078–1087, 1999.

[50] P. Ponthiaux, F. Wenger, and J.-P. Celis, "Tribocorrosion: Material Behavior Under Combined Conditions of Corrosion and Mechanical Loading, Corrosion Resistance," in *In Tech*, Dr. Shih, Ed., ISBN: 978-953-51-0467-4, 2012, http://dx.doi.org/10.5772/35634.

[51] ASTM G102-89(2015)e1, *Standard Practice for Calculation of Corrosion Rates and Related Information from Electrochemical Measurements*, ASTM International, West Conshohocken, PA, USA, 2015.

[52] I. García, D. Drees, and J. P. Celis, "Corrosion-wear of passivating materials in sliding contacts based on a concept of active wear track area," *Wear*, vol. 249, no. 5-6, pp. 452–460, 2001.

[53] M. Stemp, S. Mischler, and D. Landolt, "The effect of mechanical and electrochemical parameters on the tribocorrosion rate of stainless steel in sulphuric acid," *Wear*, vol. 255, no. 1-6, pp. 466–475, 2003.

[54] S. Barril, S. Mischler, and D. Landolt, "Electrochemical effects on the fretting corrosion behaviour of Ti6Al4V in 0.9% sodium chloride solution," *Wear*, vol. 259, no. 1-6, pp. 282–291, 2005.

[55] M. Favero, P. Stadelmann, and S. Mischler, "Effect of the applied potential of the near surface microstructure of a 316L steel submitted to tribocorrosion in sulfuric acid," *Journal of Physics D: Applied Physics*, vol. 39, no. 15, article no. S07, pp. 3175–3183, 2006.

[56] J. Stojadinović, D. Bouvet, M. Declercq, and S. Mischler, "Effect of electrode potential on the tribocorrosion of tungsten," *Tribology International*, vol. 42, no. 4, pp. 575–583, 2009.

[57] A. Bidiville, M. Favero, P. Stadelmann, and S. Mischler, "Effect of surface chemistry on the mechanical response of metals in sliding tribocorrosion systems," *Wear*, vol. 263, no. 1-6, pp. 207–217, 2007.

[58] M. N. F. Ismail, T. J. Harvey, J. A. Wharton, R. J. K. Wood, and A. Humphreys, "Surface potential effects on friction and abrasion of sliding contacts lubricated by aqueous solutions," *Wear*, vol. 267, no. 11, pp. 1978–1986, 2009.

[59] N. Espallargas and S. Mischler, "Tribocorrosion behaviour of overlay welded Ni-Cr 625 alloy in sulphuric and nitric acids: Electrochemical and chemical effects," *Tribology International*, vol. 43, no. 7, pp. 1209–1217, 2010.

[60] I. Golvano, I. Garcia, A. Conde, W. Tato, and A. Aginagalde, "Influence of fluoride content and pH on corrosion and tribocorrosion behaviour of Ti13Nb13Zr alloy in oral environment," *Journal of the Mechanical Behavior of Biomedical Materials*, vol. 49, pp. 186–196, 2015.

[61] S. Radice and S. Mischler, "Effect of electrochemical and mechanical parameters on the lubrication behaviour of Al2O3 nanoparticles in aqueous suspensions," *Wear*, vol. 261, no. 9, pp. 1032–1041, 2006.

[62] K. L. Dahm, "Direct observation of the interface during sliding tribo-corrosion," *Tribology International*, vol. 40, no. 10-12, pp. 1561–1567, 2007.

[63] N. Espallargas, C. Torres, and A. I. Muñoz, "A metal ion release study of CoCrMo exposed to corrosion and tribocorrosion conditions in simulated body fluids," *Wear*, vol. 332-333, 2017.

[64] Y. Liao, E. Hoffman, M. Wimmer, A. Fischer, J. Jacobs, and L. Marks, "CoCrMo metal-on-metal hip replacements," *Physical Chemistry Chemical Physics*, vol. 15, no. 3, pp. 746–756, 2013.

[65] S. Hiromoto and S. Mischler, "The influence of proteins on the fretting-corrosion behaviour of a Ti6Al4V alloy," *Wear*, vol. 261, no. 9, pp. 1002–1011, 2006.

[66] M. T. Mathew, M. J. Runa, M. Laurent, J. J. Jacobs, L. A. Rocha, and M. A. Wimmer, "Tribocorrosion behavior of CoCrMo alloy for hip prosthesis as a function of loads: A comparison between two testing systems," *Wear*, vol. 271, no. 9-10, pp. 1210–1219, 2011.

[67] K. Sadiq, M. M. Stack, and R. A. Black, "Wear mapping of CoCrMo alloy in simulated bio-tribocorrosion conditions of a hip prosthesis bearing in calf serum solution," *Materials Science and Engineering C: Materials for Biological Applications*, vol. 49, pp. 452–462, 2015.

[68] M. J. Runa, M. T. Mathew, and L. A. Rocha, "Tribocorrosion response of the Ti6Al4V alloys commonly used in femoral stems," *Tribology International*, vol. 68, pp. 85–93, 2013.

[69] R. Bayón, A. Igartua, J. J. González, and U. Ruiz De Gopegui, "Influence of the carbon content on the corrosion and tribocorrosion performance of Ti-DLC coatings for biomedical alloys," *Tribology International*, vol. 88, article no. 3593, pp. 115–125, 2015.

[70] S-I.. Pyun, H-C. Shin, J-W. Lee, and J-Y. Go, "Electrochemistry of Insertion Materials for Hydrogen and Lithium," in *Chapter 2: Electrochemical Methods*, ISBN: 978-3-642-29463-1, pp. 11–32, Springer, 2012.

[71] A. Lasia, *Electrochemical Impedance Spectroscopy and its Applications*, ISBN: 978-1-4614-8932-0, Springer, 2014.

[72] F. Scholz, *Electroanalytical Methods, Guide to Experiments and Applications*, Springer, Guide to Experiments and Applications, 2010.

[73] L. Mendizabal, A. Lopez, R. Bayón, P. Herrero-Fernandez, J. Barriga, and J. J. Gonzalez, "Tribocorrosion response in biological environments of multilayer TaN films deposited by HPPMS," *Surface and Coatings Technology*, vol. 295, pp. 60–69, 2015.

[74] R. Bayón, A. Igartua, X. Fernández et al., "Corrosion-wear behaviour of PVD Cr/CrN multilayer coatings for gear applications," *Tribology International*, vol. 42, no. 4, pp. 591–599, 2009.

[75] M. J. Runa, M. T. Mathew, M. H. Fernandes, and L. A. Rocha, "First insight on the impact of an osteoblastic layer on the bio-tribocorrosion performance of Ti6Al4V hip implants," *Acta Biomaterialia*, vol. 12, no. 1, pp. 341–351, 2015.

[76] V. S. De Viteri, R. Bayón, A. Igartua et al., "Structure, tribocorrosion and biocide characterization of Ca, P and I containing TiO2 coatings developed by plasma electrolytic oxidation," *Applied Surface Science*, vol. 367, pp. 1–10, 2016.

[77] N. Papageorgiou, A. Von Bonin, and N. Espallargas, "Tribocorrosion mechanisms of NiCrMo-625 alloy: An electrochemical modeling approach," *Tribology International*, vol. 73, pp. 177–186, 2014.

[78] B. Alemón, M. Flores, W. Ramírez, J. C. Huegel, and E. Broitman, "Tribocorrosion behavior and ions release of CoCrMo alloy coated with a TiAlVCN/CNx multilayer in simulated body fluid plus bovine serum albumin," *Tribology International*, vol. 81, pp. 159–168, 2015.

[79] M. Fazel, H. R. Salimijazi, M. A. Golozar, and M. R. Garsivaz Jazi, "A comparison of corrosion, tribocorrosion and electrochemical impedance properties of pure Ti and Ti6Al4V alloy treated by micro-arc oxidation process," *Applied Surface Science*, vol. 324, pp. 751–756, 2015.

[80] S. Rossi, L. Fedrizzi, M. Leoni, P. Scardi, and Y. Massiani, "(Ti,Cr)N and Ti/TiN PVD coatings on 304 stainless steel substrates: Wear-corrosion behaviour," *Thin Solid Films*, vol. 350, no. 1, pp. 161–167, 1999.

[81] S. A. Alves, R. Bayón, V. S. de Viteri et al., "Tribocorrosion Behavior of Calcium- and Phosphorous-Enriched Titanium Oxide Films and Study of Osteoblast Interactions for Dental Implants," *Journal of Bio- and Tribo-Corrosion*, vol. 1, no. 3, article no. 23, 2015.

[82] M. Buciumeanu, A. Bagheri, J. C. M. Souza, F. S. Silva, and B. Henriques, "Tribocorrosion behavior of hot pressed CoCrMo alloys in artificial saliva," *Tribology International*, vol. 97, pp. 423–430, 2016.

[83] Consejo Superior de Investigaciones Científicas (CSIC). Ciencia e ingeniería de la superficie de los materiales metálicos. Raycar (2000). ISBN: 84-00-07920-5.

[84] J. Botana Pedemonte, M. Marcos Bárcena, A. Aballe Villero. Ruido electroquímico: Métodos de análisis. Septem Ediciones (2002) ISBN: 84-95687-33-X.

[85] J. M. Sánchez-Amaya, M. Bethencourt, L. Gonzalez-Rovira, and F. J. Botana, "Medida de ruido electroquímico para el estudio de procesos de corrosión de aleaciones metálicas," *Revista de Metalurgia*, vol. 45, no. 2, pp. 142–156, 2009.

[86] P.-Q. Wu and J.-P. Celis, "Electrochemical noise measurements on stainless steel during corrosion-wear in sliding contacts," *Wear*, vol. 256, no. 5, pp. 480–490, 2004.

[87] R. J. K. Wood, J. A. Wharton, A. J. Speyer, and K. S. Tan, "Investigation of erosion-corrosion processes using electrochemical noise measurements," *Tribology International*, vol. 35, no. 10, pp. 631–641, 2002.

[88] F. Galliano, E. Galvanetto, S. Mischler, and D. Landolt, "Tribocorrosion behavior of plasma nitrided Ti-6Al-4V alloy in neutral NaCl solution," *Surface and Coatings Technology*, vol. 145, no. 1-3, pp. 121–131, 2001.

[89] A. de Frutos, M. A. Arenas, G. G. Fuentes et al., "Tribocorrosion behaviour of duplex surface treated AISI 304 stainless steel," *Surface and Coatings Technology*, vol. 204, no. 9-10, pp. 1623–1630, 2010.

[90] M. Salasi, G. Stachowiak, and G. Stachowiak, "Triboelectrochemical behaviour of 316L stainless steel: The effects of contact configuration, tangential speed, and wear mechanism," *Corrosion Science*, vol. 98, pp. 20–32, 2015.

[91] A. K. Basak, P. Matteazzi, M. Vardavoulias, and J.-P. Celis, "Corrosion-wear behaviour of thermal sprayed nanostructured FeCu/WC-Co coatings," *Wear*, vol. 261, no. 9, pp. 1042–1050, 2006.

[92] J. de Damborenea, C. Navas, J. A. García, M. A. Arenas, and A. Conde, "Corrosion-erosion of TiN-PVD coatings in collagen and cellulose meat casing," *Surface and Coatings Technology*, vol. 201, no. 12, pp. 5751–5757, 2007.

[93] E. Gracia-Escosa, I. García, J. C. Sánchez-López et al., "Tribocorrosion behavior of TiBxCy/a-C nanocomposite coating in strong oxidant disinfectant solutions," *Surface and Coatings Technology*, vol. 263, pp. 78–85, 2015.

[94] Z. Quan, P.-Q. Wu, L. Tang, and J.-P. Celis, "Corrosion-wear monitoring of TiN coated AISI 316 stainless steel by electrochemical noise measurements," *Applied Surface Science*, vol. 253, no. 3, pp. 1194–1197, 2006.

[95] A. Berradja, F. Bratu, L. Benea, G. Willems, and J.-P. Celis, "Effect of sliding wear on tribocorrosion behaviour of stainless steels in a Ringer's solution," *Wear*, vol. 261, no. 9, pp. 987–993, 2006.

[96] R. C. C. Silva, R. P. Nogueira, and I. N. Bastos, "Tribocorrosion of UNS S32750 in chloride medium: Effect of the load level," *Electrochimica Acta*, vol. 56, no. 24, pp. 8839–8845, 2011.

[97] N. Espallargas, R. Johnsen, C. Torres, and A. I. Muñoz, "A new experimental technique for quantifying the galvanic coupling effects on stainless steel during tribocorrosion under equilibrium conditions," *Wear*, vol. 307, no. 1-2, pp. 190–197, 2013.

[98] M. P. Licausi, A. I. Muñoz, V. A. Borrás, and N. Espallargas, "Tribocorrosion Mechanisms of Ti6Al4V in Artificial Saliva by Zero-Resistance Ammetry (ZRA) Technique," *Journal of Bio- and Tribo-Corrosion*, vol. 1, no. 1, article no. 8, 2015.

[99] M. Bryant, R. Farrar, R. Freeman, K. Brummitt, J. Nolan, and A. Neville, "Galvanically enhanced fretting-crevice corrosion of cemented femoral stems," *Journal of the Mechanical Behavior of Biomedical Materials*, vol. 40, pp. 275–286, 2014.

[100] M. Godet, "The third-body approach: A mechanical view of wear," *Wear*, vol. 100, no. 1-3, pp. 437–452, 1984.

[101] J.-P. Celis, P. Ponthiaux, and F. Wenger, "Tribo-corrosion of materials: Interplay between chemical, electrochemical, and mechanical reactivity of surfaces," *Wear*, vol. 261, no. 9, pp. 939–946, 2006.

[102] D. Shakhvorostov, B. Gleising, R. Büscher, W. Dudzinski, A. Fischer, and M. Scherge, "Microstructure of tribologically induced nanolayers produced at ultra-low wear rates," *Wear*, vol. 263, no. 7-12, pp. 1259–1265, 2007.

[103] K. Y. Kim, V. Agarwala, and S. Bhattacharyya, "An electrochemical polarization technique for evaluation of wear-corrosion in moving components under stress," in *Wear of materials*, Ludema, Ed., pp. 772–778, ASME, New York, 1981.

[104] A. W. Batchelor and G. W. Stachowiak, "Predicting synergism between corrosion and abrasive wear," *Wear*, vol. 123, no. 3, pp. 281–291, 1988.

[105] K. C. Barker and A. Ball, "Synergistic abrasive-corrosive wear of chromium containing steels," *British Corrosion Journal*, vol. 24, no. 3, pp. 222–228, 1989.

[106] D. Kotlyar, C. H. Pitt, and M. E. Wadsworth, "Simultaneous corrosion and abrasion measurements under grinding conditions," *Corrosion*, vol. 44, no. 4, pp. 221–228, 1988.

[107] B. W. Madsen, "Measurement of wear and corrosion rates using a novel slurry wear test," *Materials Performance*, vol. 26, no. 1, pp. 21–28, 1987.

[108] B. W. Madsen, "Measurement of erosion-corrosion synergism with a slurry wear test apparatus," *Wear*, vol. 123, no. 2, pp. 127–142, 1988.

[109] S. Yin and D. Y. Li, "Effects of prior cold work on corrosion and corrosive wear of copper in HNO3 and NaCl solutions," *Materials Science and Engineering: A Structural Materials: Properties, Microstructure and Processing*, vol. 394, no. 1-2, pp. 266–276, 2005.

[110] S. Yin, D. Y. Li, and R. Bouchard, "Effects of the strain rate of prior deformation on the wear-corrosion synergy of carbon steel," *Wear*, vol. 263, no. 1-6, pp. 801–807, 2007.

[111] M. S. Jellesen, T. L. Christiansen, L. R. Hilbert, and P. Møller, "Erosion-corrosion and corrosion properties of DLC coated low temperature gas-nitrided austenitic stainless steel," *Wear*, vol. 267, no. 9-10, pp. 1709–1714, 2009.

[112] D. López, N. Alonso Falleiros, and A. Paulo Tschiptschin, "Effect of nitrogen on the corrosionerosion synergism in an austenitic stainless steel," *Tribology International*, vol. 44, no. 5, pp. 610–616, 2011.

[113] J. A. Alegría-Ortega, L. M. Ocampo-Carmona, F. A. Suárez-Bustamante, and J. J. Olaya-Flórez, "Erosion-corrosion wear of Cr/CrN multi-layer coating deposited on AISI-304 stainless steel using the unbalanced magnetron (UBM) sputtering system," *Wear*, vol. 290-291, pp. 149–153, 2012.

[114] Ç. Albayrak, I. Hacisalihoğlu, S. Yenal vangölü, and A. Alsaran, "Tribocorrosion behavior of duplex treated pure titanium in Simulated Body Fluid," *Wear*, vol. 302, no. 1-2, pp. 1642–1648, 2013.

[115] R. Priya, C. Mallika, and U. K. Mudali, "Wear and tribocorrosion behaviour of 304L SS, Zr-702, Zircaloy-4 and Ti-grade2," *Wear*, vol. 310, no. 1-2, pp. 90–100, 2014.

[116] C. T. Kwok, P. K. Wong, and H. C. Man, "Laser surface alloying of copper with titanium: Part I. Electrical wear resistance in dry condition. Part II. Electrical wear resistance in wet and corrosive condition," *Surface and Coatings Technology*, vol. 297, pp. 58–73, 2016.

[117] F. Ma, J. Li, Z. Zeng, and Y. Gao, "Structural, mechanical and tribocorrosion behaviour in artificial seawater of CrN/AlN nano-multilayer coatings on F690 steel substrates," *Applied Surface Science*, vol. 428, pp. 404–414, 2018.

[118] A. Neville and X. Hu, "Mechanical and electochemical interactions during liquid-solid impingement on high-alloy stainless steels," *Wear*, vol. 250-251, no. 2, pp. 1284–1294, 2001.

[119] A. J. Gant, M. G. Gee, and A. T. May, "The evaluation of tribo-corrosion synergy for WC-Co hardmetals in low stress abrasion," *Wear*, vol. 256, no. 5, pp. 500–516, 2004.

[120] M. R. Thakare, J. A. Wharton, R. J. K. Wood, and C. Menger, "Exposure effects of alkaline drilling fluid on the microscale abrasion-corrosion of WC-based hardmetals," *Wear*, vol. 263, no. 1-6, pp. 125–136, 2007.

[121] B. T. Lu, J. F. Lu, and J. L. Luo, "Erosion-corrosion of carbon steel in simulated tailing slurries," *Corrosion Science*, vol. 53, no. 3, pp. 1000–1008, 2011.

[122] M. Abedini and H. M. Ghasemi, "Synergistic erosion-corrosion behavior of Al-brass alloy at various impingement angles," *Wear*, vol. 319, no. 1-2, pp. 49–55, 2014.

[123] J. O. Bello, R. J. K. Wood, and J. A. Wharton, "Synergistic effects of micro-abrasion-corrosion of UNS S30403, S31603 and S32760 stainless steels," *Wear*, vol. 263, no. 1-6, pp. 149–159, 2007.

[124] M. M. Stack, M. T. Mathew, and C. Hodge, "Micro-abrasion-corrosion interactions of Ni-Cr/WC based coatings: Approaches to construction of tribo-corrosion maps for the abrasion-corrosion synergism," *Electrochimica Acta*, vol. 56, no. 24, pp. 8249–8259, 2011.

[125] S. Akonko, D. Y. Li, and M. Ziomek-Moroz, "Effects of cathodic protection on corrosive wear of 304 stainless steel," *Tribology Letters*, vol. 18, no. 3, pp. 405–410, 2005.

[126] V. Sáenz de Viteri, G. Barandika, R. Bayón et al., "Development of Ti-C-N coatings with improved tribological behavior and antibacterial properties," *Journal of the Mechanical Behavior of Biomedical Materials*, vol. 55, pp. 75–86, 2015.

[127] F. B. Saada, Z. Antar, K. Elleuch, and P. Ponthiaux, "On the tribocorrosion behavior of 304L stainless steel in olive pomace/tap water filtrate," *Wear*, vol. 328-329, pp. 509–517, 2015.

[128] F. B. Saada, Z. Antar, K. Elleuch, P. Ponthiaux, and N. Gey, "The effect of nanocrystallized surface on the tribocorrosion behavior of 304L stainless steel," *Wear*, vol. 394-395, pp. 71–79, 2018.

[129] S. Mischler, S. Debaud, and D. Landolt, "Wear-accelerated corrosion of passive metals in tribocorrosion systems," *Journal of The Electrochemical Society*, vol. 145, no. 3, pp. 750–758, 1998.

[130] J. Jiang, M. M. Stack, and A. Neville, "Modelling the tribocorrosion interaction in aqueous sliding conditions," *Tribology International*, vol. 35, no. 10, pp. 669–679, 2002.

[131] J. Jiang and M. M. Stack, "Modelling sliding wear: From dry to wet environments," *Wear*, vol. 261, no. 9, pp. 954–965, 2006.

[132] C. B. von der Ohe, R. Johnsen, and N. Espallargas, "Modeling the multi-degradation mechanisms of combined tribocorrosion interacting with static and cyclic loaded surfaces of passive metals exposed to seawater," *Wear*, vol. 269, no. 7-8, pp. 607–616, 2010.

[133] N. Papageorgiou and S. Mischler, "Electrochemical simulation of the current and potential response in sliding tribocorrosion," *Tribology Letters*, vol. 48, no. 3, pp. 271–283, 2012.

[134] S. Cao, S. Guadalupe Maldonado, and S. Mischler, "Tribocorrosion of passive metals in the mixed lubrication regime: Theoretical model and application to metal-on-metal artificial hip joints," *Wear*, vol. 324-325, pp. 55–63, 2015.

[135] S. Guadalupe, S. Cao, M. Cantoni, W.-J. Chitty, C. Falcand, and S. Mischler, "Applicability of a recently proposed tribocorrosion model to CoCr alloys with different carbides content," *Wear*, vol. 376-377, pp. 203–211, 2017.

[136] S. Cao and S. Mischler, "Assessment of a recent tribocorrosion model for wear of metal-on-metal hip joints: Comparison between model predictions and simulator results," *Wear*, vol. 362-363, pp. 170–178, 2016.

[137] A. Dalmau, A. R. Buch, A. Rovira, J. Navarro-Laboulais, and A. I. Muñoz, "Wear model for describing the time dependence of the material degradation mechanisms of the AISI 316L in a NaCl solution," *Wear*, vol. 394-395, pp. 166–175, 2018.

[138] A. López, R. Bayón, F. Pagano et al., "Tribocorrosion behaviour of mooring high strength low alloy steels in synthetic seawater," *Wear*, vol. 338-339, pp. 1–10, 2015.

[139] A. López-Ortega, R. Bayón, J. Arana, A. Arredondo, and A. Igartua, "Influence of temperature on the corrosion and tribocorrosion behaviour of High-Strength Low-Alloy steels used in offshore applications," *Tribology International*, vol. 121, pp. 341–352, 2018.

10

Monitoring the Interaction of Two Heterocyclic Compounds on Carbon Steel by Electrochemical Polarization, Noise, and Quantum Chemical Studies

Vinod P. Raphael,[1] Shaju K. Shanmughan,[1] and Joby Thomas Kakkassery[2]

[1]*Department of Chemistry, Government Engineering College, Thrissur, Kerala 680009, India*
[2]*Research Division, Department of Chemistry, St. Thomas' College (Autonomous), Thrissur, Kerala 680001, India*

Correspondence should be addressed to Shaju K. Shanmughan; shaju5699@gmail.com

Academic Editor: Flavio Deflorian

A heterocyclic phenylhydrazone 2-[(E)-(2-phenylhydrazinylidene)methyl]pyridine (P2APH) and its reduced form 2-[(2-phenylhydrazinyl)methyl]pyridine (RP2APH) were synthesized, characterized, and subjected to corrosion inhibition investigation on carbon steel (CS) in 1 M HCl using gravimetric, polarization, electrochemical noise, quantum chemical, and surface studies. P2APH showed more inhibition capacity than RP2PPH. But RP2PPH was very stable in acid medium and showed pronounced corrosion inhibition efficacy for days. Energy of HOMO and LUMO, their difference, number of electrons transferred, electronegativity, chemical hardness, and so forth were evaluated by quantum chemical studies. Agreeable correlation was observed between the results of quantum chemical calculations and other corrosion monitoring techniques.

1. Introduction

Acid corrosion is a serious problem in many metal industries. Metal surface cleaning processes like pickling and descaling will escalate the rate of corrosion appreciably since these practices consume large amount of acids. In oil industry, acidizing of oil wells with strong acids enhances the rate of corrosion of oil pipes. Generally HCl or H_2SO_4 is used for the above industrial purposes. The only solution to combat corrosion in these circumstances is the application of corrosion inhibitors [1–6]. Investigation of corrosion inhibition response of various molecules is very much alive in contemporary research. Many organic molecules possessing aromatic rings and heterocyclic atoms act as corrosion inhibitors in acidic media [7–10]. Structure and geometry of a molecule have a definite role in preventing the corrosion of metal in acid medium. Sometimes inhibitor molecules react with the aggressive medium and this leads to alteration in the inhibition response. For instance, the corrosion inhibition behaviour of Schiff base N,N'-bis(salicylidene)-1,2-ethylenediamine (Salen), mixture of its parent compounds, and reduced form of Salen on carbon steel in 1 M HCl solution was studied by da Silva et al. [11] using corrosion potential measurements, polarization studies, EIS, and spectrophotometric measurements. Results obtained in the presence of Salen were similar to those obtained in the presence of the parent mixture, implying that in acid medium the Salen molecule undergoes hydrolysis, regenerating its precursor molecules. Reduced Salen exhibited more inhibition efficiency than Salen.

To explore the mechanism of corrosion inhibition, contemporary investigators rely on computational quantum chemical and molecular simulation techniques [12–18]. Industries are ever in search of soluble, highly stable, economical, and potential corrosion inhibitors. In the present course of study, we synthesized two highly water soluble heterocyclic compounds derived from pyridine-2-aldehyde. This article reveals the corrosion inhibition behaviour of two heterocyclic molecules on carbon steel in HCl medium, which was explored by gravimetric, electrochemical, quantum chemical, and surface analytical techniques.

2. Materials and Method

2.1. Synthesis and Characterization.
The phenylhydrazone (P2APH) was synthesized by the condensation of equimolar mixtures of pyridine-2-aldehyde and phenylhydrazine hydrochloride in ethanol medium under reflux condition for 4 h. The mixture was cooled in ice and the precipitated pale yellow coloured solid was filtered, washed with ethanol-water (1:1) mixture, and dried [19].

5 mmol of P2APH was dissolved in ethanol-water mixture (1:1). The mixture was cooled in ice bath and ethanolic solution of $NaBH_4$ (1 g in 10 mL) was added dropwise with constant stirring until the yellow colour of the reaction mixture disappeared. This mixture was kept for 10 h and the precipitated product (RP2APH) was filtered, washed with small quantities of water-ethanol mixture (1:1), and dried [20].

The products were characterized by elemental (Vario EL III Element Analyzer) analysis analysis and mass (Shimadzu, QP 2010 GCMS), nmr (Bruker Avance III HD, dmso-d_6 solvent), and IR (Shimadzu Affinity-1, KBr pellet method) spectroscopic analyses.

2.2. Metal Specimen and Corrosive Medium.
Carbon steel coupons were abraded ($1 \times 1 \times 0.15$ cm) with various grades of SiC papers (120, 400, 600, 800, 1000, and 1200), washed with soap solution, and degreased by acetone. The approx. composition of the steel specimen was determined by EDAX technique (0.58% C, 0.07% Mn, 0.02% P, 0.015% S, 0.02% Si, and Fe the rest, Hitachi SU6600 model SEM). A stock solution of P2APH and RP2APH (1.0 mM) in 1 M HCl was prepared and diluted with 1 M HCl solution to get aggressive solutions in different concentrations (0.2-1.0 mM).

2.3. Gravimetric Studies.
The metal specimens were immersed in the aggressive medium for 4 days at $29 \pm 0.2°C$ with periodical evaluation of the corrosion rate. The rate of corrosion [21, 22] was determined by

$$v = \frac{KW}{DSt}, \tag{1}$$

where v = corrosion rate (mmy^{-1}), W = weight loss (g), S = surface area of metal specimen (cm^2), t = time of treatment (h), D = density of specimen (gcm^{-3}), and K = a constant (8.76×10^4).

The inhibition efficiency (η_w%) was obtained [21] by the following equation:

$$\eta_w\% = \frac{v - v'}{v} \times 100, \tag{2}$$

where v & v' are the corrosion rate of the metal specimen in the absence and presence of the inhibitor, respectively.

2.4. Electrochemical Corrosion Studies.
Three-electrode system was used for polarization studies in which saturated calomel electrode (SCE) acted as reference electrode. Platinum and metal specimen (both having 1 cm^2 area) acted as counter and working electrodes, respectively. The working electrode was allowed to contact with the aggressive solution to attain steady-state open-circuit potential (OCP). Each metal specimen was immersed in the aggressive medium for a period of 30 min prior to the experiment at 29°C. Polarization studies were carried out in the potential range +250 to −250 mV with a sweep rate of 1 mV/s [23–26]. The inhibition efficiencies were calculated from the corrosion current densities using the following equation:

$$\eta_{pol}\% = \frac{I_{corr} - I'_{corr}}{I_{corr}} \times 100. \tag{3}$$

Electrochemical noise (ECN) experiments were performed in a three-electrode cell assembly, which consisted of two carbon steel electrodes (1 cm^2) and SCE. All ECN analyses were conducted for a period of 1200 s [27, 28]. Ivium Compactstate electrochemical system controlled by iviumsoft software was employed for the electrochemical investigations.

2.5. Surface Morphological Studies.
SEM analyses were conducted using Hitachi SU6600 model scanning electron microscope. Analysis of surface film deposited on CS was done by IR spectroscopy.

3. Results and Discussion

3.1. Structure of Molecules.
The structures of molecules were confirmed by CHN and spectroscopic techniques. The results are summarized as follows. P2APH: CHN, found: (calc.), C%; 72.69 (73.09), H%; 4.98 (5.58), N%; 21.08 (21.32). Mass spectrum; M+ peak m/z = 196.75 (base peak) ($[C_{12}H_{11}N_3]^+$), ^{1}Hnmr; 8.17δ(s) (CH=N), 12.08(s) (NH), ^{13}Cnmr; 148.97 ppm (C-2 pyr), 143.35 ppm (C-1 ph), 144.41 ppm (CH=N).

RP2APH: CHN, found: (calc.), C%; 71.69 (72.36), H%; 6.66 (6.53), N%; 20.87 (21.11). Mass spectrum; M+ peak m/z = 198.84 (base peak) ($[C_{12}H_{13}N_3]^+$), ^{1}Hnmr; 6.8(t) (CH$_2$), 8.5δ(d) (NH-1), 10.67δ(d) (NH-2), ^{13}Cnmr; 111.39 ppm (CH$_2$), 153.78 ppm (C-2 pyr), 148.18 ppm (C-1 ph). The structures and optimized geometries of molecules P2APH and RP2APH are given in Figure 1.

3.2. Gravimetric Corrosion Inhibition Studies.
The corrosion inhibition efficiencies of two molecules for 24 h at ($29 \pm 0.1°C$) are listed in Table 1. Both molecules displayed very high inhibition efficiency on CS surface. For a period of 24 h, the P2APH molecule showed little bit higher inhibition efficiency than its reduced form RP2APH at all concentrations. This can be attributed to the presence of strongly coordinating azomethine linkage in P2APH and the complete planar geometry of the molecule. Figure 2 represents the variation of corrosion inhibition efficiency with time for the molecules at 1.0 mM. The plot reveals that decrease of η_w% for P2APH was very much steeper than that of RP2APH as the days went on. This may be due to the slow hydrolysis of P2APH molecules in acidic medium. The azomethine linkage (C=N) present in the molecule is susceptible to hydrolysis [11]. The

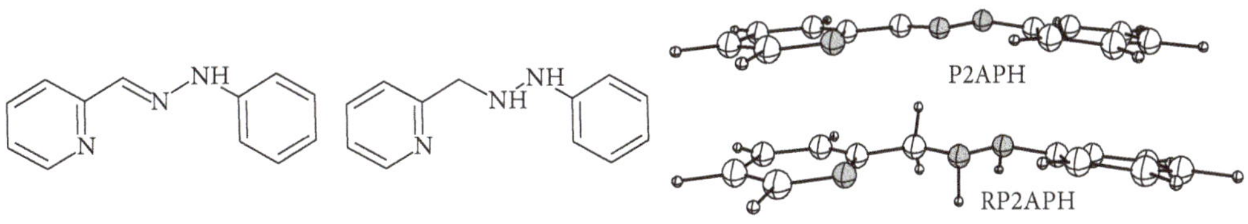

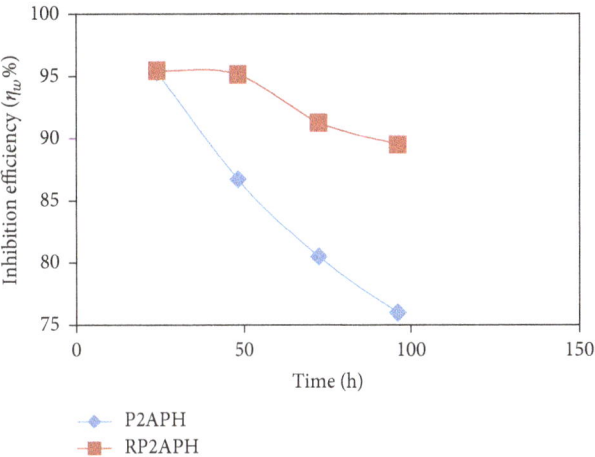

FIGURE 1: Structure of P2APH and RP2APH and their optimized geometries.

TABLE 1: Gravimetric corrosion inhibition efficiencies ($\eta_w\%$) of P2APH and RP2APH on CS in 1 M HCl for 24 h at 29 ± 0.2°C.

C (mM)	P2APH	RP2APH
0.2	92.14	91.64
0.4	92.98	91.91
0.6	94.42	92.57
0.8	94.82	93.80
1.0	95.50	95.40

reduced form of this molecule (RP2APH) is free from the azomethine linkage and is not vulnerable to hydrolysis and showed higher inhibition efficiency for days. The stability of the inhibitor in the aggressive medium is very important if it is to be recommended for a long time use. From the results of gravimetric corrosion studies one can reach into the conclusion that RP2APH molecule acts as very good corrosion inhibitor for a long time.

To get a more insight into the mechanism of corrosion, adsorption isotherms were plotted. Among the various isotherm models tried, the most suitable one was selected with the help of correlation coefficient. The most fit isotherm model for P2APH and RP2APH on CS in 1 M HCl was Langmuir adsorption isotherm equation (4). Isotherm plots having correlation coefficients close to one are given in Figure 3.

$$\frac{C}{\theta} = \frac{1}{K_{ads}} + C, \qquad (4)$$

where C is the concentration of the inhibitor, θ is the fractional surface coverage, and K_{ads} is the adsorption equilibrium constant [1]. Standard free energy of adsorption (ΔG^0_{ads}) is related to equilibrium constant of adsorption (K_{ads}) by the following equation:

$$\Delta G^0_{ads} = -RT \ln\left(55.5\, K_{ads}\right), \qquad (5)$$

where 55.5 is the molar concentration of water, R is the universal gas constant, and T is the temperature in Kelvin. Both molecules exhibited free energies of adsorption approximately 38.0 kJ/mol, which established that both the molecules interact with the metal surface by strong chemical forces [29].

3.3. Polarization Studies.
To get an idea about the electrokinetic reactions and the involvement of molecules in the process polarization studies have been performed. Table 2 provides the Tafel polarization parameters of CS in the presence and absence of the inhibiting molecules. Figures 4(a) and 4(b)

FIGURE 2: Variation of corrosion inhibition efficiencies ($\eta_w\%$) of 1.0 mM P2APH and RP2APH with time on CS in 1 M HCl.

show the Tafel polarization curves for metal specimens in 1 M HCl with and without P2APH and RP2APH. The corrosion current density of the metal specimens markedly decreased with the concentration of inhibitors. A significant reduction of the active surface area of the metal specimen exposed to the corrosive medium occurs due to the increase of the thickness of the adsorbed layer with concentration [14]. The notable rise in $\eta_{pol\%}$ was observed by these inhibitors with concentration. At 1.0 mM concentration, P2APH displayed 92.3% inhibition efficiency. Since the cathodic slopes of Tafel lines (Figure 4) were deviated more than anodic slopes, one can assume that these two molecules delay the metal corrosion by inhibiting the cathodic process of corrosion more. In other words, the hydrogen evolution process was considerably hindered by both molecules compared to the oxidation of Fe into Fe^{2+}.

3.4. Electrochemical Noise Studies.
Electrochemical noise experiments were performed with identical CS electrodes and reference electrode (SCE) immersed in the test solution (1 M HCl) in the presence and absence of P2APH and RP2APH for a period of 1200 s at 29±0.2°C. Figure 5 represents the current noise for blank metal specimen and metal treated with acid in the presence of 1.0 mM P2APH and RP2APH. From the noise data it is understandable that the mean values of the current noise follow the order blank > RP2APH > P2APH, which reveals that the protective power of the molecules is in the order P2APH > RP2APH. The ratio of standard deviation of potential noise to the standard deviation of current noise gave the noise resistance R_n [30]. The noise resistance for

TABLE 2: Potentiodynamic polarization parameters of CS in the presence and absence of P2APH and RP2APH in 1 M HCl at 29 ± 0.2°C for immersion period of 30 min.

	P2APH					RP2APH				
C (mM)	$-E_{corr}$ (mV/SCE)	I_{corr} (μA/cm^2)	$-b_c$ (mV/dec)	b_a (mV/dec)	$\eta_{pol\%}$	$-E_{corr}$ (mV/SCE)	I_{corr} (μA/cm^2)	$-b_c$ (mV/dec)	b_a (mV/dec)	$\eta_{pol\%}$
0	531	639.31	102	152	—	531	639.31	102	152	
0.2	511	383.20	110	137	40.03	433	375.61	126	161	41.22
0.4	497	219.93	85	123	66.68	517	249.71	122	175	60.92
0.6	488	108.08	67	126	83.08	511	206.93	94	162	67.62
0.8	501	55.67	52	126	91.23	509	105.76	119	168	83.45
1.0	484	49.40	112	136	92.27	512	77.55	146	167	87.86

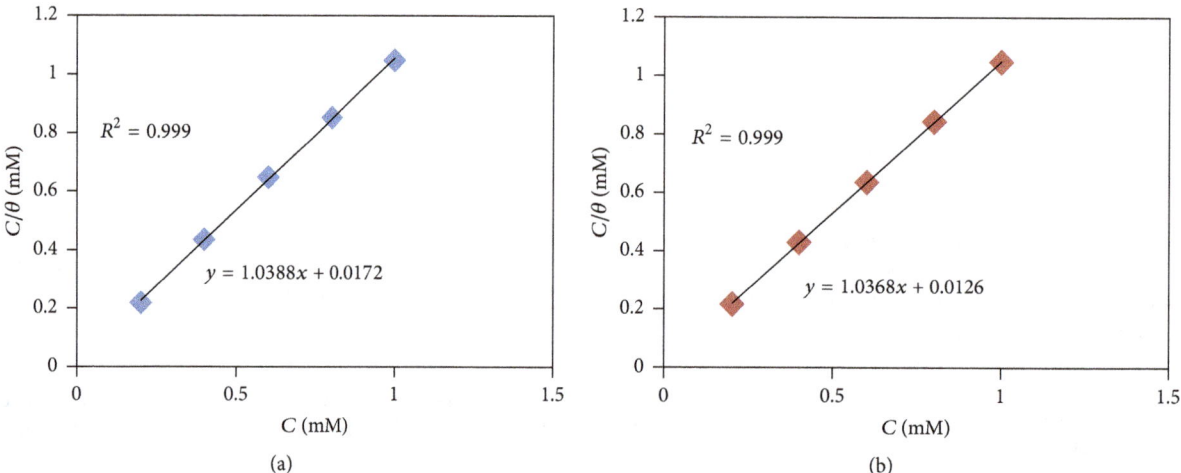

(a) (b)

FIGURE 3: Langmuir adsorption isotherm for (a) P2APH and (b) RP2APH on CS in 1 M HCl for 24 h.

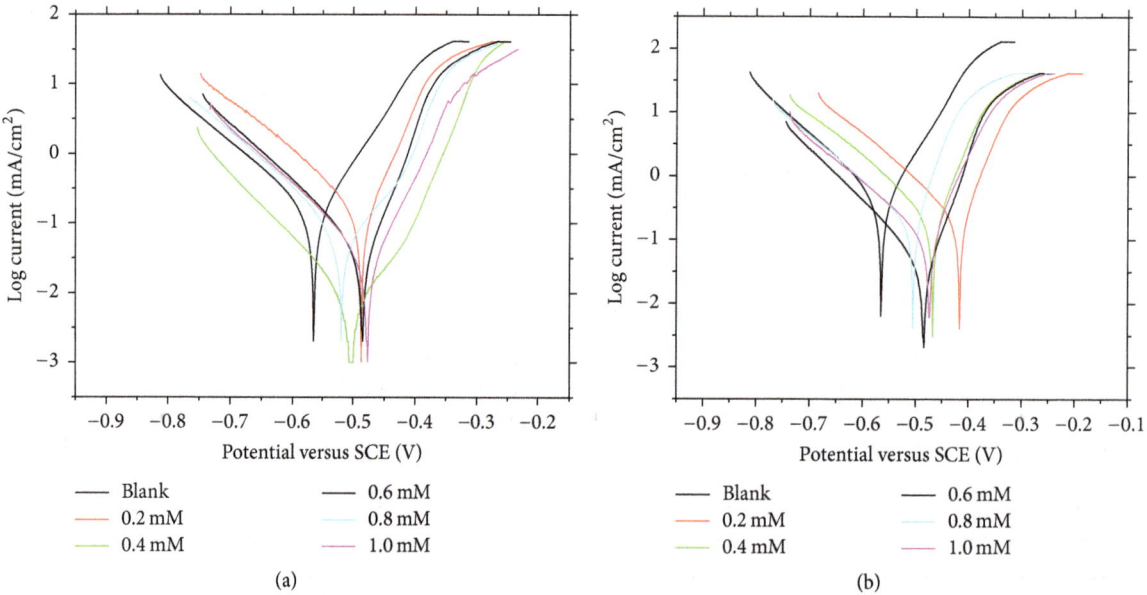

(a) (b)

FIGURE 4: Tafel lines for CS in the absence and presence of (a) P2APH and (b) RP2APH in 1 M HCl at 29 ± 0.2°C.

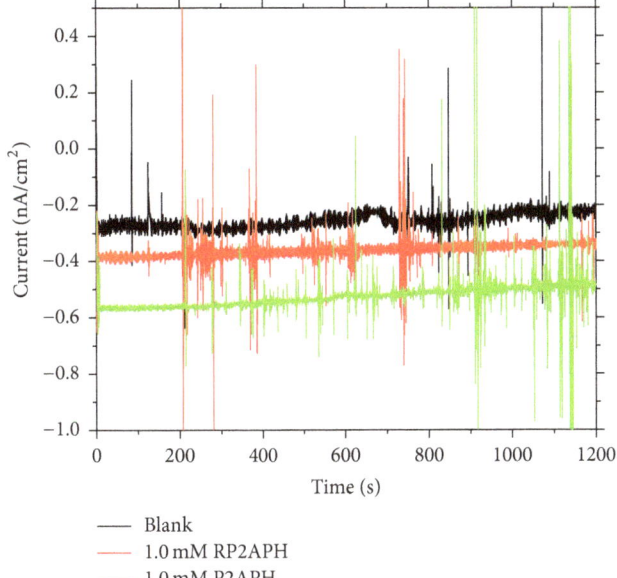

FIGURE 5: Current noise for CS in the absence and presence of P2APH and RP2APH (1.0 mM) in 1 M HCl 29 ± 0.2°C.

the blank metal specimen was 2.8, which was increased appreciably by the addition of inhibitors. CS treated with P2APH and RP2APH displayed noise resistances 4.22 and 3.92, respectively, which obviously reflects the involvement of molecules in the metal dissolution process.

The frequency domain analysis of noise data gave the Power spectral density (PSD) of various systems (Figure 6). The advantage of PSD is that it is independent of time and the signal statistics do not alter with time [31–33]. Fast Fourier Transformation (FFT) was adopted to convert time dependent noise data into PSD plots using ivium software. The frequency response of the noise signals can be monitored with the help of PDS plots. To improve the spectral resolution, Burg introduced maximum entropy method (MEM) for short time records [34]. Mathematically this method makes some assumptions about unmeasured data which are consistent with the existing data [35]. The red lines shown in the FFT spectrum of various CS specimens depicts MEM curves.

Current and potential noise for the blank specimen were comparatively higher than the specimens treated with inhibiting molecules. The magnitude of noise was significantly lowered with the frequency. Higher amplitude of potential noise signal is an indication of appreciable localized metallic corrosion [36]. On analyzing the PSD plots and MEM curves, it is clear that the magnitude of potential noise values is very much higher for blank metal specimen than the specimens immersed in acid in the presence of P2APH and RP2APH (1.0 mM). This is the indication of the considerable localized corrosion on CS surface in the absence of inhibitors. At all frequencies potential noise and current noise of CS treated with P2APH are lower than those of specimen treated with RP2APH, suggesting that the former molecule inhibits well the dissolution of CS in 1 M HCl.

The slope of the PSD plot is important parameter which depicts the response of the inhibiting molecules on the metal surface. A considerable change in the magnitude of slope of a PSD plot is a measure of the vulnerability for the metal dissolution [37]. Though the slopes of the PSD_v curves almost remain unchanged for CS treated with 1.0 mM P2APH and RP2APH, significant change in the slopes of PSD_i plots was noticed. Slopes of PSD_i spectral curves of CS specimen treated with 1.0 mM inhibitor molecules roughly changed to zero from −0.19. This result establishes that carbon steel in the presence of P2APH and RP2APH is less vulnerable to corrosion in 1 M HCl.

3.5. IR Spectral Studies. The surface film deposited on the CS specimen treated with 1.0 mM inhibitor in the aggressive medium for 24 h was removed mechanically and analyzed using IR spectroscopy. Figures 7(a) and 7(b) denote the overlay IR spectra of the P2APH and RP2APH and their surface film spectra on CS surface, respectively. The phenylhydrazone P2APH displayed a peak at 3330 cm^{-1} which was due to the stretching vibration of N-H group. This peak in the surface film spectrum shifted to higher frequency (3420 cm^{-1}) with very strong absorption. When we consider the structure of P2APH, the ample chance for the intramolecular hydrogen bonding between the heterocyclic nitrogen atom and N-H group is very obvious. The enhancement of N-H stretching frequency in the surface film spectrum may be due to the rupturing of hydrogen bond during the adsorption of molecule on CS surface through the N-atom of the pyridine ring which will lead to the free vibration of N-H bond. The CH=N stretching frequency appearing at 1600 cm^{-1} in the IR spectrum of P2APH changed to 1620 cm^{-1} in the spectrum of surface film. This is due to the interaction of this group with the surface metal atoms. The C-N(pyr) stretching vibration of P2APH was also shifted from 1540 cm^{-1} to 1520 cm^{-1} in surface film spectrum which is a clear indication of the interaction of the compound with metal atoms through pyridine nitrogen. Strong absorption below 900 cm^{-1} in the spectrum of surface film can be attributed to the formation of metal-nitrogen and metal-carbon bonds. Due to the lack of complete coplanarity of RP2APH, it is difficult to propose intramolecular hydrogen bonding. The N-H stretching frequency appearing at 3400 cm^{-1} in the spectrum of RP2APH is not appreciably altered in the spectrum of surface film. The peak observed at 1580 cm^{-1} was changed to 1600 cm^{-1} in the IR spectrum of surface film, which can be attributed to the interaction of pyridine nitrogen with Fe atoms. Below 1000 cm^{-1}, the spectrum of surface film of RP2APH exhibited very intense IR absorption indicating the formation of strong Fe-N and Fe-C bonds.

3.6. SEM Studies. Surface morphology of the metal surfaces was evaluated with the aid of SEM. All micrographs were recorded at a magnification of 200x. Figures 8(a)– 8(d), respectively, show the magnified surface images of bare metal, metal immersed in acid, and metal treated with acid in the presence of P2APH and RP2APH (1.0 mM) for

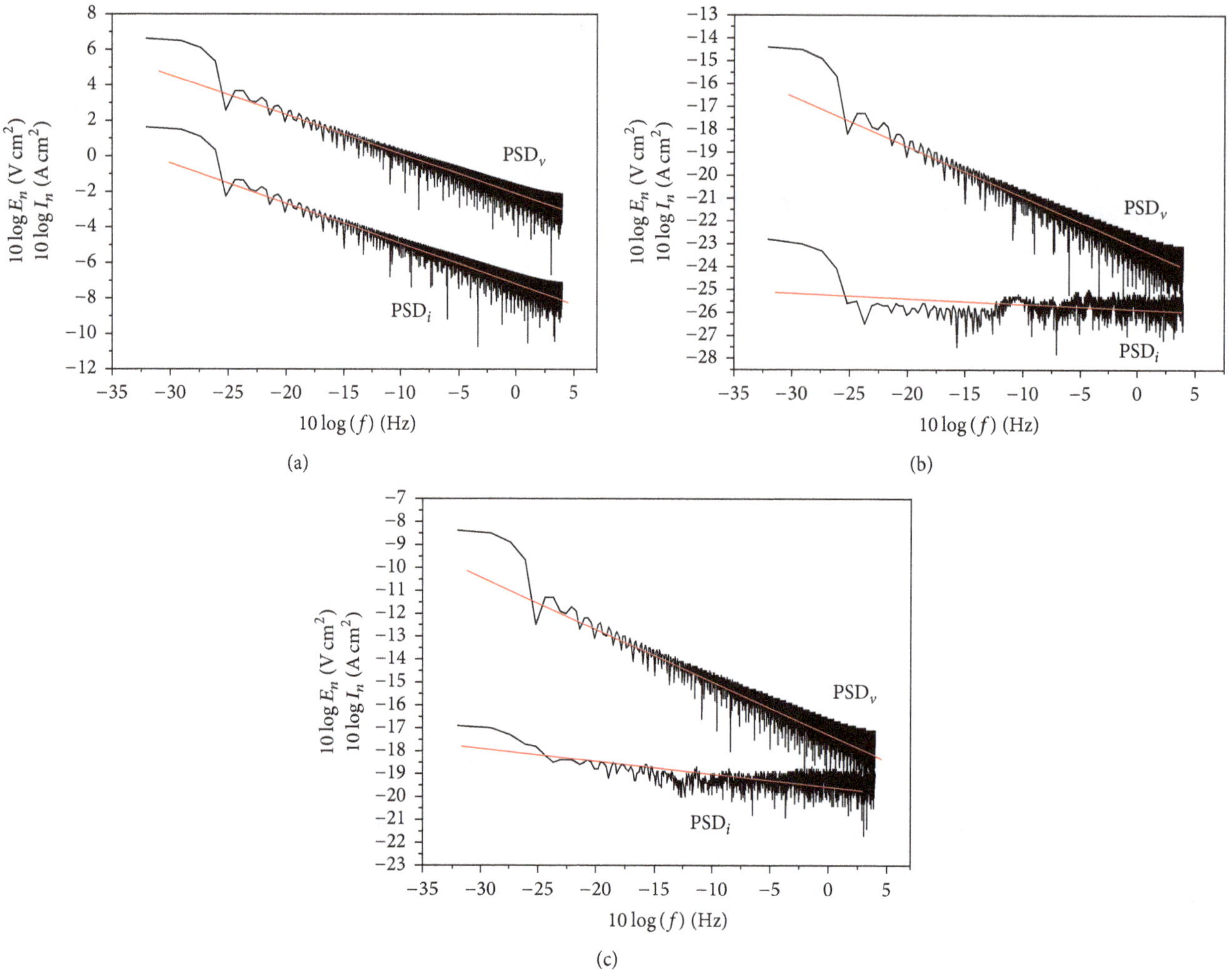

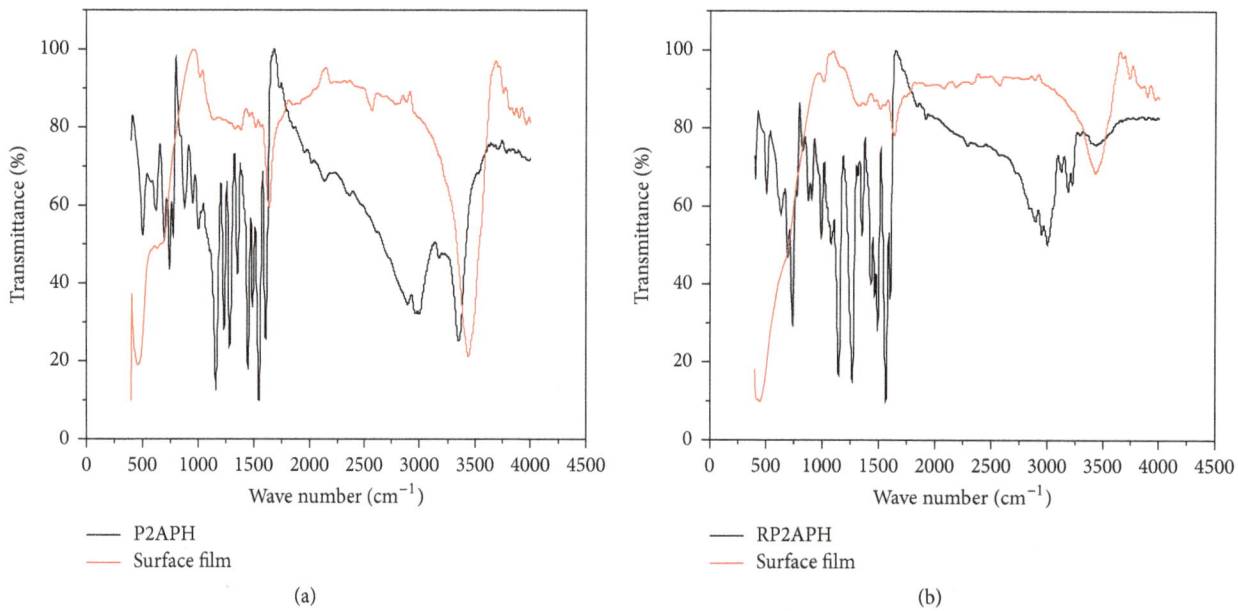

FIGURE 6: Power spectral density (current and voltage) of CS (a) and blank (b) in presence of P2APH and (c) RP2APH (1.0 mM) in 1 M HCl.

FIGURE 7: IR spectrum of (a) P2APH and surface film on CS; (b) RP2APH and surface film on CS.

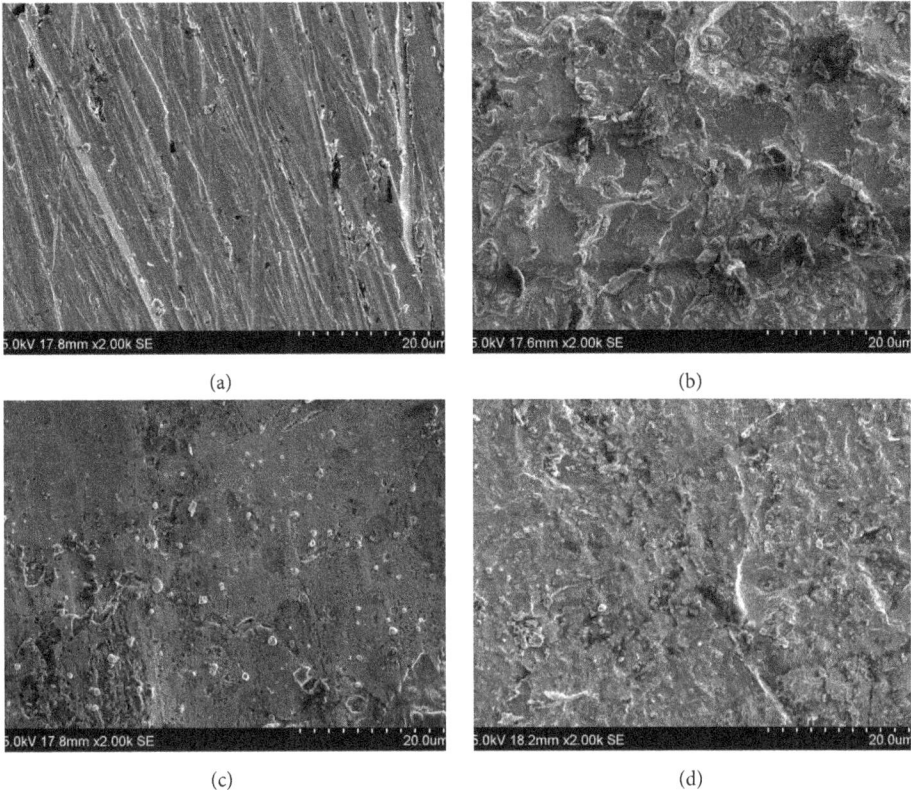

FIGURE 8: SEM images of (a) bare metal; (b) CS treated with 1.0 M HCl; (c) with P2APH (1.0 mM); (d) with RP2APH (1.0 mM) for 24 h.

24 h. The textures of all surface images showed considerable differences. The pits and cracks appearing in the SEM image of the bare metal specimen were due to the polishing effects. The micrograph of blank specimen appeared to be very rough due to severe corrosion. Comparatively smoother images were displayed by metal specimens immersed in HCl in the presence of P2APH (1.0 mM) and RP2APH (1.0 mM). This can be attributed to the prevention of CS acid corrosion in acid by the adsorption of molecules.

3.7. Quantum Chemical Calculations. The corrosion inhibition response of organic molecules can be correlated with the energy of frontier molecular orbitals. The donor-acceptor interactions (HSAB concept) between the filled molecular orbitals of the inhibitor molecules and the vacant d-orbitals of Fe atoms on the surface are very important in the prevention of metal dissolution. High value of E_{HOMO} and the lowest value of $E_{LUMO} - E_{HOMO}$ (ΔE) are the important quantum chemical parameters which facilitate the strong binding of the molecules on the metal surface [34, 38]. Optimization of geometry of molecules and quantum chemical calculations were performed by DFT method using GAMMES software. A combination of Beck's three-parameter exchange functional and Lee–Yang–Parr nonlocal correlation functional (B3LYP) was employed in DFT calculations [34]. Estimated quantum mechanical parameters like E_{HOMO}, E_{LUMO}, and ΔE for P2APH and RP2APH are provided in Table 3. The HOMO and LUMO of the molecules are P2APH and RP2APH represented in Figure 9. Approximate HSAB parameters like electronegativity (χ) and chemical hardness (η) of the molecules were evaluated by the following equations [39] and are also reported in Table 3:

$$\chi \approx -\frac{1}{2}\left(E_{HOMO} + E_{LUMO}\right),$$
$$\eta \approx \frac{1}{2}\left(E_{HOMO} - E_{LUMO}\right). \qquad (6)$$

The E_{HOMO} values of the two molecules were almost the same, but the energy separation between HOMO and LUMO was considerably lower for P2APH than RP2APH, which indicate that P2APH is a finer inhibitor than RP2APH. The energy required to render electrons from HOMO of P2APH to the vacant d-orbitals of Fe is low compared to that of RP2APH. Lowest value of E_{LUMO} is an indication of the higher probability of the molecule to accept electrons from the metal surface. On comparing the E_{LUMO} values of the two molecules, it is obvious that P2APH molecule possesses very low value than RP2APH. This originates a strong backdonation of electrons from the filled orbital of Fe to the unoccupied molecular orbital of P2APH. Quantum chemical parameters provide a room for the calculation of number of electrons transferred (ΔN) from donor to acceptor molecules. As an approximation, the chemical hardness of Fe bulk metal is assumed as zero and the approximate electronegativity of bulk Fe is taken as 7 eV [39]. Equation (7) gives the approximate quantity of electrons transferred from the inhibitor molecule to the Fe atoms.

$$\Delta N = \frac{\chi_{Fe} - \chi_{inhib}}{2\left(\eta_{Fe} + \eta_{inhib}\right)}. \qquad (7)$$

TABLE 3: Quantum chemical parameters of P2APH and RP2APH.

Molecule	E_{HOMO} (eV)	E_{LUMO} (eV)	ΔE (eV)	χ	η	ΔN
P2APH	−3.4640	0.7429	4.2069	1.36055	2.10345	1.340524
RP2APH	−3.4613	1.4803	4.941	0.9905	2.4708	1.216104

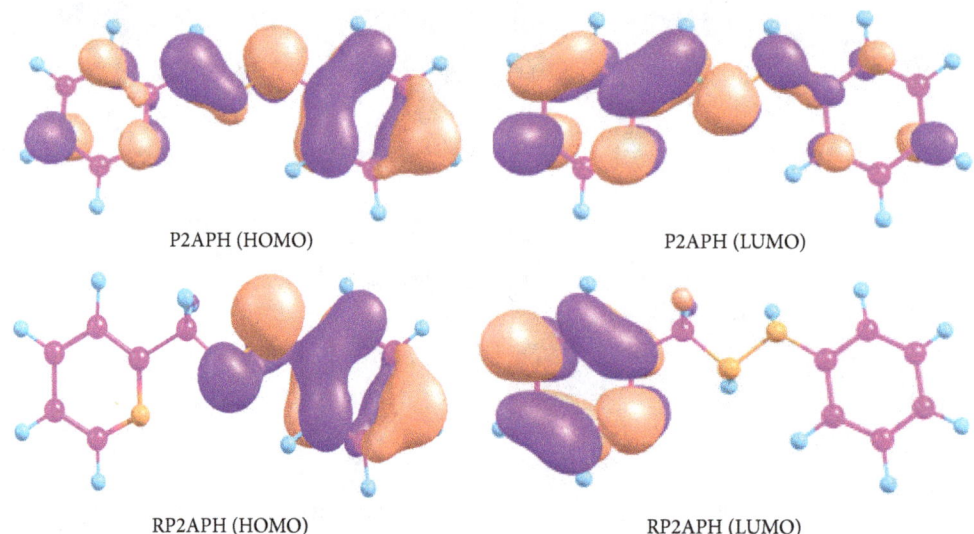

P2APH (HOMO)   P2APH (LUMO)

RP2APH (HOMO)   RP2APH (LUMO)

FIGURE 9: HOMO and LUMO of P2APH and RP2APH.

The number of electrons transferred from the inhibitor to the acceptor atom is higher for P2APH than RP2APH, which suggests that P2APH can make a strong coordinate type interaction with the metal atoms.

Perturbation of electron cloud occurs in molecules during chemical reaction. Chemical hardness of a molecule is a measure of resistance against the polarization of electron cloud. Soft molecules generally have low value of chemical hardness, low HOMO-LUMO energy gap, and high polarizability [40]. They are also associated with high chemical reactivity. If we consider the interaction of Fe with two organic molecules as soft-soft type, it is evident that P2APH has a great tendency to adsorb on Fe surface and hence prevent the metal dissolution appreciably, since it possesses low value of chemical hardness than RP2APH.

4. Conclusions

(1) P2APH and its reduced form RP2APH were very effective corrosion inhibitors for CS in HCl medium.

(2) All corrosion monitoring studies revealed that inhibition efficiency of P2APH is greater than that of RP2APH at all concentrations, which may be attributed to the presence of azomethine linkage present in P2APH.

(3) Gravimetric corrosion studies showed that the prolonged corrosion inhibition efficiency of molecules follows the order RP2APH > P2APH.

(4) Two molecules obeyed Langmuir adsorption isotherm on CS surface during the inhibition process.

(5) The inhibitors delay the process of corrosion by hindering the cathodic process mainly.

(6) The participation of P2APH and RP2APH in preventing the localized corrosion on CS in HCl was verified by electrochemical noise studies.

(7) Quantum chemical studies differentiated the protective power of molecules.

(8) Surface analysis using SEM and IR clearly portrayed the anticorrosive mechanism of inhibitors.

Competing Interests

The authors declare that they have no competing interests.

Acknowledgments

Authors are grateful to TEQIP (phase II), Government Engineering College, Thrissur, Kerala, India, for the financial assistance in the form of seed money project.

References

[1] V. P. Raphael, K. J. Thomas, K. S. Shaju, and A. Paul, "Corrosion inhibition investigations of 3-acetylpyridine semicarbazone on carbon steel in hydrochloric acid medium," *Research on Chemical Intermediates*, vol. 40, no. 8, pp. 2689–2701, 2014.

[2] S. John, B. Joseph, K. V. Balakrishnan, K. K. Aravindakshan, and A. Joseph, "Electrochemical and quantum chemical

study of 4-[(E)-[(2,4-dihydroxy phenyl) methylidine] amino]-6-methyl-3-sulphanylidine-2,3,4,5-tetra hydro-1,2,4-triazin-5-one [DMSTT]," *Materials Chemistry and Physics*, vol. 123, no. 1, pp. 218–224, 2010.

[3] F. Bentiss, M. Traisnel, L. Gengembre, and M. Lagrenée, "Inhibition of acidic corrosion of mild steel by 3,5-diphenyl-4H-1,2,4-triazole," *Applied Surface Science*, vol. 161, no. 1, pp. 194–202, 2000.

[4] A. K. Singh, S. K. Shukla, M. Singh, and M. A. Quraishi, "Inhibitive effect of ceftazidime on corrosion of mild steel in hydrochloric acid solution," *Materials Chemistry and Physics*, vol. 129, no. 1-2, pp. 68–76, 2011.

[5] S. V. Ramesh and A. V. Adhikari, "N′-[4-(diethylamino)benzylidine]-3-[8-(trifluoromethyl) quinolin-4-yl]thiopropano hydrazide) as an effective inhibitor of mild steel corrosion in acid media," *Materials Chemistry and Physics*, vol. 115, no. 2-3, pp. 618–627, 2009.

[6] A. Fiala, A. Chibani, A. Darchen, A. Boulkamh, and K. Djebbar, "Investigations of the inhibition of copper corrosion in nitric acid solutions by ketene dithioacetal derivatives," *Applied Surface Science*, vol. 253, no. 24, pp. 9347–9356, 2007.

[7] A. K. Singh and M. A. Quraishi, "Inhibiting effects of 5-substituted isatin-based Mannich bases on the corrosion of mild steel in hydrochloric acid solution," *Journal of Applied Electrochemistry*, vol. 40, no. 7, pp. 1293–1306, 2010.

[8] N. Khalil, "Quantum chemical approach of corrosion inhibition," *Electrochimica Acta*, vol. 48, no. 18, pp. 2635–2640, 2003.

[9] J. M. Costa and J. M. Lluch, "The use of quantum mechanics calculations for the study of corrosion inhibitors," *Corrosion Science*, vol. 24, no. 11-12, pp. 929–933, 1984.

[10] E. E. Oguzie, "Inhibiting effect of crystal violet dye on aluminum corrosion in acidic and alkaline media," *Chemical Engineering Communications*, vol. 196, no. 5, pp. 591–601, 2009.

[11] A. B. da Silva, E. D'Elia, and J. A. da Cunha Ponciano Gomes, "Carbon steel corrosion inhibition in hydrochloric acid solution using a reduced Schiff base of ethylenediamine," *Corrosion Science*, vol. 52, no. 3, pp. 788–793, 2010.

[12] H.-L. Wang, H.-B. Fan, and J.-S. Zheng, "Corrosion inhibition of mild steel in hydrochloric acid solution by a mercapto-triazole compound," *Materials Chemistry and Physics*, vol. 77, no. 3, pp. 655–661, 2003.

[13] S. John and A. Joseph, "Electro analytical, surface morphological and theoretical studies on the corrosion inhibition behavior of different 1,2,4-triazole precursors on mild steel in 1 M hydrochloric acid," *Materials Chemistry and Physics*, vol. 133, no. 2-3, pp. 1083–1091, 2012.

[14] A. Pourghasemi Hanza, R. Naderi, E. Kowsari, and M. Sayebani, "Corrosion behavior of mild steel in H_2SO_4 solution with 1,4-di [1′-methylene-3′-methyl imidazolium bromide]-benzene as an ionic liquid," *Corrosion Science*, vol. 107, pp. 96–106, 2015.

[15] N. O. Obi-Egbedi, I. B. Obot, and M. I. El-Khaiary, "Quantum chemical investigation and statistical analysis of the relationship between corrosion inhibition efficiency and molecular structure of xanthene and its derivatives on mild steel in sulphuric acid," *Journal of Molecular Structure*, vol. 1002, no. 1–3, pp. 86–96, 2011.

[16] K. Babic-Samardzija, C. Lupu, N. Hackerman, and A. R. Barron, "Inhibitive properties, adsorption and surface study of butyn-1-ol and pentyn-1-ol alcohols as corrosion inhibitors for iron in HCl," *Journal of Materials Chemistry*, vol. 15, no. 19, pp. 1908–1916, 2005.

[17] I. Ahamad, R. Prasad, and M. A. Quraishi, "Thermodynamic, electrochemical and quantum chemical investigation of some Schiff bases as corrosion inhibitors for mild steel in hydrochloric acid solutions," *Corrosion Science*, vol. 52, no. 3, pp. 933–942, 2010.

[18] E. E. Ebenso, M. M. Kabanda, L. C. Murulana, A. K. Singh, and S. K. Shukla, "Electrochemical and quantum chemical investigation of some azine and thiazine dyes as potential corrosion inhibitors for mild steel in hydrochloric acid solution," *Industrial and Engineering Chemistry Research*, vol. 51, no. 39, pp. 12940–12958, 2012.

[19] S. Erdemir, "Synthesis of novel chiral Schiff base and amino alcohol derivatives of calix[4]arene and chiral recognition properties," *Journal of Molecular Structure*, vol. 1007, pp. 235–241, 2012.

[20] V. K. Aghera and P. H. Parsania, "A cleaner approach for reduction of some symmetric diimines using $NaBH_4$," *Indian Journal of Chemistry—Section B Organic and Medicinal Chemistry*, vol. 48, no. 3, pp. 438–442, 2009.

[21] N. Kuriakose, J. T. Kakkassery, V. P. Raphael, and S. K. Shanmughan, "Electrochemical impedance spectroscopy and potentiodynamic polarization analysis on anticorrosive activity of thiophene-2-carbaldehyde derivative in acid medium," *Indian Journal of Materials Science*, vol. 2014, Article ID 124065, 6 pages, 2014.

[22] ASTM, "Standard recommended practice for the laboratory immersion corrosion testing of metals," AST G-31-72, ASTM, Philadelphia, Pa, USA, 1990.

[23] H. H. Hassan, E. Abdelghani, and M. A. Amin, "Inhibition of mild steel corrosion in hydrochloric acid solution by triazole derivatives. Part I. Polarization and EIS studies," *Electrochimica Acta*, vol. 52, no. 22, pp. 6359–6366, 2007.

[24] A. Dermaj, N. Hajjaji, S. Joiret et al., "Electrochemical and spectroscopic evidences of corrosion inhibition of bronze by a triazole derivative," *Electrochimica Acta*, vol. 52, no. 14, pp. 4654–4662, 2007.

[25] H. Ashassi-Sorkhabi, T. A. Aliyev, S. Nasiri, and R. Zareipoor, "Inhibiting effects of some synthesized organic compound on the corrosion of St-3 in 0.1N H_2SO_4 solution," *Electrochimica Acta*, vol. 52, no. 16, pp. 5238–5241, 2007.

[26] F. B. Growcock and R. J. Jasinski, "Time-resolved impedance spectroscopy of mild steel in concentrated hydrochloric acid," *Journal of the Electrochemical Society*, vol. 136, no. 8, pp. 2310–2314, 1989.

[27] J. A. Edward and M. M. Fitelson, "Notes on maximum-entropy processing (Corresp.)," *IEEE Transactions on Information Theory*, vol. 19, no. 2, pp. 232–234, 1973, Reprinted as 'Modern Spectrum Analysis' (IEEE), 1978.

[28] G. C. Barker, "Noise connected with electrode processes," *Journal of Electroanalytical Chemistry and Interfacial Electrochemistry*, vol. 21, no. 1, pp. 127–136, 1969.

[29] S. A. Umoren, "Synergistic inhibition effect of polyethylene glycol-polyvinyl pyrrolidone blends for mild steel corrosion in sulphuric acid medium," *Journal of Applied Polymer Science*, vol. 119, no. 4, pp. 2072–2084, 2011.

[30] M. G. Mahjani, M. Sabzali, M. Jafarian, and J. Neshati, "An investigation of the effects of inorganic inhibitors on the corrosion rate of aluminum alloy using electrochemical noise measurements and electrochemical impedance spectroscopy," *Anti-Corrosion Methods and Materials*, vol. 55, no. 4, pp. 208–216, 2008.

[31] R. A. Cottis and C. A. Loto, "Electrochemical noise generation during SCC of a high-strength carbon steel," *Corrosion*, vol. 46, no. 1, pp. 12–19, 1990.

[32] U. Bertocci, J. Frydman, C. Gabrielli, F. Huet, and M. Keddam, "Analysis of electrochemical noise by power spectral density applied to corrosion studies: Maximum entropy method or fast Fourier transform?" *Journal of the Electrochemical Society*, vol. 145, no. 8, pp. 2780–2786, 1998.

[33] C. H. Chen, *Non-Linear Max. Entropy Spectral Analysis Methods for Signal Recognition*, Research Studies Press, 1982.

[34] R. M. Issa, M. K. Awad, and F. M. Atlam, "Quantum chemical studies on the inhibition of corrosion of copper surface by substituted uracils," *Applied Surface Science*, vol. 255, no. 5, pp. 2433–2441, 2008.

[35] B. P. Markhali, R. Naderi, M. Mahdavian, M. Sayebani, and S. Y. Arman, "Electrochemical impedance spectroscopy and electrochemical noise measurements as tools to evaluate corrosion inhibition of azole compounds on stainless steel in acidic media," *Corrosion Science*, vol. 75, pp. 269–279, 2013.

[36] D. J. Mills, G. P. Bierwagen, B. Skerry, and D. Tallman, "Investigation of anticorrosive coatings by the electrochemical noise method," *Materials Performance*, vol. 34, no. 5, pp. 33–38, 1995.

[37] R. A. Cottis, M. A. A. Al-Awadhi, H. Al-Mazeedi, and S. Turgoose, "Measures for the detection of localized corrosion with electrochemical noise," *Electrochimica Acta*, vol. 46, no. 24-25, pp. 3665–3674, 2001.

[38] V. S. Sastri and J. R. Perumareddi, "Molecular orbital theoretical studies of some organic corrosion inhibitors," *Corrosion*, vol. 53, no. 8, pp. 617–622, 1997.

[39] R. G. Pearson, "Absolute electronegativity and hardness: application to inorganic chemistry," *Inorganic Chemistry*, vol. 27, no. 4, pp. 734–740, 1988.

[40] S. Xia, M. Qiu, L. Yu, F. Liu, and H. Zhao, "Molecular dynamics and density functional theory study on relationship between structure of imidazoline derivatives and inhibition performance," *Corrosion Science*, vol. 50, no. 7, pp. 2021–2029, 2008.

11

Study on the Electrochemical Performance of Sacrificial Anode Interfered by Alternating Current Voltage

Qingmiao Ding[ID]**, Xiao Chu**[ID]**, Tao Shen**[ID]**, and Xiaoxiao Yu**

Airport College, Civil Aviation University of China, Tianjin, China

Correspondence should be addressed to Xiao Chu; xchu@cauc.edu.cn

Academic Editor: Ramazan Solmaz

The effect of alternating current (AC) voltage of 0V, 1V, 3V, and 5V on magnesium alloy sacrificial anode electrochemical properties was studied by open circuit potential (OCP) analysis, electrochemical impedance spectroscopy (EIS), and polarization curve measurements. The results demonstrate that the AC voltage has a great effect on the magnesium alloy sacrificial anode. The corrosion control is anode control in the first two days with no AC interference. The stray current accelerates the transmission and diffusion of oxygen, so the corrosion rate under AC interference is higher than that with no AC interference. And the corrosion control becomes cathodic control under AC interference. The corrosion rate of the sacrificial anode is faster and faster as the AC interference voltage increases in the range of 0~5V, while the corrosion inclination is weakened.

1. Introduction

It has been known that the AC stray current has a great influence on the potential and current of the buried metallic pipeline when it is close to high voltage AC transmission lines or rail transit systems [1, 2]. Moreover, the presence of AC has caused serious damage on interfered metallic structures even when cathodic protection is applied [2–5]. L. Y. Xu [6] found that the presence of AC interference decreased the CP effectiveness to protect the steel from corrosion. Only when CP potential was sufficiently negative, the steel was under a complete protection even when the AC current density was 400 A/m^2. The cathodic protection system of airport apron pipe network generally adopts magnesium alloy sacrificial anode cathodic protection [7], and it also inevitably has an adverse effect on the sacrificial anode performance [8, 9].

As early as 1978, Pookote [10] pointed out that AC could lead to the potential shift of Mg sacrificial anode and accelerate its dissolution. Freiman [11] found that the potential of Mg sacrificial anode shifted positively under AC and the current efficiency of Mg was decreased significantly. For pipelines protected by sacrificial anodes, excessive AC interference voltage could degrade the performance of the sacrificial anode, shorten its service life, and even "polarity reversal" which could accelerate the corrosion of the pipeline [12]. Bruchner [13] found that the polarity of pipeline-Mg sacrificial anode turned to be the cathode while the protected pipeline turned to be the anode when the applied AC current density reached 39A/m^2. Dezhi Tang [14] found that the output cathodic protection current of the Mg sacrificial anode decreased at alternating current density of 50 A/m^2. Polarity reversal occurred to Mg sacrificial anode as soon as alternating current interference of 100 A/m^2 (or larger) was applied. Yin Kehua et al. [15] studied the effects of AC interference on the properties of Mg, Al, and Zn sacrificial anodes in soil. The results showed that with the increase of AC interference, the output negative current density of Mg and Al would decrease continuously, while output current density of Zn was relatively stable. At the same time, the "polar reversal" behavior of Mg anode under AC interference was also observed.

In summary, AC voltage can interfere in the normal operation of cathodic protection system [9, 16–21], or cause polarity reversal in pipelines with the sacrificial anode,

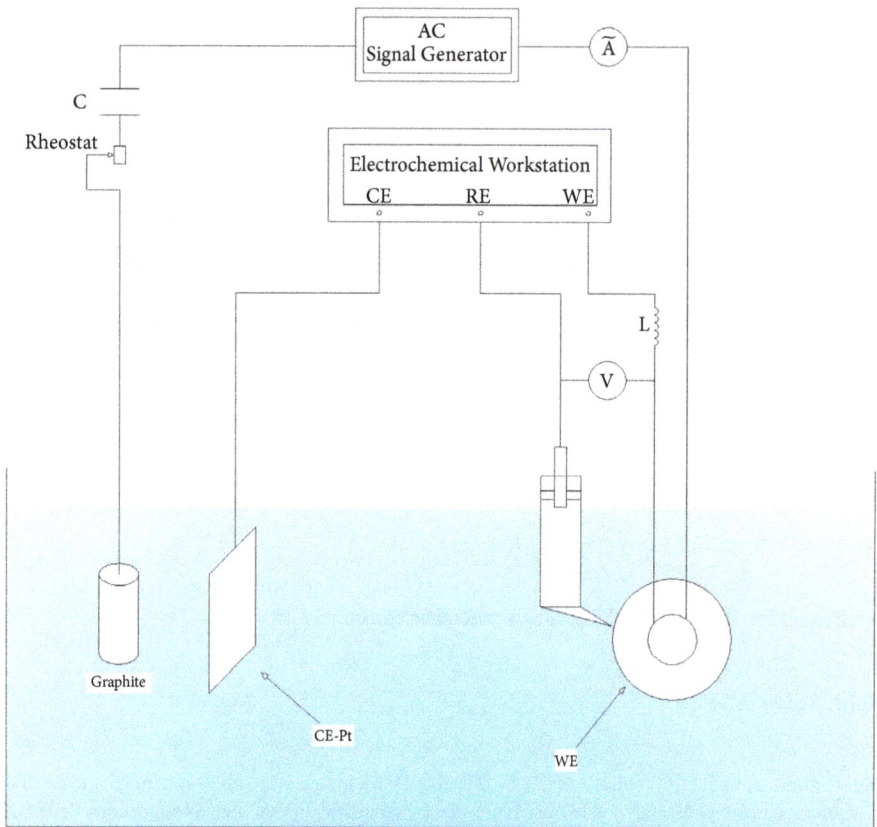

FIGURE 1: Schematic diagram of the experimental.

leading to the pipeline lack of effective protection. Therefore, it is very necessary to study the influence of AC stray current on the cathodic protection system.

2. Experimental Procedure

2.1. Specimen and Solution. Sacrificial anode specimens used in this work were magnesium alloy rods with diameter of 10mm. The chemical composition of magnesium alloy rods is as follows: 6.5 wt% Al, 3.5 wt% Zn, 0.6 wt% Mn, 0.005 wt% Fe, 0.003 wt% Ni, 0.02 wt% Cu, 0.1 wt% Si, and bal. wt% Mg. The magnesium alloy sacrificial anode rods were welded with wires and put into the electrician tubes. We used epoxy resin and curing agent mixture with the ratio of 3:1 to pour magnesium alloy sacrificial anode rods, leaving an exposure area as the working face. The electrode was ground sequentially to 2000grit emery paper and then cleaned by acetone, absolute ethanol, and distilled water.

The test solution was brought from the soil along the apron with the chemical composition of Na_2CO_3 0.1599g/L, NaCl 0.5124 g/L, Na_2SO_4 0.1712g/L, $NaHCO_3$ 0.0864g/L. All measurements were conducted at ambient temperature of 20°C and open to air.

2.2. Electrochemical Equivalent Circuit for the Experimental. Figure 1 shows the schematic diagram of the experimental set-up for studies of AC interference corrosion under

different chloride ion concentration or test temperature. The experimental device uses a three-electrode system. The magnesium alloy sacrificial anode rod was used as the working electrode (WE), the platinum electrode was used as the counter electrode (CE), and a saturated calomel electrode (SCE) was used as a reference electrode (RE). The distance between WE and RE was about 2mm, in order to reduce the ohmic drop in potential measurements. The electric circuit was specially designed to supply and measure AC and DC signals independently. Working electrode, AC power source, capacitor, and carbon rod form an AC interference system. Reference electrode, high resistance voltmeter, and working electrode compose voltage test system. Auxiliary electrode, reference electrode, working electrode, and electrochemical workstation consist of electrochemical test system. In this work, the sinusoidal AC signal with a frequency of 50 Hz was supplied between two graphite electrodes. Within the AC mesh, an electrolytic capacitor (50V, 470μF) [1] in series was used to prevent DC circulation. And within the DC mesh, an inductor [2] was used to prevent the flow of AC current into the electrochemical measurement system.

2.3. Electrochemical Measurements. The electrochemical workstation (CHI660D) used a three-electrode cell system. Specimens used for electrochemical measurements were coated with epoxy resin, leaving an exposure area of 3.925 cm^2 as the working surface. The electrode was ground

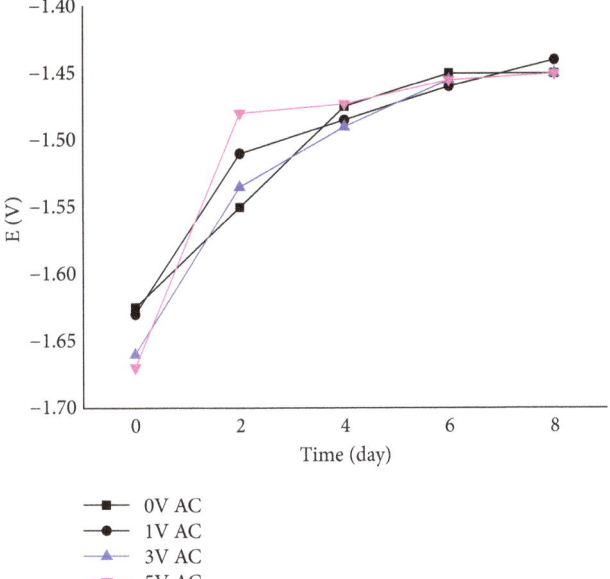

FIGURE 2: The open circuit potential of magnesium sacrificial anode under different AC interference voltages at an ambient temperature of 20°C in the test solution.

sequentially to 2000 grit emery paper and then cleaned by acetone, anhydrous alcohol, and distilled water.

The stable open potential, electrochemical impedance spectroscopy, and polarization curves of magnesium alloy sacrificial anode were measured by the electrochemical work-station every day. The frequency range of EIS was 100 kHz to 0.1Hz while the amplification of sinusoidal AC signal was 3mV. The data analysis software was ZSimpWin3.21. According to the test results of open circuit potential, it was determined that the range of polarization curve was from -0.8V to -0.3V, and the scanning rate was 0.1667mV/s.

3. Results

Figure 2 shows the open circuit potential of sacrificial anode under various AC voltages at an ambient temperature of 20°C in the test solution for eight days. The corrosion potential of the magnesium sacrificial anode was observed to shift positively under all experimental conditions. It indicated that the activity in the sacrificial anode surface decreased as time went by with or without AC interference. The open circuit potential's positive migration under AC interference of 5V voltage was the maximum while that of 0V voltage was the minimum.

Table 1 shows the corrosion electrochemical parameters fitted by polarization curves of the sacrificed anode under different AC interference voltages of 0V, 1V, 3V, or 5V. The polarization resistance increased gradually and the corrosion current decreased with the progress of the experiment. It indicated that the corrosion tendency and rate of the sacrificial anode decreased with the increasing of the experimental time. Compared with the polarization resistance and corrosion current at the same experiment time, it could be seen that the polarization resistance was smaller under AC interference

than that without AC interference, and the corrosion current was larger than that without AC interference. It indicated that AC interference voltage could reduce the corrosion tendency of the sacrificial anode while it would accelerate the dissolution rate of magnesium anodic. The larger the AC interference voltage was, the stronger the promotion effect was.

Sacrificial anode control could be observed by the slope trend of the cathode and anode polarization curves. At the beginning of the experiment, the slope of cathodic polarization curve was larger than that of anodic polarization curve. After a certain period of experiment, the slope of anodic polarization curve was gradually higher than that of cathodic polarization curve. It indicated that the cathode control gradually transformed into anode control. It showed that AC played an important role in anodic polarization of magnesium alloy sacrificial anode.

Figure 3 shows the EIS under different AC in the original soil simulation solution and at ambient temperature. It could be seen that there were two impedance arcs in the shape of EIS under the four environments of 0V, 1V, 3V, and 5V AC interference voltages. The change of the impedance arc was the largest on the second day of experiment. The variation of impedance arc became smaller with the lapse of time. We could find that the radius of impedance arc represented by the sacrificial anode surface corrosion products was the largest with no AC. And the larger the AC interference was, the larger the radius of impedance arc was. It indicated that AC interference accelerated the corrosion of the sacrificial anode and promoted the corrosion products on the surface of sacrificial anode. The greater the AC interference was, the faster the rate of corrosion was.

In order to analyze the sacrificial anodic corrosion process, the EIS of Figure 2 is fitted with the equivalent circuits. Figure 4 shows the equivalent circuit of the EIS of the magnesium alloy sacrificial anode during the experiment. The elements represented different physical meanings: Rs represented the resistance of the solution, CPEdl represented the electric double layer capacitor of the metal oxide film and the solution, CPE represented the capacitance due to the ion through surface oxide film, Rt represented the charge transfer resistance, and Rc represented the oxidation resistance. When the experiment started, the oxide film on the surface of the sacrificial anode was destroyed, and the exposing metal reacted with oxygen. The corrosion products began to accumulate on the surface of the sacrificial anode. The physical meaning of the element representation changed: Rs still represented the resistance of the solution and CPEdl still represented the electric double layer capacitor of the metal oxide film and the solution, and CPE no longer expressed the capacitance due to the ion through surface oxide film, but on behalf of the capacitance due to the ion through corrosion scale. Rt still represented the charge transfer resistance. Rc represented the corrosion product film resistance.

Figure 5 shows the process of the change of corrosion product film resistance Rc under different AC interference voltages. It could be seen that the value of Rc increased gradually as experiment time went by. The larger the AC voltage was, the greater the change of Rc was.

Table 1: The corrosion electrochemical parameters from polarization curves of the sacrificed anode under different AC voltages.

AC voltage	Electrochemical parameters	0d	2d	4d	6d	8d
0V	Polarization resistance (Ω)	659.3	731.1	820.4	940.8	1018.2
	Corrosion current (A)	$6.530*10^{-5}$	$5.825*10^{-5}$	$5.167*10^{-5}$	$4.072*10^{-5}$	$3.876*10^{-5}$
	Cathodic Tafel slope	6.340	5.813	5.082	4.458	5.049
	Anodic Tafel slope	3.760	4.396	5.175	5.174	5.97
1V	Polarization resistance (Ω)	243.2	346.4	592.4	724.8	785
	Corrosion current (A)	$1.814*10^{-4}$	$1.279*10^{-4}$	$7.558*10^{-5}$	$5.858*10^{-5}$	$5.602*10^{-5}$
	Cathodic Tafel slope	5.021	4.788	4.621	4.413	4.774
	Anodic Tafel slope	4.834	5.027	5.091	5.826	5.113
3V	Polarization resistance (Ω)	331.7	488.8	510.3	595.8	623.5
	Corrosion current (A)	$1.254*10^{-4}$	$8.871*10^{-5}$	$7.968*10^{-5}$	$7.5*10^{-5}$	$7.253*10^{-5}$
	Cathodic Tafel slope	5.376	4.775	4.829	4.027	4.017
	Anodic Tafel slope	4.894	5.211	5.864	5.703	5.597
5V	Polarization resistance (Ω)	228.5	245.3	298.7	378.2	431.6
	Corrosion current (A)	$1.914*10^{-4}$	$1.818*10^{-4}$	$1.476*10^{-4}$	$1.178*10^{-4}$	$9.838*10^{-5}$
	Cathodic Tafel slope	5.238	4.694	4.701	4.814	3.880
	Anodic Tafel slope	4.708	5.052	5.222	4.951	6.359

4. Discussion

During the corrosion of magnesium alloy sacrificial anode in the soil simulated solution, the anodic and cathodic reactions include Mg oxidation and the reduction of hydrogen ions, respectively:

$$Mg \longrightarrow Mg^{2+} + 2e \tag{1}$$

$$2H^+ + 2e \longrightarrow H_2 \tag{2}$$

Generally, the presence of alternating current in the soil around the buried pipelines is prevalent. And it has been acknowledged that buried pipelines corrode at an accelerated rate in the presence of AC interference. Magnesium alloy is a more reactive metal than carbon steel, and AC should have a greater impact on the sacrificial anode. So when

AC interference is applied (on the 0th day), the potential is negatively shifted. And the greater the AC interference voltage is, the greater the potential offset is. Because of the AC interference, magnesium dissolution is accelerated. The corrosion product Mg^{2+} does not diffuse in time; this would cause a local saturation of Mg^{2+} ions and OH^- to exceed the solubility of magnesium hydroxide ($Mg(OH)_2$), favoring the formation of $Mg(OH)_2$ scale:

$$Mg^{2+} + 2OH^- \longrightarrow Mg(OH)_2 \tag{3}$$

Then the open circuit potential of magnesium alloy has a sudden increase in the initial experiment stage, the potential offset is the largest under AC interference voltage of 5V, and the potential offset is the least under no AC interference. Then the open circuit potential continues to shift

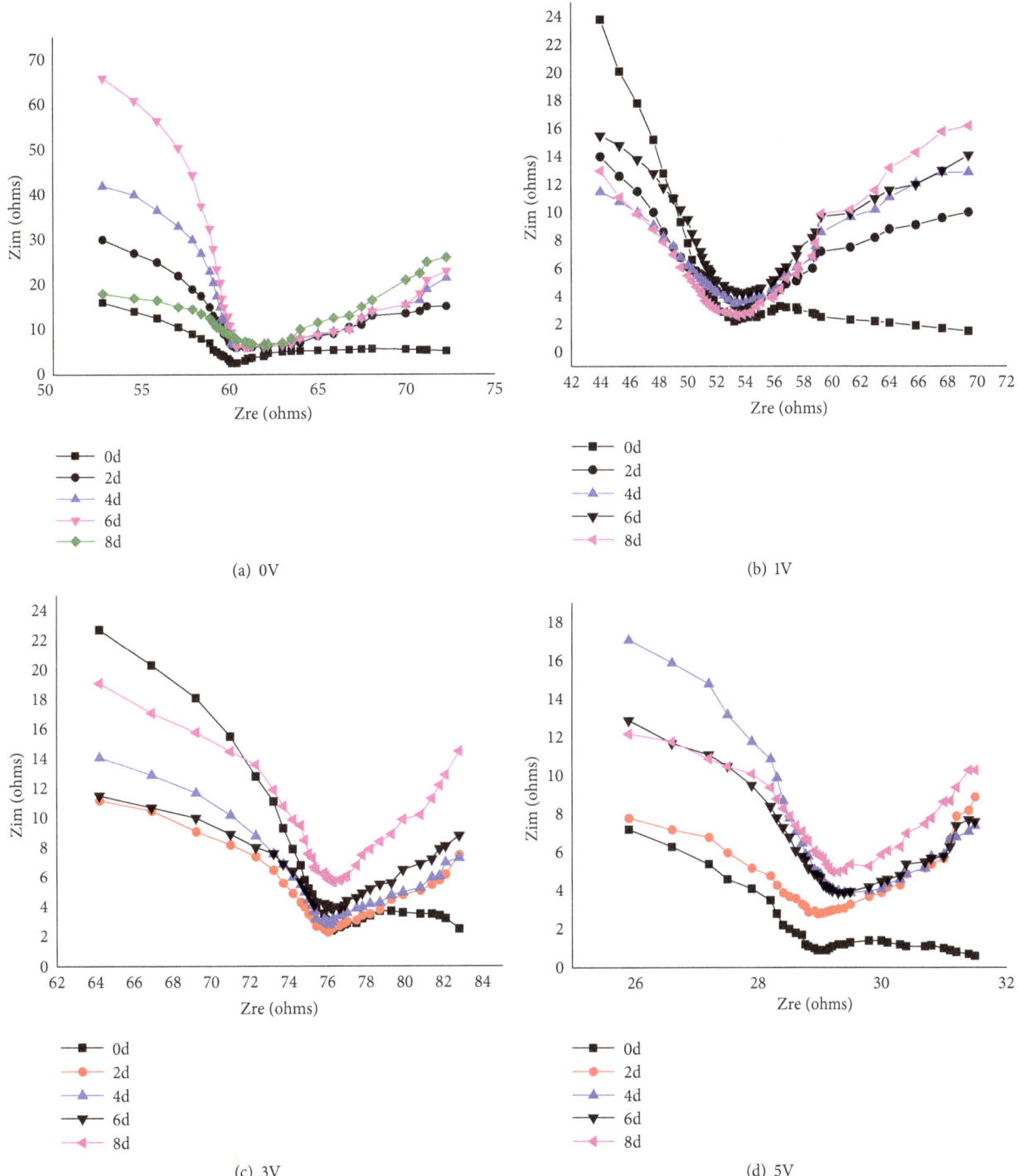

FIGURE 3: The EIS of magnesium sacrificial anode under different AC interference voltages at an ambient temperature of 20°C in the test solution.

in a positive migration, although the degree of the offset is small. It indicates that the surface corrosion is serious and the sacrificial anode surface has the largest changes because of the rapid generating of corrosion product in the first two days of the experiment. And the inclination of sacrificial anode corrosion is smaller relatively after two days of the experiment because the sacrificial anode surface is covered with corrosion

products. And then the influence of the corrosion inclination of AC voltage becomes small after two days of the experiment.

In the process of corrosion and dissolution, the cathodic reaction of magnesium sacrificial anode is the oxygen absorption reaction which is mainly due to the reaction between O_2 and the surface of magnesium alloy sacrificial anode. The corrosion of magnesium sacrificial anode with

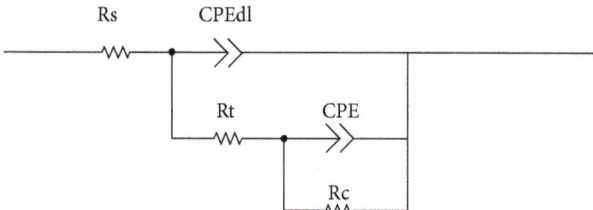

FIGURE 4: Equivalent circuit of the EIS.

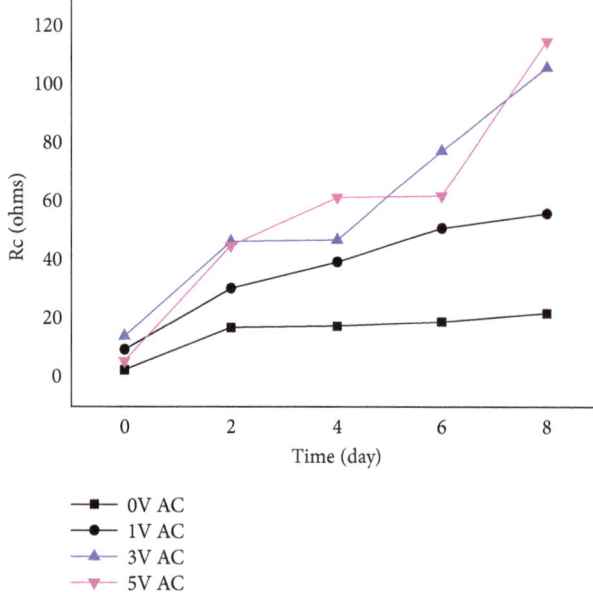

FIGURE 5: The Rc value of the sacrificial anode under different AC interference voltages at an ambient temperature of 20°C in the test solution.

most of the rest is attached to the surface of sacrificial anode. The corrosion products gradually accumulated and became thicker. The electrochemical impedance spectroscopy was the same mode shown in Figure 3. The corrosion products attached to the sacrificial anode surface accumulates and the resistance of the corrosion reaction becomes larger. With no AC interference and 1V AC interference, the increasing rate of Rc becomes slower during the experiment. It indicates that the earlier corrosion products hinder the oxygen reaction of sacrificial anode surface seriously, which makes the corrosion products increase fewer. However, the increasing rate of Rc becomes slower and then becomes faster during the experiment when the AC interference voltages are 3V and 5V. It indicates that the larger AC stray current changes the corrosion state of magnesium sacrificial anode surface. There is still a great corrosion rate although the reaction resistance caused by corrosion products increases.

5. Conclusion

Effects of AC on the performance of Mg sacrificial anode were studied in the simulated soil solution along the apron pipeline at ambient temperature (20°C). The analysis of the experimental results led to the following conclusions:

(1) AC interference is critical to the corrosion of magnesium alloy sacrificial anode. AC accelerated the dissolution of Mg sacrificial anode.

(2) The DC potential of Mg sacrificial shifted positively under AC at the same time. With the passage of time, the DC potential of the Mg sacrificial anode shifted positively in the presence or absence of AC interference, and the positive shift was more obvious in the case of AC interference than that in the absence of AC interference.

(3) As the AC interference voltage increases in the range of 0~5V, the corrosion rate of the sacrificial anode is faster and faster, but the corrosion inclination is weakened. Moreover, the corrosion products accumulate in the magnesium alloy sacrificial anode surface, and it affects the releasing of quantity of electric charge.

Conflicts of Interest

The authors declare that there are no conflicts of interest regarding the publication of this article.

Acknowledgments

The research work was supported by the Central College Foundation of CAUC (3122017038) and Airport Engineering Research Base Open Fund (study on the optimization of cathodic protection model in the apron area based on BEASY).

AC interference is biased towards cathodic control in the first two days because the Cathodic Tafel Slope is bigger than the Anode Tafel Slope. So the diffusion rate of oxygen affects the rate of corrosion. And the stray current accelerates the transmission and diffusion of oxygen, so the corrosion rate under AC interference is bigger than that with no AC interference; the bigger the AC interference voltage is, the faster the corrosion rate is. In the process of experiment, the longer the time of the oxygen absorption reaction on the surface of the anode is, the more the surface corrosion products accumulate. It would be difficult for O_2 to contact with the sacrificial anode surface, the reaction speed would gradually reduce and converge in all the experiments (as shown in Table 1), and the change tendency of impedance arc would be smaller (as shown in Figure 4). It indicates that the corrosion of sacrificial anode surface is relatively serious in the initial period. After that, although the corrosion is still aggravated, the rate of corrosion gradually slows down. The corrosion products on the sacrificial anode surface have accumulated very seriously, which hinders the reaction rate between O_2 and sacrificial anode surface.

In the process of experiment, the corrosion products of magnesium alloy sacrificial anode would drop a little and

References

[1] D. Kuang and Y. F. Cheng, "AC corrosion at coating defect on pipelines," *Corrosion*, vol. 71, no. 3, pp. 267–276, 2015.

[2] L. Wang, L. Cheng, J. Li, Z. Zhu, S. Bai, and Z. Cui, "Combined effect of alternating current interference and cathodic protection on pitting corrosion and stress corrosion cracking behavior of X70 Pipeline Steel in near-neutral pH environment," *Materials* , vol. 11, no. 4, 2018.

[3] S. Goidanich, L. Lazzari, and M. Ormellese, "AC corrosion. Part 2: Parameters influencing corrosion rate," *Corrosion Science*, vol. 52, no. 3, pp. 916–922, 2010.

[4] L. Y. Xu, X. Su, and Y. F. Cheng, "Effect of alternating current on cathodic protection on pipelines," *Corrosion Science*, vol. 66, pp. 263–268, 2013.

[5] D.-K. Kim, S. Muralidharan, T.-H. Ha et al., "Electrochemical studies on the alternating current corrosion of mild steel under cathodic protection condition in marine environments," *Electrochimica Acta*, vol. 51, no. 25, pp. 5259–5267, 2006.

[6] A. Q. Fu and Y. F. Cheng, "Effect of alternating current on corrosion and effectiveness of cathodic protection of pipelines," *Canadian Metallurgical Quarterly*, vol. 51, no. 1, pp. 81–90, 2012.

[7] H. Zhao, P. Bian, and D. Ju, "Electrochemical performance of magnesium alloy and its application on the sea water battery," *Journal of Environmental Sciences*, vol. 21, no. 1, pp. S88–S91, 2009.

[8] I. A. Metwally, H. M. Al-Mandhari, A. Gastli, and Z. Nadir, "Factors affecting cathodic-protection interference," *Engineering Analysis with Boundary Elements*, vol. 31, no. 6, pp. 485–493, 2007.

[9] N. Kouloumbi, G. Batis, N. Kioupis, and P. Asteridis, "Study of the effect of AC-interference on the cathodic protection of a gas pipeline," *Anti-Corrosion Methods and Materials*, vol. 49, no. 5, pp. 335–345, 2002.

[10] S. R. Pookote and D.-T. Chin, "Effect of alternating current on the underground corrosion of steels," *Materials Performance*, vol. 17, no. 3, pp. 9–15, 1978.

[11] L. l. Freiman and M. Yunovich, "Special behavior of steel cathode in soil and protection assessment of underground pipe with a buried coupon," *Protection of Metals*, vol. 27, pp. 437–447, 1991.

[12] D. Tang, Y. Du, M. Lu, L. Dong, and Z. Jiang, "Progress in the mutual effects between AC interference and the cathodic protection of buried pipelines," *Journal of the Chinese Society of Corrosion and Protection*, vol. 33, no. 5, pp. 351–356, 2013.

[13] H. Bruckner W, "Electrochemical methods for the study of corrosion lead- encased water pipe and cables," *Corrosion*, vol. 8, pp. 135-136, 1965.

[14] D. Tang, Y. Du, X. Li, Y. Liang, and M. Lu, "Effect of alternating current on the performance of magnesium sacrificial anode," *Materials and Corrosion*, vol. 93, pp. 133–145, 2016.

[15] K. Yin H, M. Tang H, and J. Xiong X, "Corrosion of buried steel structure under effect of electrical field with industry frequency," *Journal of Chinese Society for Corrosion and protection*, vol. 2, no. 3, pp. 33–41, 1982.

[16] X. Wang, G. Yang, H. Huang, Z. Chen, and L. Wang, "Study on AC stray current corrosion law of buried steel pipelines," *Applied Mechanics and Materials*, vol. 263-266, no. 1, pp. 448–451, 2013.

[17] M. Ormellese, S. Goidanich, and L. Lazzari, "Effect of AC interference on cathodic protection monitoring," *Corrosion Engineering, Science and Technology*, vol. 46, no. 5, pp. 618–623, 2011.

[18] Y. T. Li, X. Li, G. W. Cai, and L. H. Yang, "Influence of AC interference to corrosion of Q235 carbon steel," *Corrosion Engineering, Science and Technology*, vol. 48, no. 5, pp. 322–326, 2013.

[19] A. Q. Fu and Y. F. Cheng, "Effects of alternating current on corrosion of a coated pipeline steel in a chloride-containing carbonate/bicarbonate solution," *Corrosion Science*, vol. 52, no. 2, pp. 612–619, 2010.

[20] D. Kuang and Y. F. Cheng, "Effects of alternating current interference on cathodic protection potential and its effectiveness for corrosion protection of pipelines," *Corrosion Engineering, Science and Technology*, vol. 52, no. 1, pp. 22–28, 2017.

[21] M. Büchler and H.-G. Schöneich, "Investigation of alternating current corrosion of cathodically protected pipelines: Development of a detection method, mitigation measures, and a model for the mechanism," *Corrosion*, vol. 65, no. 9, pp. 578–586, 2009.

12

The Discrete Wavelet Transform and Its Application for Noise Removal in Localized Corrosion Measurements

Rogelio Ramos,[1] **Benjamin Valdez-Salas,**[1] **Roumen Zlatev,**[1]
Michael Schorr Wiener,[1] **and Jose María Bastidas Rull**[2]

[1]*Engineering Institute, Autonomous University of Baja California, Boulevard Benito Juarez, Insurgentes Este,*
21280 Mexicali, BC, Mexico
[2]*National Center of Metallurgical Research (CENIM) Madrid, The Spanish State Council for Scientific Research (CSIC),*
Madrid, Spain

Correspondence should be addressed to Rogelio Ramos; rramosi@uabc.edu.mx

Academic Editor: Flavio Deflorian

The present work discusses the problem of induced external electrical noise as well as its removal from the electrical potential obtained from Scanning Vibrating Electrode Technique (SVET) in the pitting corrosion process of aluminum alloy A96061 in 3.5% NaCl. An accessible and efficient solution of this problem is presented with the use of virtual instrumentation (VI), embedded systems, and the discrete wavelet transform (DWT). The DWT is a computational algorithm for digital processing that allows obtaining electrical noise with Signal to Noise Ratio (SNR) superior to those obtained with Lock-In Amplifier equipment. The results show that DWT and the threshold method are efficient and powerful alternatives to carry out electrical measurements of potential signals from localized corrosion processes measured by SVET.

1. Introduction

Currently, VI offers novel and efficient alternatives in digital measurement and processing of electrical signals in corrosion studies, with significant benefits in the cost of VI systems, compared to the costs of traditional instrumentation, whose function is defined by the manufacturer and not by the end user, allowing the replacement of traditional measurement and control instruments with modern computerized instruments.

1.1. Electrode Scanning Techniques. The electrochemical electrode scanning techniques are represented by scanning reference and scanning vibrating reference electrode (SRET and SVET) [1], since SVET is the best SNR, if we compare the minimum measurement ranges for the SRET technique (that precedes the SVET), which are 200 μV and are limited by the noise level of the signal [2, 3].

In the SVET the displacement of the vibrating electrode is small, in the range of 1 to 100 μm; this displacement provides a small ion flux corresponding to the electric potential at the tip of the electrode, in the order of microvolts [4].

Having electrical signals in the range of microvolts implies that these are vulnerable to electrical noise, which interferes considerably in the electrical signal making it strongly dependent on the SNR, such as the potential curve of the SVET. Under these conditions, measuring the potential difference with the vibrating electrode causes an uncertainty in the potential values of the electrical signal.

1.2. Sources of Electrical Noise. There are two main types of noise sources in SVET system: internal noise and external noise. Internal noise is generated by the components of the system as a consequence of its properties and nature (thermal noise, ripple noise, and fluctuating noise). The internal noise of the electrochemical system appears in the solid-liquid interface known as electrochemical noise (EN) whose sources are several types of fluctuations: fluctuation in the concentration of the species in the metal/electrolyte interface, fluctuation in surface morphology, fluctuation in electrode activity, and fluctuation of activity on the electrode surface [5].

External noise is the main identified source of electrical noise also known as white or Gaussian noise which is considered as a disturbing variable to our measurement system. This comes from external electrical or magnetic sources that affect all the components of the electrical system including the electrochemical part, electrodes, wiring, and instrumentation. The interferences that are presented can be stationary, nonstationary, or random, where the main sources of electrical noise are the current of the conventional electrical line, fluorescent lamps, computers, and electric motors.

1.3. Virtual Instrumentation. VI is an indispensable tool of great utility for the development of measurement and control instruments in the laboratory or field and its application of measurement and control of chemical and electrochemical systems applied to corrosion is not the exception [6–8]. When talking about VI, we are referring to an instrument found in the hardware of a computer and an interface that allows the physical and functional connection between the VI and the system to be measured and controlled. VIs are recyclable and easily scaled, can be connected to the outside world, and have simple configuration and low cost per acquisition (measurement) channel. Currently VIs are programmable in embedded systems, with the main advantage that their functionality is defined by the end user; that is, we design them according to our needs unlike a traditional rigid instrument whose functionality is defined by the manufacturer and is designed to meet general or very specific needs. The background for VI development dates back to 1986 when National Instruments introduced the LabVIEW® virtual instrumentation platform which was made to operate on MAC computers.

One of the main features of LabVIEW is the ease of use, valid for both professional programmers and people with little programming knowledge. LabVIEW is a graphical programming tool, meaning that programs are not written but drawn, facilitating their understanding. By having already predesigned a lot of programming blocks, it facilitates the programmer creating the project, allowing him to spend less time on programming.

1.4. Embedded Systems. Embedded systems are computers designed to have a specific function, unlike a general purpose computer which, as its name implies, is designed to perform general computing functions. The embedded systems have a great diversity of applications, some of daily use as the control of operation of a domestic appliance or being more complex as is the case of data acquisition and control in industry, aerospace, automotive, and so on. With an embedded system it is possible to carry out programmable measurement and control operations in real time. The main advantage of an embedded system is to do Digital Signal Processing (DSP) regardless of computer usage, making the embedded system a fully recyclable, portable, efficient, and economical VI with end-user defined functionality.

1.5. The Wavelet Transform. The wavelet transform (WT) has broad application in the analysis of stationary and nonstationary signals. These applications include the removal of electrical noise from the signals, detection of abrupt discontinuities, and compression of large amounts of data. The use of WT in corrosion studies is not the exception, as shown by the works published in the literature [9–12].

With the WT, it is possible to decompose a signal into a group of constituent signals, known as wavelets, each with a well defined, dominant frequency, similar to the Fourier transform (FT) in which the representation of a signal is by sine and cosine functions of unlimited duration. In WT, wavelets are transient functions of short duration, that is, limited duration centered around a specific time. The problem of the FT is that when passing from the time domain to the frequency domain the information of what happens in time is lost. Observing the frequency spectrum obtained using the FT is simple to distinguish the frequency content of the signal being analyzed but it is not possible to deduce in what time the components of the signal of the frequency spectrum appear or disappear. Unlike the FT, the WT allows an analysis in both time and frequency domains giving information on the evolution of the frequency content of a signal over time [13].

As in the case of FT, the WT has been discretized and is known as a discrete wavelet transform (DWT) and represents an important advantage over traditional FT methods. The WT decomposes a signal into several scales representing different frequency bands, and, at each scale, the position of the WT can be determined at the important time characteristic with which the electrical noise can be identified and effectively removed. Short-time wavelets allow information to be extracted from high-frequency components. This is important information to eliminate electrical noise since electrical noise is more likely to exhibit high-frequency fluctuations [14]. Long-term wavelets allow you to extract information from low frequencies. With the information of the high and low frequencies, we can define a threshold and zero the frequencies below the undesired threshold of the electric noise [15].

The pioneering work of removing electrical noise from signals using the WT has its origin in the works of Donoho and Johnstone [16, 17], which propose the use of a threshold for the removal of Gaussian white electrical noise in the signals. Similar works but with the use of the discrete wavelet transform undecimated (UWT) are given later allowing the removal of the electrical noise by a nonlinear method suggested by Coifman and Donoho [18].

The DWT is considered a suitable tool for the elimination of electric noise as a novel alternative that replaces procedures of attenuation of electrical noise with the use of low-pass filters of the systems Lock-In Amplifier or fast Fourier transform (FFT) that alone can be used in the circumstances in which the electrical noise has a very small overlap of bands or completely different and separated from the signal and the noise to be able to use the method of filtering, this being an important limitation in the moment of processing digitally signals not stationary whose content changes over time.

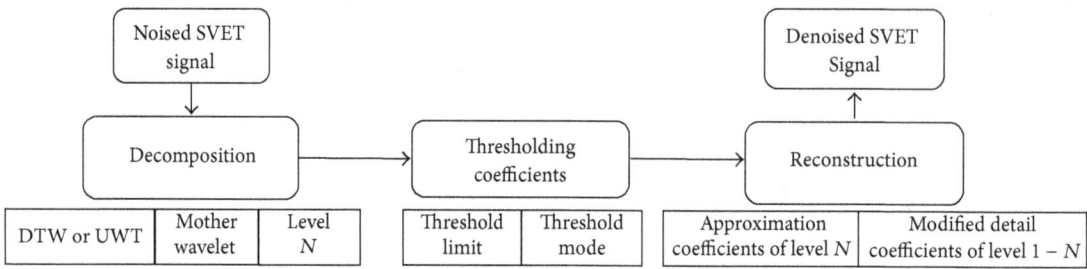

FIGURE 1: General thresholding procedure.

1.6. Noise Attenuation Procedure with the DWT Using the Threshold Method. The threshold method is the most used method by the wavelet transform for the attenuation of electrical noise [19], in which the main theoretical basis is to determine, in the time and frequency domain, the coefficients of the signals that throughout the space concentrate the energy of the SVET signal as well as the coefficients and energy of the noise that is generally distributed in the time domain and the whole frequency whose energy is less than the energy of the SVET signal.

By decomposing the SVET signal with noise by the DWT, we find that the resulting WT coefficients of the SVET potential signal are larger than the wavelet coefficients of the noise signal; this identification of coefficients allows defining a threshold and making the coefficients below the threshold equal to zero, resulting in noise removal from the potential SVET signal.

1.7. Stages in Noise Removal by the Threshold Method. Figure 1 shows a scheme for the general method of the threshold method for the removal of electrical noise. We show the three main stages, decomposition, threshold application, and reconstruction, as well as their respective substages. In LabVIEW the decomposition, threshold, and reconstruction are implicit in the Denoise function as follows:

(i) Decomposition: it consists of determining the coefficients of the low and high frequencies and the level to which the threshold would be applied using the DWT of the SVET potential curve contaminated with white electric noise. In the Denoise function of LabVIEW, we can choose the mother wavelet and a level to decompose the potential signal.

(ii) Application of the threshold: subsequently, once the SVET potential signal coefficients have been determined, a threshold is chosen to eliminate all threshold values below the threshold. In the Denoise function we can choose between soft thresholding or hard thresholding, type of rescaling, and threshold rule.

(iii) Reconstruction: finally, after the threshold is applied and the noise is eliminated, the resulting coefficients, the level, and the wavelet used are part of the reconstruction. With the reconstruction, we finally obtain a signal of SVET potential free of electrical noise.

The objective of the present investigation is to obtain a potential SVET signal with a high SNR for aluminum alloy A96061 in contact with an aqueous solution of 3.5% NaCl.

2. Materials and Methods

2.1. Metals and Specimens. In the experiment, samples of aluminum alloy A96061 protected with chromium conversion coating were used, applying chromium oxide as an anticorrosion pretreatment for aerospace applications. For the pitting corrosion formation process, the aluminum specimens were exposed in a saline mist chamber. The solution used in the fog chamber was 3.5% NaCl at 20°C; tests were carried out following the procedure recommended in standard ASTM B117.

2.2. System Employed. Here are four main devices that integrate the configuration of the system utilized to determine and measure the potential signal with the SVET.

The first one is a homemade SVET device that utilized a vibrating microelectrode with 10 to 20 μm spherical platinum black tip and vibrated with 20 μm amplitude at an average distance of 100 μm above the surface of the sample, with a vibration frequency of 60 Hz.

The second one is a mighty scope near infrared 10x–200x 5.0 MP video inspection system, Color CMOS, 6-LED light microscope operating at 850 nm, and USB 2.0 infrared digital microscope with a homemade lineal stage focus controlling mechanism in the range from 8.5 mm to 112 mm, with stepper motor NPM PF35-24CL.

The third one is homemade SVET device with XY stages, stepper motors VEXTA model PX245M-01, and a gain controlled instrumentation amplifier with high input impedance (1013 Ω/100 fA).

The fourth one is an embedded data acquisition FPGA (Field Programmable Gate Array) NI RIO NI myRIO with a dual-core Xilinx Z-7010 processor, 667 MHz processing speed 256 MB nonvolatile memory, and USB ports. This includes a 12-bit CAD and a sample rate of 500 kS/s connected to a computer whose main function is to display the results and store the information as well as to execute the virtual instrument that controls the SVET systems and computational vision of the optical preanalysis of the specimen under study. It is important to clarify that the data acquisition system employed has a dual-core processor and a FPAG array of components allowing the VI program to run on the NI

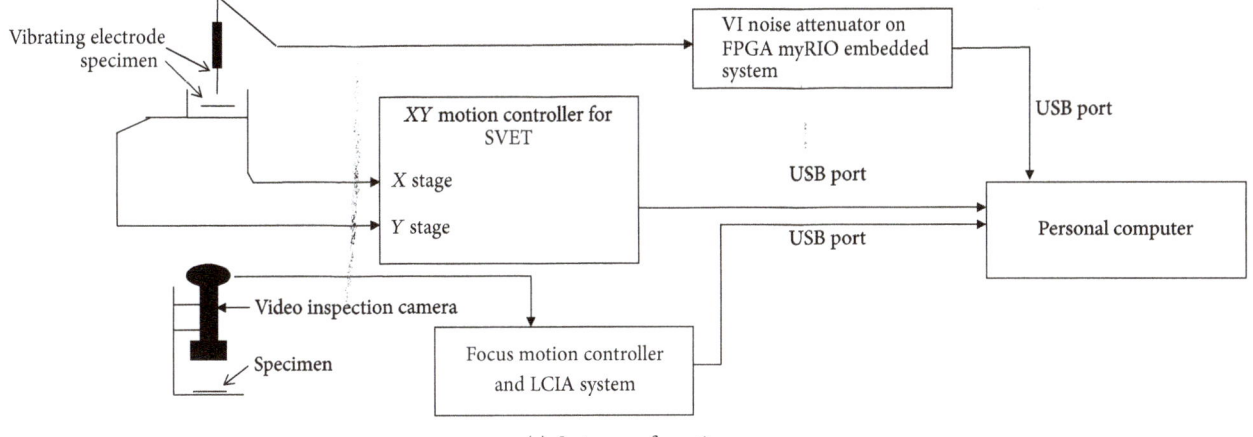

(a) System configuration

(b) System employed

FIGURE 2

myRIO making the background noise attenuation process in real time. Figure 2(a) shows the complete diagram of the system and Figure 2(b) shows the system employed.

2.3. Method. The pitting found in aluminum A96061 surface after being exposed to the saline fog is preanalyzed and identified optically using a virtual computer vision instrument developed in the laboratory and VI Localized Corrosion Image Analyzer (LCIA). The LCIA determines the location of pits in Cartesian coordinates. Subsequently, the coordinates are sent by electronic file to a workstation for electrochemical scanning that operates with the SVET. With the application of LCIA an effective sweep is made, only in the bites identified by computational vision, saving a considerable time of scanning.

The exploration of the pits that appear after the fog chamber test in 3.5% NaCl previously detected by the computer vision system was validated with the use of SVET.

VI employed in electrical noise attenuation includes the digitization of the potential signal obtained by SVET under background noise testing and the application of the Denoise and digital filter design functions of LabVIEW to the testing signal, to suppress the background noise.

3. Results

The localized corrosion pits were generated under saline fog conditions, the procedure recommended in the ASTM standard B117, the SVET technique, the application of the DWT,

digital filtering implemented, in an embedded FPGA myRIO system, and the use of a VI for the optical preprocessing of specimens with localized corrosion.

Figure 3 shows images of the optical preprocessing. (a) corresponds to the digitized image of the observed specimen and (b) shows the preoptical processing. The red, green, blue, and yellow colors indicate possible stings.

The SVET potential curve with respect to the distance shown in Figure 4 corresponds to the bite identified with the green color in the optical preprocessing of Figure 2. The result of the measurement is an electrical signal immersed in noise. In the potential curve of Figure 3, a SNR of 9.03 dB was measured.

By digitally processing the VI potential signal, the discrete wavelet transform, and the LabVIEW Denoise function implemented in an embedded FPGA myRIO system, a potential curve with a measured SNR of 178.56 dB was obtained. Figure 5 shows the potential amplitude SVET with respect to the distance of the potential curve digitally processed with the DWT. The parameters used were a Daubechies (db02), Level 15, threshold method, and soft thresholding wavelet.

By digitally processing the potential signal with the Lab-VIEW digital filter design (DFD) functions implemented in the embedded FPGA myRIO system, the resulting potential curve with the use of DFD is shown in Figure 6 presenting a SNR of 38.59 dB. For the SVET anode potential curve in Figure 3, the digital filter used to obtain these results is a Dolph-Chebyshev Window low-pass type FIR filter.

(a)

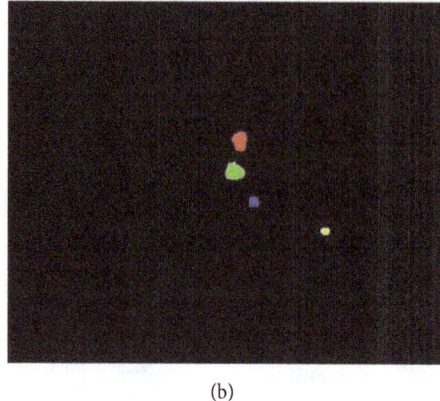

(b)

FIGURE 3: Identification of potential pitting by computational vision: (a) digitized image; (b) preprocessing of colors red, green, blue, and yellow identifying potential stings.

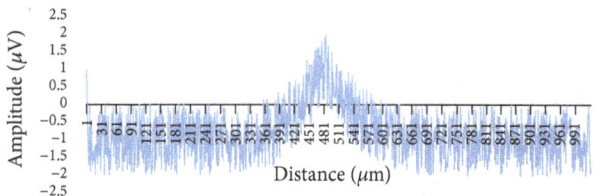

FIGURE 4: Signal of potential SVET immersed in a noise environment with a SNR of 9.03 dB.

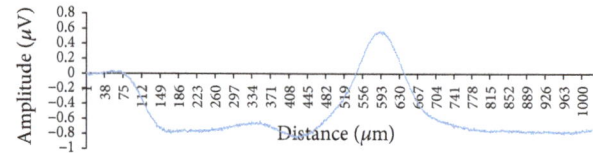

FIGURE 6: Electrical noise suppression with SNR digital filtering of 38.59 dB.

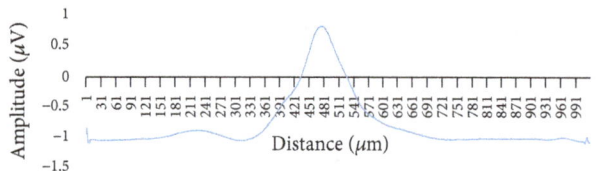

FIGURE 5: Electrical noise attenuation with Wavelet Denoise with SNR of 178.56 dB.

Table 1 summarizes the values obtained in the potential measurements on the SVET.

4. Discussion

The results were obtained by comparing two VIs implemented in an embedded data acquisition system FPGA myRIO. The first one uses the Wavelet Denoise function of the advanced signal processing toolkit and the second with the tools and functions for LabVIEW digital filter design. In addition, information on optical preprocessing is presented as an alternative preanalysis method for A96061 aluminum specimens.

With the preoptical processing it is possible to detect in real time potential pitting and Cartesian coordinates of the pitting, in order to perform a sweep only in the incipient bites. The immediate benefit of this procedure is a reduction of scanning time with the SVET technique.

Figure 4 shows a SVET potential curve immersed in electrical noise making it impossible to measure the electrical potential due to corrosion without the application of noise suppression.

The graphs of Figures 5 and 6 and Table 1 demonstrate the electrical noise attenuation with the DWT and digital filtering as well as a summary of the electrical measurements made by SVET, respectively.

It is evident that DTW increases significantly the SNR to 178.56 dB corresponding to a noise removal of 169.53 dB compared to 9.03 dB of the SVET potential at the electrode tip, shown in Figure 4.

Using a digital filter, a SNR of 38.59 dB was achieved, representing an electrical noise attenuation of 31.3 dB, with a significant reduction in the amplitude of the potential and a displacement in the location distance of the potential peak of the bite of 100 μm regarding the time of the potential waveform, and it is due mainly to the transient response characteristics or filters instability zone.

The use of embedded systems presents an important advantage in the implementation of virtual instrumentation which makes it possible to develop portable instruments of specific function, recyclable and scalable for studies of localized corrosion.

If we compare the obtained results of the DTW, we find a noise removal 169.53 dB higher than that provided by the digital filtering of 31.31 dB. On the other hand, with the DWT the peak of the anodic potential waveform of the pit in its original position of 481 micrometers is maintained which does not happen if we use digital filtering.

TABLE 1: Results of measurement and noise attenuation with wavelet transform and digital filtering.

Signal	Amplitude (μV)	SNR (dB)	Noise Suppression (dB)	Position of the peak potential (μm)
Electrode SVET	−2 to 2	9.03	0	481
Electrode SVET, DWT denoised	−1.10 to 0.80	178.56	169.53	481
Electrode SVET, digital filter	−0.90 to 0.60	38.59	31.31	581

5. Conclusions

VI implemented in an embedded system was developed with application in the electrical potential measurement and electric noise removal in SVET results. VI and embedded systems are recyclable, scalable, and end-user-defined for specific applications and are an alternative to traditional electrochemical measurements.

It was experimentally verified that the DWT and the threshold method are efficient and powerful alternatives to carry out electrical measurements of potential signals from localized corrosion processes measured by SVET. The application of the LabVIEW functions for the DWT can effectively remove electrical noise with a SNR of 169.53 dB or higher.

Conflicts of Interest

The authors declare that there are no conflicts of interest regarding the publication of this paper.

Acknowledgments

The authors acknowledge the Materials Corrosion Laboratory of the UABC Institute of Engineering for the economic support and facilities provided for the development of electronic instrumentation and testing of the experimental processes. They are grateful for the international collaboration with the National Center of Metallurgical Research (CENIM) Madrid, the Spanish State Council for Scientific Research (CSIC), for the support provided in the research and experimentation process in this work.

References

[1] R. Akid and M. Garma, "Scanning vibrating reference electrode technique: a calibration study to evaluate the optimum operating parameters for maximum signal detection of point source activity," Electrochimica Acta, vol. 49, no. 17-18, pp. 2871–2879, 2004.

[2] H. S. Isaacs and Y. Ishikawa, "Current and Potential Transients during Localized Corrosion of Stainless Steel," Journal of The Electrochemical Society, vol. 132, no. 6, pp. 1288–1293, 1985.

[3] S. Fujimoto and T. Shibata, Denki Kagaku, 64, 967 (1996).

[4] R. Ramos, R. K. Zlatev, M. S. Stoytcheva, B. Valdez, S. Flores, and A. M. Herrera, "Pitting corrosion characterization by SVET applying a synchronized noise suppression technique," ECS Transactions, vol. 29, no. 1, pp. 33–42, 2010, The lectrochemical Society.

[5] P. Marcus and F. Mansfeld, Analytical Methods in Corrosion Science and Engineering, 2006.

[6] R. Ramos, R. Zlatev, B. Valdez, M. Stoytcheva, M. Carrillo, and J.-F. García, "LabVIEW 2010 computer vision platform based virtual instrument and its application for pitting corrosion study," Journal of Analytical Methods in Chemistry, vol. 2013, Article ID 193230, 8 pages, 2013.

[7] H. Meng, J.-Y. Li, and Y.-H. Tang, "Virtual instrument for determining rate constant of second-order reaction by pX based on LabVIEW 8.0," Journal of Automated Methods and Management in Chemistry, vol. 2009, Article ID 849704, 7 pages, 2009.

[8] W.-B. Wang, J.-Y. Li, and Q.-J. Wu, "The design of a chemical virtual instrument based on LabVIEW for determining temperatures and pressures," Journal of Automated Methods and Management in Chemistry, vol. 2007, Article ID 68143, 7 pages, 2007.

[9] R. Moshrefi, M. G. Mahjani, and M. Jafarian, "Application of wavelet entropy in analysis of electrochemical noise for corrosion type identification," Electrochemistry Communications, vol. 48, pp. 49–51, 2014.

[10] H. Men, C. Wang, Y. Peng, S. Yang, and Z. Xu, "Wavelet analysis and its application in the field of microbial corrosion," in Proceedings of the 3rd International Symposium on Intelligent Information Technology Application, IITA 2009, IEEE Computer Society, NanChang, China, November 2009.

[11] J. A. Wharton, R. J. K. Wood, and B. G. Mellor, "Wavelet analysis of electrochemical noise measurements during corrosion of austenitic and superduplex stainless steels in chloride media," Corrosion Science, vol. 45, no. 1, pp. 97–122, 2003.

[12] X. Wang, J. Wang, C. Fu, and Y. Gao, "Determination of corrosion type by wavelet-based fractal dimension from electrochemical noise," International Journal of Electrochemical Science, vol. 8, pp. 7211–7222, 2013.

[13] L. A. Montejo and L. E. Suárez, "Aplicaciones de la Transformada Ondícula ("Wavelet") en Ingeniería Estructural," in Mecanica Computacional, vol. XXVI, pp. 2742–2753, Córdoba, Argentina, October 2007.

[14] M. Bitenc, D. S. Kieffer, and K. Khoshelham, "Evaluation of wavelet denoising methods for small-scale joint rougness estimation using terrestrial laser scanning," in Proceedings of the ISPRS Annals of Photogrammetry, Remote Sensing and Spatial Information Sciences, vol. II-3/W5, ISPRS Geospatial Week 2015, La Grade Motte, France, 28 Sep–03 Oct 2015, 28 Sep–03 Oct 2015.

[15] S. G. Chang, B. Yu, and M. Vetterli, "Adaptive wavelet thresholding for image denoising and compression," IEEE Transactions on Image Processing, vol. 9, no. 9, pp. 1532–1546, 2000.

[16] D. L. Donoho and J. M. Johnstone, "Ideal spatial adaptation by wavelet shrinkage," Biometrika, vol. 81, no. 3, pp. 425–455, 1994.

[17] D. L. Donoho, "De-noising by soft-thresholding," IEEE Transactions on Information Theory, vol. 41, no. 3, 1995.

[18] R. R. Coifman and D. L. Donoho, *Translation-Invariant De-Noising*, Spriger, New York, USA, 1995.

[19] B. Wu and C. Cai, "Wavlet denoising and its implementation in LabVIEW," in *Proceedings of the 2009 2nd International Congress on Image and Signal Processing, CISP '09*, IEEE, Tianjin, China, October 2009.

13

Corrosion of Reinforced Concrete Structures Submerged by the 2004 Tsunami in West Aceh, Indonesia

Herdi Susanto,[1] Syifaul Huzni ⓘ,[2] and Syarizal Fonna ⓘ[2]

[1]*Department of Mechanical Engineering, Faculty of Engineering, Teuku Umar University, Meulaboh 23681 Aceh Barat, Indonesia*
[2]*Department of Mechanical and Industrial Engineering, Faculty of Engineering, Syiah Kuala University, Darussalam, Banda Aceh 23111, Indonesia*

Correspondence should be addressed to Syarizal Fonna; syarizal.fonna@unsyiah.ac.id

Academic Editor: Flavio Deflorian

The earthquake and tsunami of 26 December 2004 caused the infrastructure in Aceh's West Coast region to be submerged by seawater and to require the rehabilitation and reconstruction. The infrastructure that was submerged in the tsunami might experience a decrease in strength due to corrosion attack and would unexpectedly collapse if an earthquake occurs even on a small scale. This study was conducted to examine the corrosion risk level of the infrastructures in Aceh's West Coast region, Indonesia, which submerged by the 2004 tsunami. Three locations were chosen for the study, i.e., Suak Ribee, Ujong Kalak, and Padang Seurahet. The assessments were carried out in 2014 and 2015. Three to four columns in each of the buildings were selected for the assessment. The half-cell potential technique method which refers to ASTM C876 was used to obtain and analyze the assessment data. The results of the assessment show that the electrical potentials on the surface of concrete for the buildings which submerged by the tsunami were range between -100 and -450 mV (vs. $Cu/CuSO_4$) and categorized into low to high corrosion risk level. Meanwhile, the electrical potentials for new buildings range between (-100) and (-350) mV which indicated low to medium corrosion risk. Hence, the corrosion actively occurred in the areas having medium to high corrosion risk. Also, it was found that the corrosion risk level for the building tends to increase by increasing time. Therefore, the prevention and/or rehabilitation is necessary for stopping the corrosion, and so the premature failure of the building might be avoided.

1. Introduction

The West Aceh region is vulnerable to catastrophic earthquakes [1]. Meanwhile, there are a large number of infrastructures which were submerged by the 2004 tsunami in the region. The infrastructures were repaired and reused again. Therefore, the infrastructures might be susceptible to corrosion attack [2, 3] that can cause a decrease in strength [4], and sudden collapse may happen when an earthquake occurs even on a small scale [5, 6]. This is certainly not desirable because, besides the material loss, it can also lead to loss of life of people.

To avoid the possibility of a sudden failure of the infrastructures, especially in the regions hit by the 2004 tsunami in West Aceh, corrosion assessment must be done to determine the level of corrosion risk of reinforced concrete infrastructures, and it is important to be performed periodically.

Hence, this study aims to conduct an assessment in order to determine corrosion risk level of reinforced concrete buildings in West Aceh region. For this study, the half-cell potential technique is used by obtaining electrical potential on the surface of the concrete.

2. Half-Cell Potential Technique

The half-cell potential technique is a nondestructive test widely used to monitor and determine the corrosion of reinforced concrete infrastructures [7]. Many researchers have performed and justified the method to evaluate the reinforcing steel corrosion in concrete [3, 5–8]. Figure 1 shows the comparison of half-cell potential technique result to the actual condition of corrosion from the previous study [5] in which the half-cell potential technique result was consistent with the actual corrosion. This method measures the value

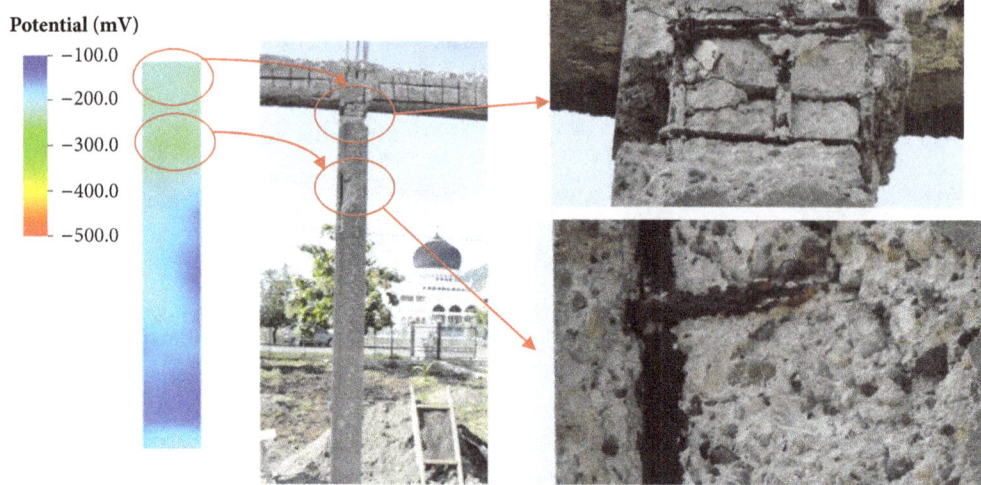

FIGURE 1: Comparison of half-cell potential technique result to the actual corrosion in the previous result [5].

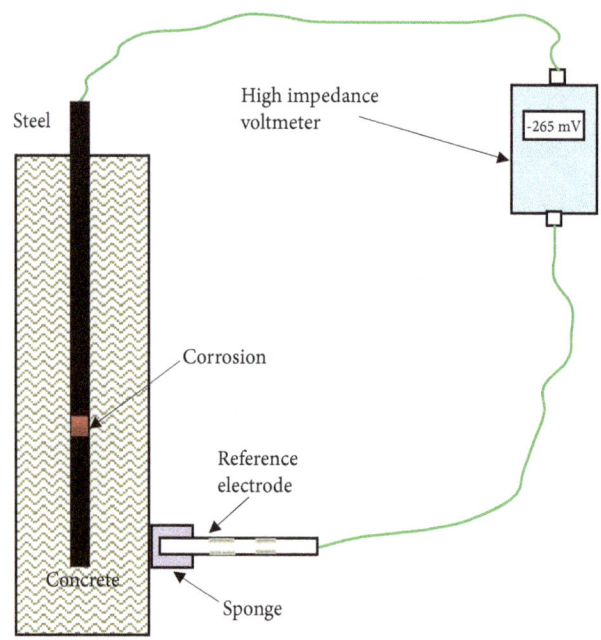

FIGURE 2: Electrical potential measurement with half-cell potential technique.

of electrical potential on the concrete surface in order to determine the level of corrosion risk [4, 9]. ASTM C876 is used as guidance in performing the half-cell potential technique and a criterion to analyze the electrical potential data [4]. Figure 2 shows the schematic of electrical potential measurement on the concrete surface which performed using the methods and Table 1 shows the criteria for the levels of corrosion risk.

3. Corrosion Assessment of the Reinforced Concrete Structures

This assessment was carried out in the region hit by the 2004 tsunami, especially West Aceh Regency. Three locations were selected in the regions, i.e., Suak Ribee (Objects I and III), Padang Seurahet (Object II), and Ujung Kalak (Object IV).

The first study was conducted in August 2014 and the second in June 2015. Two reinforced concrete buildings that submerged by the tsunami (Objects I and II) and two new buildings built after the 2004 tsunami (Objects III and IV) were selected in this study. The locations of the buildings/objects can be seen in Figure 3. The main equipment for the study was half-cell meter and rebar locator as shown in Figure 4.

Then, for every building, at least three columns were selected to be measured to their corrosion potential. On each column, the location of the reinforcement was determined with the rebar locator. Then, a grid was constructed on the

TABLE 1: Criteria for corrosion risk levels with referring to ASTM C876 [4].

No	Reference Electrode (mV)				Corrosion risk level
	Cu/CuSO$_4$	Ag/AgCl	Standard Hydrogen	Calomel	
1	> (-200)	> (-100)	> (+120)	> (-80)	Low (10% risk of corrosion)
2	(-200)– (-350)	(-100)– (-250)	(+120) – (-30)	(-80) – (-230)	Intermediate corrosion risk
3	< (-350)	< (-250)	< (-30)	< (-230)	High (>90% risk of corrosion)
4	< (-500)	< (-400)	< (-180)	< (-380)	Severe corrosion

FIGURE 3: The locations of the reinforced concrete building for the study.

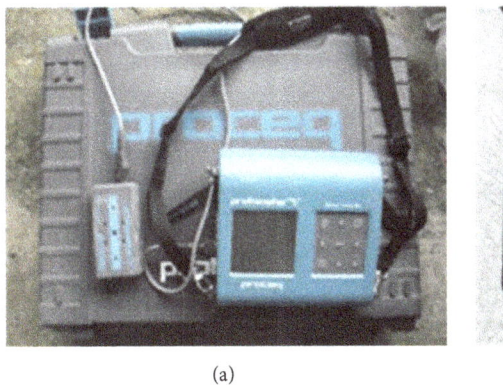

(a)

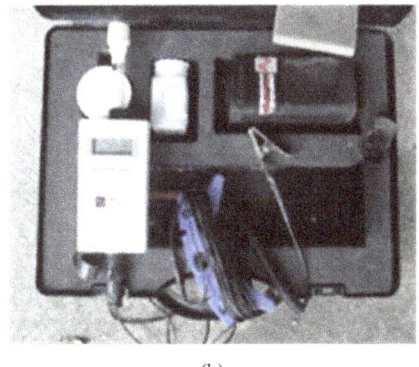

(b)

FIGURE 4: (a) Rebar locator (profometer 5$^+$) and (b) SCRIBE DHC digital half-cell meter.

surface of the column corresponding to the location of the reinforcement. The four buildings are shown in Figures 5 and 6.

Next, the four buildings were assessed for their corrosion risk levels. Every column for buildings to be assessed was shown in Figures 7 and 8. The measurement of rebar location using profometer can be seen in Figure 9(a). Meanwhile, the potential corrosion measurement was shown in Figure 9(b).

4. Result and Discussion

The potential corrosion data which were obtained on the rein-forced concrete structure were mapped using the software

VisIt-2.7.3. The software is an open source. It used to generate color contour represented potential values on the columns of the building.

Figure 10 shows the corrosion risk assessment results for Object I in Suak Ribee submerged by the 2004 tsunami. The potential distributions on four columns were presented in the figure. The potential distribution of year 2014 for column 1, column 2, column 3, and column 4 ranged between (-100) and (-350) mV, <-200 mV, (-200) and (-350) mV, and (-100) and (-450) mV, respectively, while those potential distributions of the columns ranged between (-200) and (-350) mV, (-200) and (-350) mV, (-200) and (-350) mV, and (-200) and (-350) mV, respectively, of year 2015.

FIGURE 5: Buildings submerged by the tsunami: (a) Objects I and (b) Object II.

FIGURE 6: New buildings: (a) Objects III and (b) Object IV.

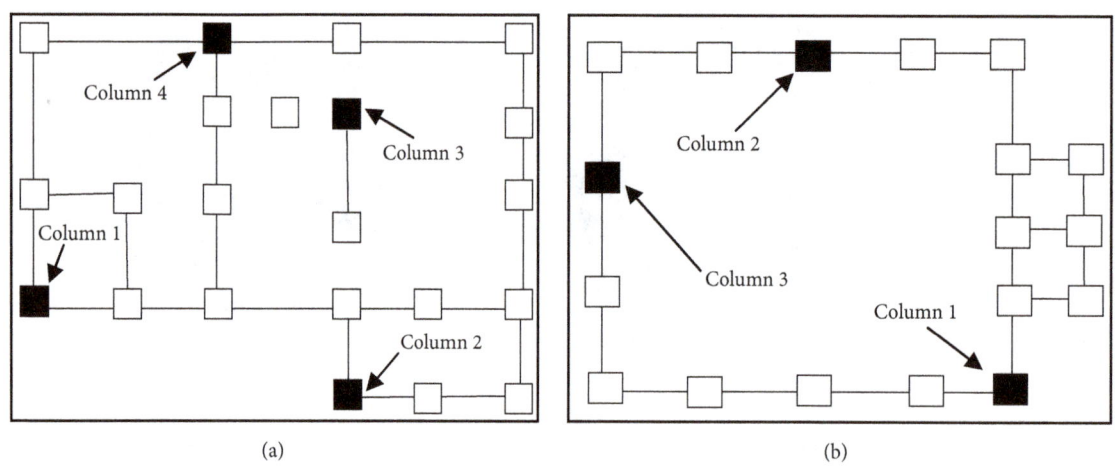

FIGURE 7: Schematic sketch for (a) Object I and (b) Object II.

Referring to Table 1, the distribution of the potentials on the columns indicated that the corrosion risk level varied from low to high-risk level. Some parts of column 1 (especially lower part), column 2, and lower part column 4 were categorized into low corrosion risk level for the year 2014. In 2015, the corrosion risk level for those parts became intermediate risk level. The corrosion risk level for column 3 was intermediate risk level in 2014 and remained at this level in 2015. The high-risk level occurs in column 4. However, it became intermediate risk level in 2015. It might happen due to repassivation on the rebar. However, further study is needed.

The increasing of corrosion risk level of the columns was consistent with the previous study conducting the half-cell potential mapping in Banda Aceh city, Indonesia [6] as seen in Figure 11(a). The indication of repassivation also occurred in the previous study that was shown in Figure 11(b) [6]. However, the corrosion risk generally tends to become higher by increasing the time for all columns in this study and also in the previous study [6].

Another assessment result for infrastructure submerged by the 2004 tsunami was given in Figure 12. The figure shows the corrosion risk assessment results for Object II in Padang

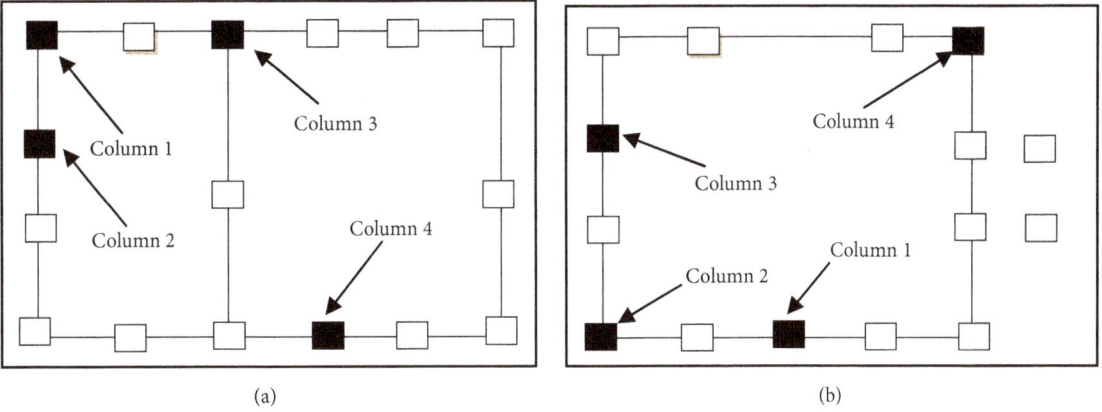

FIGURE 8: Schematic of column position of (a) Object III and (b) Object IV.

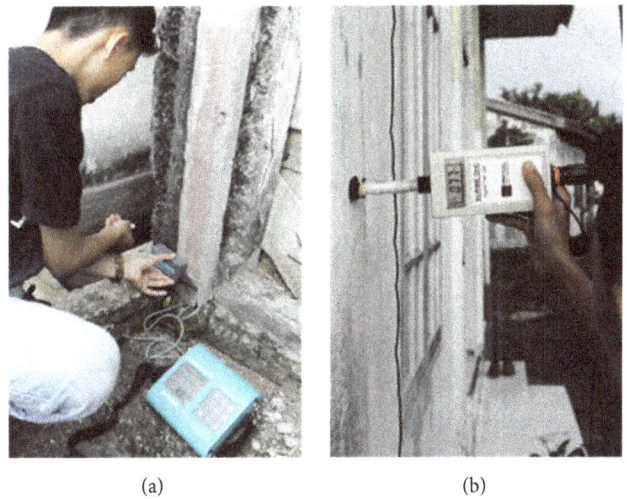

(a)   (b)

FIGURE 9: (a) Measurement of the location of steel reinforcement and (b) potential corrosion measurement of the column.

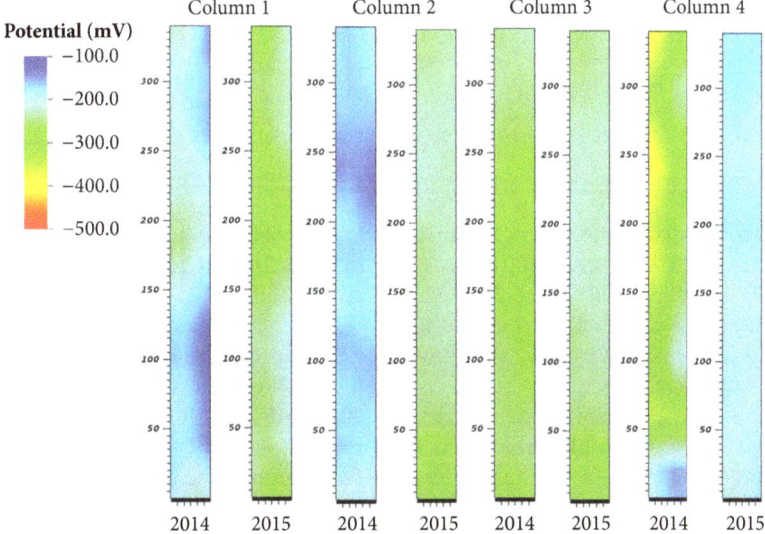

FIGURE 10: The potential distribution on each column of Object I for 2014 and 2015.

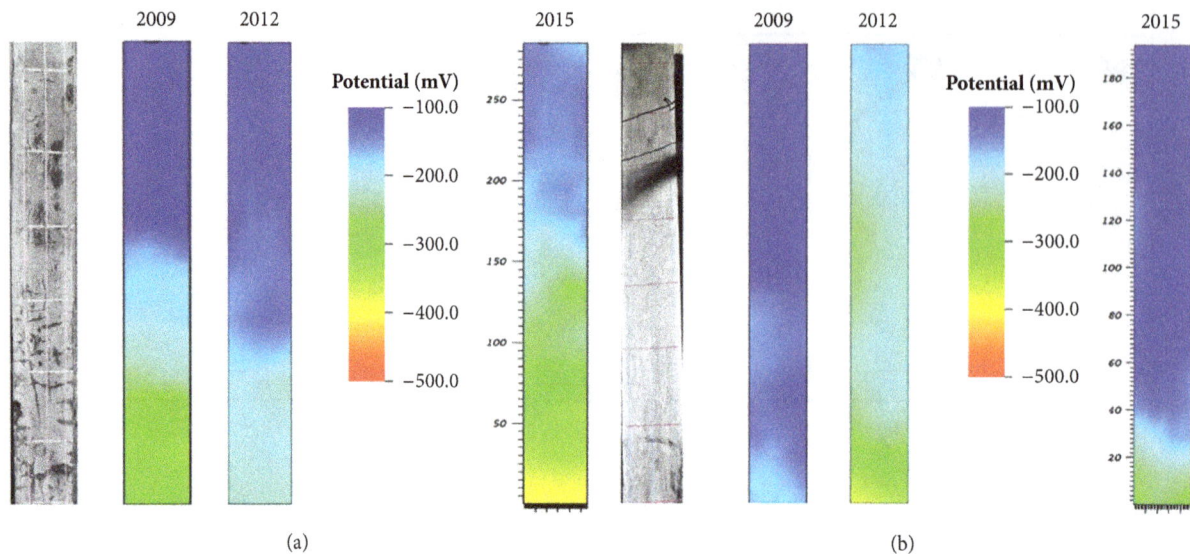

FIGURE 11: The potential distribution on the column for the previous study in Banda Aceh city, Indonesia: (a) indicating an increase of corrosion risk and (b) indicating repassivation [6].

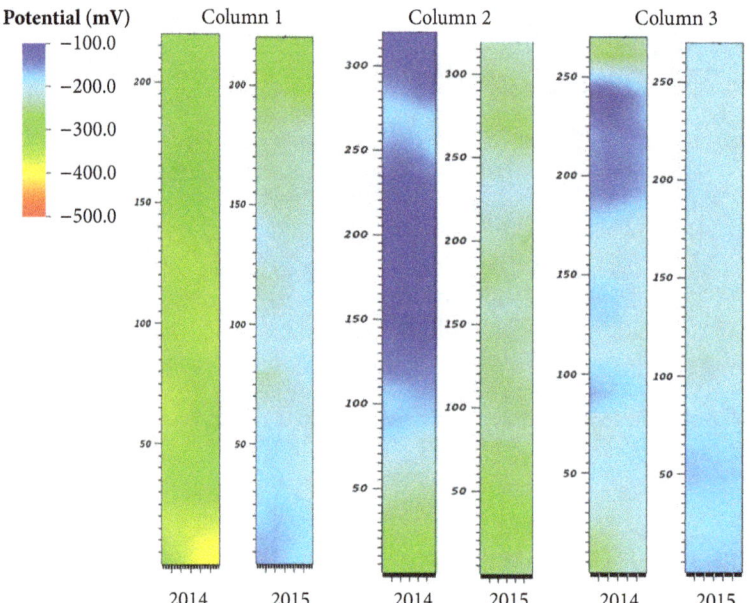

FIGURE 12: The potential distribution on each column of Object II for 2014 and 2015.

Seurahet. In 2014, the potential distribution for column 1, column 2, and column 3 ranged between (-200) and (-450) mV, (-100) and (-350) mV, and (-100) and (-350) mV, respectively. Thus, those potential distributions became in range between (-200) and (-350) mV, (-200) and (-350) mV, and (-200) and (-350) mV, respectively for 2015.

Using the criteria in Table 1, the potential on the columns of Object II was categorized into low to high corrosion risk level. Almost all of column 1 had completely fallen into intermediate corrosion risk, but the lower part was already at high corrosion risk level in 2014. However, the data in 2015 show that the potential value became less negative but is still into intermediate corrosion risk level. The corrosion risk level

for columns 2 and 3 falls into low and intermediate risk level in 2014. Thus, it completely became intermediate corrosion risk in 2015.

The phenomenon that the corrosion worsened by increasing time can be confirmed by looking at Figure 13. The figure shows the visual of a column condition of the year 2014 and 2015. Initially, there were cracks on the column in 2014, and the cracks became bigger and longer in 2015. Furthermore, the corroded rebar was also revealed in 2015. It occurred because the corrosion product will push the concrete cover with the load over the strength of concrete and so the crack might be initiated [10]. Finally, by increasing time, the concrete cover might collapse completely.

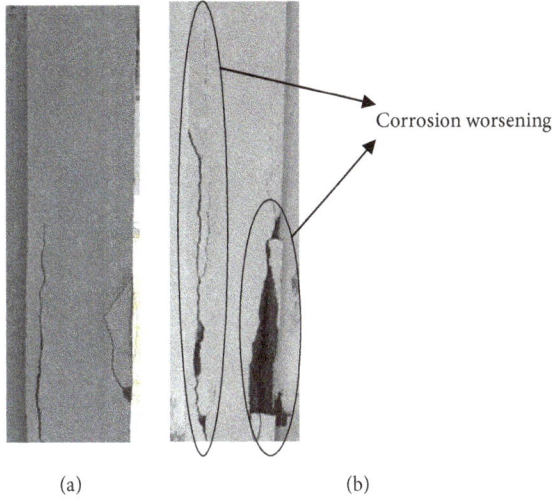

(a)   (b)

FIGURE 13: The visual of a column of Object II for (a) 2014 and (b) 2015.

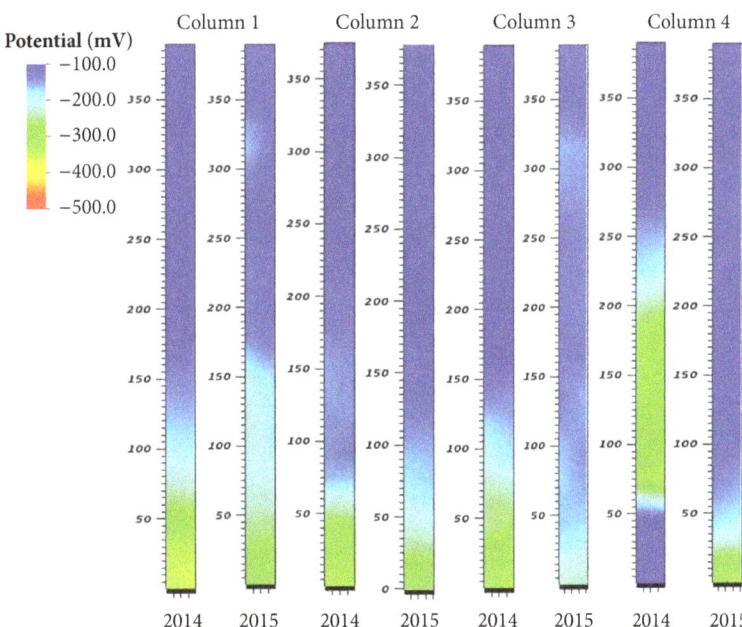

FIGURE 14: The potential distribution on each column of Object III for 2014 and 2015.

Moreover, the potential distribution for one of the infrastructures built in the tsunami-affected area (Object III) was shown in Figure 14. As mention before, this building is located in Suak Ribee. In 2014, the potential distribution for column 1, column 2, column 3, and column 4 was the same, i.e., ranging between (-100) and (-350) mV. Thus, it was found that those potential distribution values remain the same in 2015, i.e., (-100) and (-350) mV. However, there was difference in the potential distribution pattern comparing the 2014 data with the 2015 data.

The potential values on the columns of Object III might be categorized into low to intermediate corrosion risk level by referring to the criteria in Table 1. The majority of the columns lower part had completely fallen into intermediate

corrosion risk but the upper parts were at low corrosion risk level in 2014. However, it was found that an area on the lower part of column 4 is still in low-risk level. The data in 2015 show that the potential values for all columns tend the same. The corrosion risk level for all columns lower part falls into intermediate risk level and the upper parts are still at low-risk level. Thus, the lower parts of columns seem more susceptible to corrosion. This might be due to the capillarity of concrete that the lower part could obtain more water from the soil and/or rainwater than others part of the column. Hence, the possibility of corrosion to initiate becomes higher.

Furthermore, by conducting a visual inspection, it was found that there was a difference on the condition between the column directly facing the sea and the column facing the

(a)   (b)

FIGURE 15: The visual of a column of Object III for (a) directly facing to the sea and (b) facing to the opposite of sea.

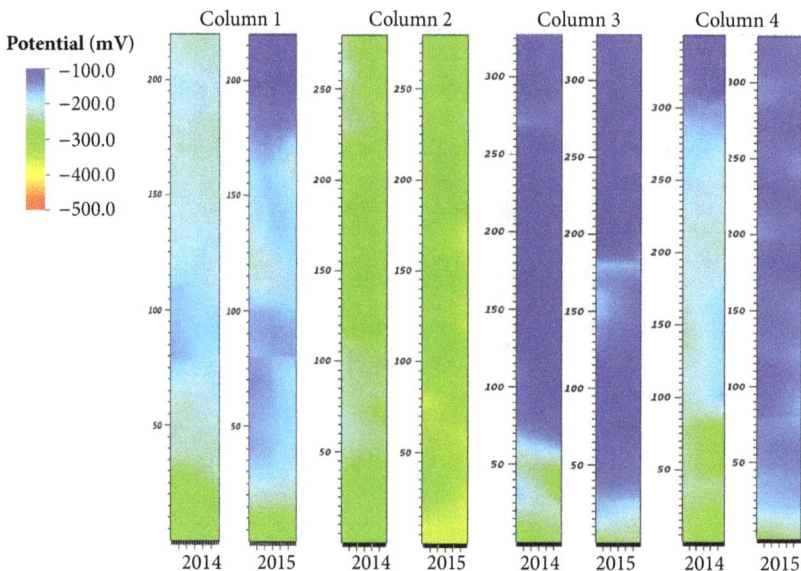

FIGURE 16: The potential distribution on each column of Object IV for 2014 and 2015.

opposite direction of the sea as seen in Figure 15. The figure shows that the column facing the sea already has a crack on it. The crack was an indication that the corrosion actively occurs [10] below the concrete surface at the region. This condition might happen due to the ion chloride more accumulated on the concrete. The ion chloride could be carried by the wind from the sea [11]. It was consistent with the previous research that the ion chloride content in the rainwater around the coastal area was relatively high (24 ppm) [12].

The potential distribution for the last infrastructures was shown in Figure 16. This building was built in the tsunami-affected area (Object IV) and located in Ujong Kalak. In 2014, the potential distribution for column 1, column 2, column 3, and column 4 ranged between (-200) and (-350) mV, (-200) and (-450) mV, (-100) and (-350) mV, and (-100) and (-350) mV, respectively. Thus, it was found that those potential distribution values for column 1 tend to become more positive

in 2015, i.e., (-100) and (-350) mV. The potential on column 2 became more negative in 2015, i.e., (-300) and (-450) mV. Lastly, the potential values for column 3 and column 4 tend to be more positive but remain in the same range in 2015, i.e., (-100) and (-350) mV.

Based on the criteria in Table 1, the corrosion risk level for the columns could be determined. The corrosion risk level for column 1 and column 2 was already at intermediate risk in 2014. In 2015, the risk level tends to become higher for column 2. The below part was at a high-risk level. This condition was suspected because of the position of column 2 directly facing the sea and, thus, it might be more exposed to ion chloride from the sea. However, the upper part of column 1 became low-risk level in 2015, while the corrosion risk level for column 3 and column 4 was relatively the same at intermediate risk for 2014 and 2015.

By comparing the assessment results of the 2014 and 2015 data, a tendency of increasing the negativity of corrosion potential values is shown. This means that the corrosion actively occurred on the reinforcing steels. Furthermore, this shows that the new structures built in the tsunami-affected area would be susceptible to corrosion attack indicated by its intermediate corrosion risk level. These results were consistent with assessment results conducted around Banda Aceh city [6]. Therefore, it is important to take action to stop the corrosion, so the sudden failure of the structures could be prevented.

5. Conclusion

The corrosion risk assessment based on ASTM C876 has been conducted for four infrastructures in West Aceh Regency, Aceh Province, Indonesia. The results show that some parts of the structures were already at intermediate to high corrosion risk level. The structures that submerged by the 2004 tsunami show the tendency at higher risk than the structures built after the 2004 tsunami. However, these structures are also already at intermediate corrosion risk level. This indicated that the corrosion actively occurs for all infrastructures. Hence, the rehabilitation and/or protection is necessary to be carried out in order to prevent the corrosion worsening and so the premature failure might be avoided.

Conflicts of Interest

The authors declare that they have no conflicts of interest.

References

[1] Pemerintah Aceh, *Rencana Pembangunan Jangka Panjang (RPJP) Aceh Tahun 2012-2032*, Indonesian, 2012, https://bappeda.acehprov.go.id/download/download/24.

[2] NACE, "Corrosion cost by industrial sector," *Supplement to Material Performance*, vol. 41, no. 7, p. 4, 2006.

[3] R. R. Hussain, "Underwater half-cell corrosion potential bench mark measurements of corroding steel in concrete influenced by a variety of material science and environmental engineering variables," *Measurement*, vol. 44, no. 1, pp. 274–280, 2011.

[4] J. P. Broomfield, *Corrosion of Steel in Concrete - Understanding, Investigation and Repair*, Taylor & Francis, London, UK, 2nd edition, 2007.

[5] M. Ridha, S. Fonna, S. Huzni, and A. K. Ariffin, "Corrosion risk assessment of public buildings affected by the 2004 tsunami in Banda Aceh," *Journal of Earthquake and Tsunami*, vol. 7, no. 1, pp. 1–22, 2013.

[6] S. Fonna, M. Ridha, S. Huzni, W. A. Walid, T. D. Mulya, and T. A. K. Ariffin, "Corrosion risk of RC buildings after ten years the , 2004, tsunami in banda aceh – Indonesia," *Procedia Engineering*, vol. 171, pp. 965–976, 2017.

[7] V. Leelalerkiet, J.-W. Kyung, M. Ohtsu, and M. Yokota, "Analysis of half-cell potential measurement for corrosion of reinforced concrete," *Construction and Building Materials*, vol. 18, no. 3, pp. 155–162, 2004.

[8] T. Parthiban, R. Ravi, and G. T. Parthiban, "Potential monitoring system for corrosion of steel in concrete," *Advances in Engineering Software*, vol. 37, no. 6, pp. 375–381, 2006.

[9] ASTM C876-91, "Standard Test Methods for Half-Cell Potentials of Uncoated Reinforcing Steel in Concrete, Wear and Erosion; Metal Corrosion," Vol. 03.02, August 2005.

[10] C. Jiang, Y.-F. Wu, and M.-J. Dai, "Degradation of steel-to-concrete bond due to corrosion," *Construction and Building Materials*, vol. 158, pp. 1073–1080, 2018.

[11] L. J. Korb and D. L. Olson, *ASM Handbook: Corrosion*, vol. 13, ASM International, 1992.

[12] I. Musalam and R. Nasoetion, "Penelitian karakteristik korosi atmosfer di daerah pantai utara jakarta," *Korosi*, vol. 14, no. 1, pp. 1–8, 2005, in Indonesian.

14

Effect of Grain Size on the Stress Corrosion Cracking of Ultrafine Grained Cu-10 wt% Zn Alloy in Ammonia

Takuma Asabe, Muhammad Rifai, Motohiro Yuasa, and Hiroyuki Miyamoto

Department of Mechanical Engineering, Doshisha University, Kyoto 610-0394, Japan

Correspondence should be addressed to Hiroyuki Miyamoto; hmiyamot@mail.doshisha.ac.jp

Academic Editor: Jerzy A. Szpunar

The effect of grain size in the *micron* to submicron range on the stress corrosion cracking (SCC) of Cu-10 wt% Zn alloys was investigated using constant-load tests in ammonia vapor. The grain size was systematically varied from 4 μm to 0.12 μm by either cold-rolling or equal-channel angular pressing (ECAP), followed by annealing. The time to fracture increased with decreasing grain size above 1 μm but then began to decrease with decreasing grain size into the submicron range. This inverse trend in the submicron range is discussed in terms of a severe plastic deformation- (SPD-) induced ultrafine grain microstructure.

1. Introduction

While the effect of grain size reduction on the strength of metallic materials has been well established, its effect on corrosion is more complicated and does not appear to be explainable *using a universal law such as the Hall-Petch law of the yield stress*. It seems that the effect of grain size reduction on corrosion resistance is mostly positive in stainless steels [1] and aluminum alloys [2, 3], whereas the effect is marginal in copper [4, 5] and titanium alloys [6–8]. However, there are several contradictory reports involving the same materials and environment [9–11]. The limited available literature on the effect of grain size on stress corrosion cracking (SCC) [12, 13] reports an increasing resistance to SCC with decreasing grain size. Edmund investigated the effect of grain size on the SCC of α-brass in an ammonia environment using a constant-load test and reported increasing fracture time with decreasing grain size [14]. Additional discussion of the effect of grain size on corrosion can be found in comprehensive review papers [15, 16].

Severe plastic deformation (SPD) enables the reduction of grain size to the submicron range in bulk metallic materials for load-carrying structural applications [17]. The so-called ultrafine grained (UFG) materials formed by SPD exhibit unique physical and mechanical properties. For example, the strength increases enormously after ECAP and the ductility also remains relatively high or even increases in some cases [17–19]. These unique properties have frequently been attributed to deformation-induced UFG structures superimposed with the dislocation structure. The advent of SPD technology has rendered the corrosion behavior of UFG materials a more pressing issue and has brought the question of whether the dependence of corrosion properties on grain size can be extrapolated into the submicron range back to the scientific community [15].

The SCC susceptibility of UFG or nanostructured materials has been far less studied and much remains to be explored [20–31]. One limitation of SCC studies of SPD materials is that structural materials that exhibit SCC are generally high-strength alloys such as austenitic stainless steels and Cu-Zn alloys, which are difficult to process by SPD. Fortunately, corrosion studies only require a UFG structure in the surface layer, and surface modification of even very hard materials is possible using methods such as shot peening [32] or surface mechanical attrition treatment (SMAT) [33]. Another reason for the lack of research on this topic is the very large difference in yield strength between UFG materials formed by SPD and conventional coarse-grained materials, as a result of which the standard SCC tests that apply either a common load or displacement inevitably impose different levels of strain or

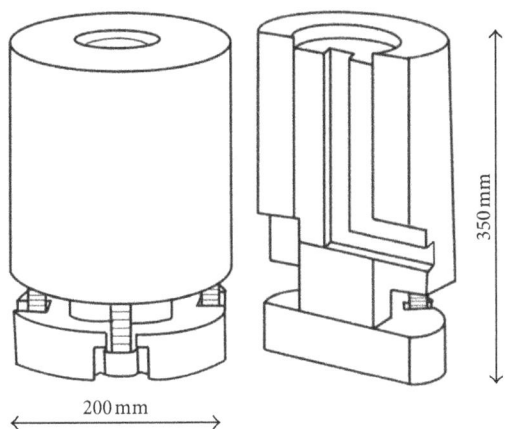

FIGURE 1: Schematic diagram and dimensions of the ECAP die.

TABLE 1: Processing recipes and grain size.

Sample	Deformation	Annealing	Grain size (μm)
1	None	None	38.0
2	CR 50%	673 K-60 min	12.6
3	CR 50%	623 K-60 min	3.30
4	ECAP 2 pass	623 K-15 min	1.10
5	ECAP 2 pass	623 K-10 min	0.71
6	ECAP 2 pass	623 K-5 min	0.44
7	ECAP 8 pass	473 K-10 min	0.15
8	ECAP 8 pass	473 K-1 min	0.12
9	ECAP 8 pass		0.12

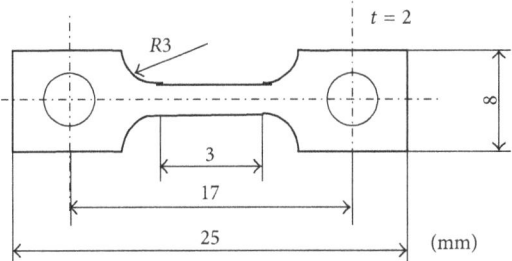

FIGURE 2: SCC specimen dimensions.

stress in these two materials. Therefore, these two approaches may yield considerable discrepancies in experimental results, for example, in the case of the time to fracture. As an alternative approach, these two very different materials have been compared by slow-strain-rate testing (SSRT), and their susceptibility to SCC was determined from the ratio of elongation to fracture in air to that in solution [20–22, 24, 26–29, 31]. However, it remains unclear whether the susceptibility to SCC determined by SSRT reflects performance in actual service environments. In this study, we investigated the effect of grain size, ranging from fine grains to submicron grains, on the susceptibility to SCC using a model Cu-10 wt% Zn-ammonia system under a common constant load. The objective of this study was not to compare the two extremes (UFG and conventional course-grained counterparts), as has been done in many previous studies, but to track the variation in SCC susceptibility in the transitional regime ranging from fine to submicron grains, to better understand the SCC of UFG materials from a physical point of view. As far as we know, this paper is the first to report the SCC susceptibility under constant-load tests for this range of grain sizes.

2. Materials and Methods

The grain sizes of commercial Cu-10 wt% Zn alloys were systematically controlled by cold-rolling and ECAP followed by heat treatment. For ECAP, 100 mm long billets with an 8 mm square cross-section were pressed through an ECAP die for either two or eight passes by the so-called Bc route, in which the sample is rotated by 90° around its longitudinal axis between passes. The ECAP die was deliberately designed so that harder materials such as brass and stainless steel can be pressed, as shown in Figure 1. Nevertheless, α-brass with a higher Zn content, for example, Cu-30 wt% Zn, is too hard for this ECAP die, even though it is more susceptible to SCC and would have been an appropriate material for the present study. Thus, the choice of Cu-10 wt% Zn was a compromise between the applicability to ECAP and susceptibility to SCC. Cold-rolling was carried out to a 50% reduction. *After*

ECAP and cold-rolling, the samples were then annealed in an electronic furnace as summarized in Table 1, leading to grain sizes ranging from 0.12 μm to 38 μm. *Grain size was estimated using the intercept method, and twins were ignored following the rule of grain size measurement of copper and copper alloys described in JIS0501.*

The susceptibility to SCC was evaluated in constant-load tests using a cantilever-type apparatus (Figure 2) for specimens with a yield stress above 290 MPa. A common constant stress of 280 MPa was applied to all SCC specimens regardless of their yield stress and was selected based on our previous studies [25], which indicated that the susceptibility to SCC was much higher in as-ECAPed UFG Cu-10 wt% Zn than in the coarse-grained same alloys. The dog-bone specimens shown in Figure 3 were placed in a chamber in which a 14% ammoniacal solution was placed in the bottom to fill the chamber with ammonia vapor. The time to fracture was recorded by a stopwatch placed under the edge of the beam. The microstructure and fracture surface were examined by transmission electron microscopy (TEM, JSM 2100F) and field-emission-type scanning electron microscopy (FE-SEM, JSM-FE7001).

3. Results and Discussion

Nominal stress-strain curves obtained from tensile tests are shown in Figure 4. ECAPed materials had typical stress-strain curves with a high tensile strength and little strain hardening capability. Thus, they exhibited necking at the onset of plastic deformation. The tensile stress exceeded 600 MPa, much greater than that of pure copper (approximately 450 MPa

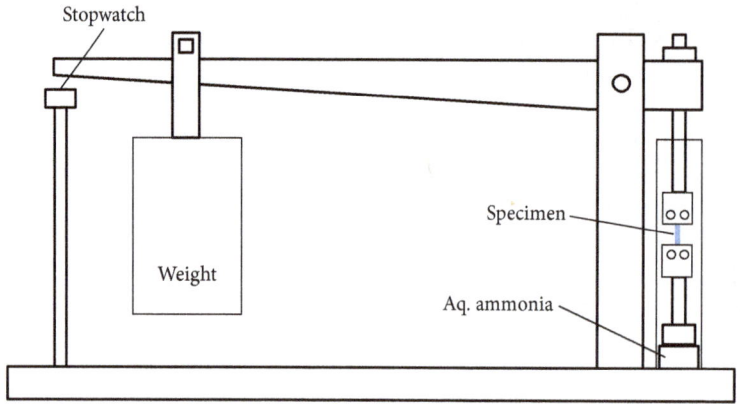

FIGURE 3: Schematic diagram of the constant-load SCC test.

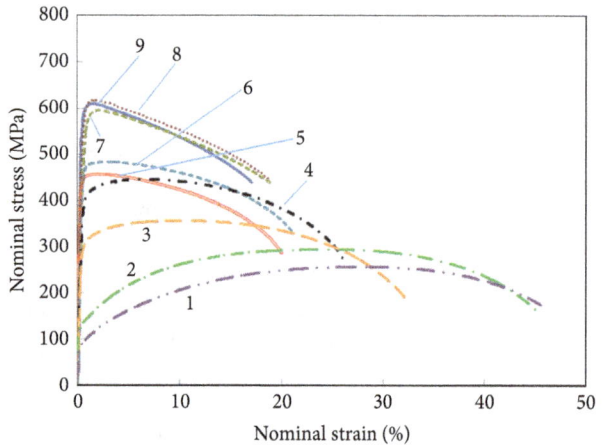

FIGURE 4: Nominal stress-strain curves of tensile tests. Numbers 1–9 correspond to the sample numbers in Table 1.

after 8 passes of ECAP). This higher tensile stress was mainly caused by solid-solution hardening and a lower stacking fault energy (SFE). Cu-10 wt% Zn has an SFE of 35 mJ/m^2, which is much lower than that of pure Cu (78 mJ/m^2) [34]. Microstructures observed by TEM and SEM are shown in Figure 5, and the processing routes and resulting grain sizes are summarized in Table 1. The minimum grain size of 0.12 μm was achieved after 8 passes of ECAP and is somewhat smaller than that achieved with pure copper [5, 31] and pure aluminum [35] in our previous studies. This smaller grain size achieved by SPD is associated with a lower SFE, and several studies have demonstrated that the minimum grain size achieved by SPD scales with the SFE [36–39]. The larger grain size shown in Figure 5 was obtained by recrystallization and/or grain growth using appropriate combinations of ECA/cold-rolling and annealing. Whereas a relatively uniform grain size was observed in samples processed by ECAP, samples processed by cold-rolling were less uniform, indicating that continuous grain growth occurred in the former case while discontinuous recrystallization occurred in the latter. Several studies have reported that UFG structures

with a high fraction of high-angle grain boundaries (HAGBs) exhibit continuous grain growth during annealing after a very high strain is imposed by SPD, whereas discontinuous grain growth or recrystallization tends to occur when the HAGB fraction is lower and less strain is imposed by SPD [40–44]. The relationship between yield stress and the reciprocal root of grain size is shown in Figure 6. Hall-Petch relationship can be divided into two regions with different slopes: Region I with grain sizes larger than 1 μm and Region II with smaller grain sizes. The slope corresponds to the constant k in the Hall-Petch relationship, represented as $\sigma_y = \sigma_o + k/\sqrt{d}$, where σ_y is the yield stress, σ_o is a constant, and d is the grain size. The constant k was estimated to be 0.40 MPam$^{1/2}$, which is not far from 0.27 MPam$^{1/2}$ reported by Armstrong et al. [45]. A transition of Hall-Petch slope with decreasing grain size in the submicron range was reported for UFG aluminum processed by ARB [46]. The lower slope for grain sizes smaller than 1 μm was estimated to be 0.07 and attributed to residual mobile dislocations inside the grains that carry plastic strain. For larger grain sizes after longer annealing times, the slope becomes steeper because of reduced density of mobile dislocations inside the grains [46]. This discussion was based on the assumption that high-angle grain boundaries act as sinks of mobile dislocations, consuming them during post-SPD annealing [47]. Similarly, in Region II in our studies, some residual mobile dislocations may carry plastic strain. However, in comparison with pure copper and pure aluminum with a higher SFE, dislocations are extended into Shockley partial dislocations with stacking faults and are difficult to be absorbed into grain boundaries, as discussed later.

In the SCC tests, the specimens were covered with a black tarnish film, which is well known to be a thick brittle copper oxide (Cu$_2$O) that causes intergranular SCC under tensile stress [48–50]. In Cu-Zn alloys with less than 20 wt% Zn, intergranular SCC tends to occur, whereas alloys with more than 20 wt% Zn tend to exhibit transgranular SCC [51, 52]. Cracks propagated in the tarnish film and formed at a higher rate on grain boundaries than within grains. Residual dislocations along the grain boundaries may enhance the

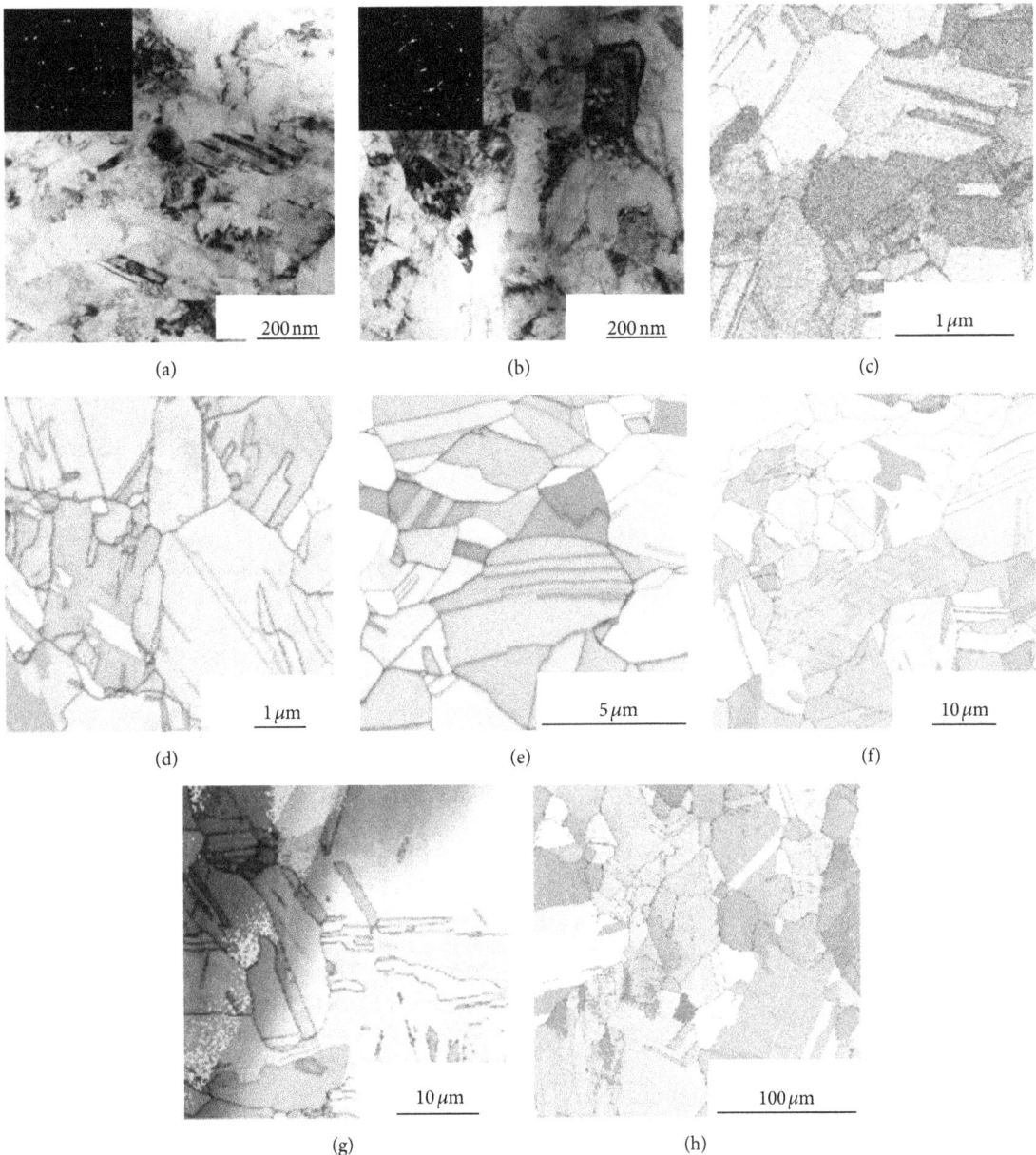

FIGURE 5: Microstructure after ECAP/cold-rolling and final annealing of (a) sample 8, (b) sample 7, (c) sample 6, (d) sample 5, (e) sample 4, (f) sample 3, (g) sample 2, and (h) sample 1.

reactivity of the grain boundaries with ammonia and the susceptibility to IGSCC. Figure 7 shows times to fracture obtained in SCC tests as a function of the inverse square root of grain size. Data for Cu-30 wt% Zn in ammonia from Edmund [14] is also plotted in the figure. As with the yield stress shown in Figure 6, the variation can be divided into two regimes, with a border at $1\,\mu m$. Considering Edmund's data for Cu-30 wt% Zn with large grains, it seems reasonable that the time to fracture increases with decreasing grain size from the conventional coarse grain regime down to $1\,\mu m$ (Region I) but then decreases with decreasing grain size in Region II. The positive trend in Region I can be explained on the basis of the assumption that an intergranular crack

initiates and propagates when local stress concentration by dislocation accumulation at the grain boundary reaches a certain critical value, so at smaller grain sizes, a higher applied stress is required to activate enough dislocations [53]. The changes in SCC susceptibility for grain sizes smaller than $1\,\mu m$ (Region II) may be closely associated with the change in the Hall-Petch slope and can be attributed to residual dislocations, mobile or immobile, in the grains. These dislocations reside along grain boundaries or are trapped at the grain boundaries in a nonequilibrium state [54] and enhance the reactivity of grain boundaries in a corrosive environment in Region II, rendering the susceptibility less dependent on grain size. In our previous studies of UFG pure copper,

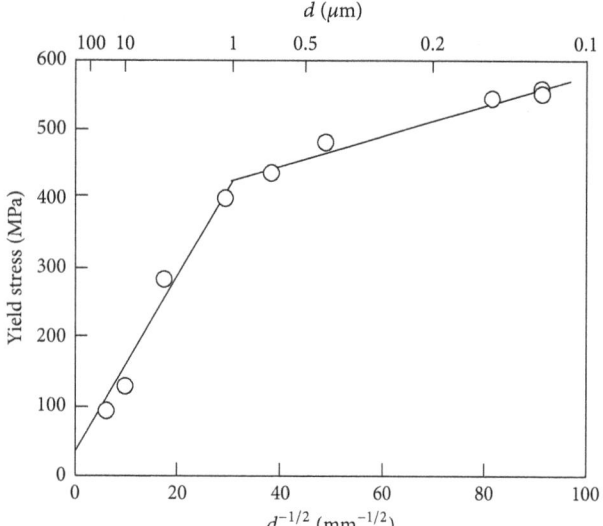

FIGURE 6: Relationship between yield stress and the inverse square root of grain size.

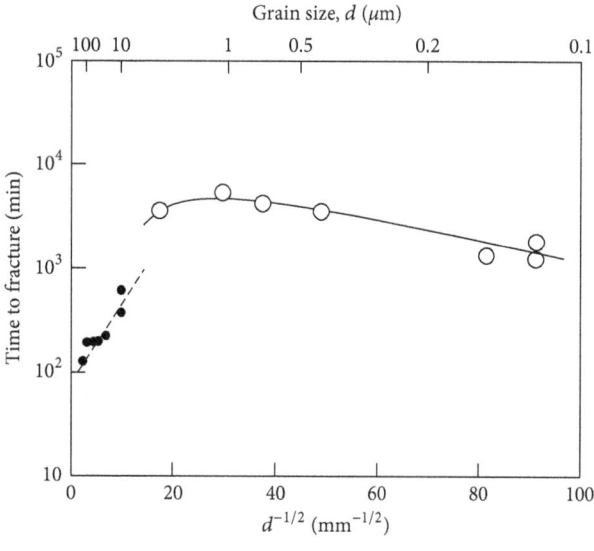

• Cu-30 wt% Zn [14]

FIGURE 7: Relationship between time to fracture and the inverse square root of grain size in the SCC tests.

very short annealing for 90 seconds at 200°C relaxed the nonequilibrium grain boundaries to their equilibrium state with little grain growth [5, 31, 55, 56]. However, the extended dislocations with stacking faults in Cu-Zn alloys are expected to be difficult to be absorbed into the grain boundaries as a result of the lower stacking energy and therefore may remain in the grains during annealing for longer times at higher temperatures. Furthermore, the nonequilibrium grain boundaries may be more stable due to Zn segregation, and some may remain in Region II [57–59]. A common constant stress of 280 MPa was applied to all specimens, so

the applied stress normalized by yield stress, σ_a/σ_{ys}, was lower in specimens with a smaller grain size, which have a higher yield strength. This means that even though the dislocation activity under macroscopic elastic deformation is lower in these materials, they are more susceptible to SCC.

Grain boundary sliding is considered to be one possible mode of plastic deformation in UFG materials [18, 60] and has been observed at room temperature in UFG pure aluminum by atomic force microscopy [61]. Film rupture facilitated by grain boundary sliding at the crack tip may enhance the susceptibility to IGSCC under a constant load. The present results are compatible with our previous studies that employed compact specimens of Cu-10 wt% Zn, in which the threshold stress was lower and the time to fracture was higher in UFG samples than in their counterparts with larger grain sizes [25].

Macroscopically, all of the SCC specimens fractured in a brittle manner with little plastic deformation. SEM fractographs of coarse-grained sample 1 and UFG sample 9 are shown in Figures 8(a) and 8(b). Microscopic observation at higher magnification revealed an intergranular mode of fracture in both samples (Figures 8(c) and 8(d)). Since the grain size was very small and the fracture surface was readily covered by a tarnish film, it was difficult to obtain clear evidence of IGSCC by SEM observation.

In addition to grain size, grain boundary structure is another important factor influencing IGSCC resistance [62–64], and several attempts have been made to alleviate SCC by controlling the grain boundary character distribution (GBCD) or by an alternative approach called grain boundary engineering (GBE) [65–68]. Ideally, the grain size and grain boundary character distribution should both be considered when characterizing the SCC of UFG materials by SPD. Therefore, SPD still has potential for high SCC resistance materials through the equilibration and control of grain boundary character distribution.

4. Conclusions

In this work, the effect of decreasing grain size into the submicron regime on the SCC of Cu-10 wt% Zn alloys in ammonia was investigated using a constant-load test, leading to the following conclusions:

(1) The yield stress obtained by tensile tests increased with decreasing grain size, but the tendency decreased for grain sizes smaller than 1 μm. Thus, the Hall-Petch relationship can be divided into two regions with different slopes, with a lower slope for grain sizes smaller than 1 μm.

(2) The time to fracture by SCC increased with decreasing grain size down to 1 μm but then decreased with further decreases in grain size into the submicron scale. In other words, there was a critical grain size above which the susceptibility began to increase, and this grain size matched the grain size that divides the two Hall-Petch relationships.

FIGURE 8: Fracture surfaces after SCC tests of (a), (c) sample 1, and (b), (d) sample 9 in (a), (b) macroscopic appearance, and (c), (d) microscopic observation.

(3) Stress corrosion cracks propagated intergranularly regardless of grain size. SPD-induced grain boundaries have a high sensitivity to chemical reaction and intergranular SCC.

Conflicts of Interest

The authors declare that there are no conflicts of interest regarding the publication of this paper.

Acknowledgments

This work was financially supported by JSPS KAKENHI Grant no. 26420748 and a grant-in-aid from the Japan Copper and Brass Association.

References

[1] A. Di Schino and J. M. Kenny, "Effects of the grain size on the corrosion behavior of refined AISI 304 austenitic stainless steels," *Journal of Materials Science Letters*, vol. 21, no. 20, pp. 1631–1634, 2002.

[2] K. D. Ralston, N. Birbilis, and C. H. J. Davies, "Revealing the relationship between grain size and corrosion rate of metals," *Scripta Materialia*, vol. 63, no. 12, pp. 1201–1204, 2010.

[3] K. D. Ralston, D. Fabijanic, and N. Birbilis, "Effect of grain size on corrosion of high purity aluminium," *Electrochimica Acta*, vol. 56, no. 4, pp. 1729–1736, 2011.

[4] B. Hadzima, M. Janeček, R. J. Hellmig, Y. Kutnyakova, and Y. Estrin, "Microstructure and corrosion behaviour of ultrafine-grained copper," *Materials Science Forum*, vol. 503-504, pp. 883–888, 2006.

[5] H. Miyamoto, K. Harada, T. Mimaki, A. Vinogradov, and S. Hashimoto, "Corrosion of ultra-fine grained copper fabricated by equal-channel angular pressing," *Corrosion Science*, vol. 50, no. 5, pp. 1215–1220, 2008.

[6] N. P. Gurao, G. Manivasagam, P. Govindaraj, R. Asokamani, and S. Suwas, "Effect of texture and grain size on bio-corrosion response of ultrafine-grained titanium," *Metallurgical and Materials Transactions A: Physical Metallurgy and Materials Science*, vol. 44, no. 12, pp. 5602–5610, 2013.

[7] H. S. Kim and W. J. Kim, "Annealing effects on the corrosion resistance of ultrafine-grained pure titanium," *Corrosion Science*, vol. 89, no. C, pp. 331–337, 2014.

[8] H. S. Kim, S. J. Yoo, J. W. Ahn, D. H. Kim, and W. J. Kim, "Ultrafine grained titanium sheets with high strength and high corrosion resistance," *Materials Science and Engineering A*, vol. 528, no. 29-30, pp. 8479–8485, 2011.

[9] O. Jilani, N. Njah, and P. Ponthiaux, "Transition from intergranular to pitting corrosion in fine grained aluminum processed by equal channel angular pressing," *Corrosion Science*, vol. 87, pp. 259–264, 2014.

[10] A. Korchef and A. Kahoul, "Corrosion behavior of commercial aluminum alloy processed by equal channel angular pressing," *International Journal of Corrosion*, vol. 2013, Article ID 983261, 11 pages, 2013.

[11] D. Song, A. B. Ma, J. H. Jiang, P. H. Lin, and J. Shi, "Improving corrosion resistance of pure Al through ECAP," *Corrosion Engineering Science and Technology*, vol. 46, no. 4, pp. 505–512, 2011.

[12] A. Morris and C. Bridgeport, "Stress-corrosion cracking of annealed brass," *Transactions of the American Institute of Mining and Metallurgical Engineering*, vol. 89, pp. 256–275, 1930.

[13] T. C. Tsai and T. H. Chuang, "Role of grain size on the stress corrosion cracking of 7475 aluminum alloys," *Materials Science and Engineering: A*, vol. 225, no. 1-2, pp. 135–144, 1997.

[14] G. Edmund, "Season cracking of brass, symposium on stress corrosion cracking of metals, A.S.T.M. and A.I.M.E. (1945) pp. 67-89," 1945.

[15] H. Miyamoto, "Corrosion of ultrafine grained materials by severe plastic deformation, an overview," *Materials Transactions*, vol. 57, no. 5, pp. 559–572, 2016.

[16] K. D. Ralston and N. Birbilis, "Effect of grain size on corrosion: a review," *Corrosion*, vol. 66, no. 7, 2010.

[17] R. Z. Valiev, R. K. Islamgaliev, and I. V. Alexandrov, "Bulk nanostructured materials from severe plastic deformation," *Progress in Materials Science*, vol. 45, no. 2, pp. 103–189, 2000.

[18] R. Z. Valiev, I. V. Alexandrov, Y. T. Zhu, and T. C. Lowe, "Paradox of strength and ductility in metals processed by severe plastic deformation," *Journal of Materials Research*, vol. 17, no. 1, pp. 5–8, 2002.

[19] C. C. Koch, "Optimization of strength and ductility in nano-crystalline and ultrafine grained metals," *Scripta Materialia*, vol. 49, no. 7, pp. 657–662, 2003.

[20] G. R. Argade, W. Yuan, K. Kandasamy, and R. S. Mishra, "Stress corrosion cracking susceptibility of ultrafine grained AZ31," *Journal of Materials Science*, vol. 47, no. 19, pp. 6812–6822, 2012.

[21] G. R. Argade, N. Kumar, and R. S. Mishra, "Stress corrosion cracking susceptibility of ultrafine grained Al-Mg-Sc alloy," *Materials Science and Engineering A*, vol. 565, pp. 80–89, 2013.

[22] Y. H. Jang, S. S. Kim, S. Z. Han, C. Y. Lim, and C. J. Kim, "Corrosion and stress corrosion cracking behavior of equal channel angular pressed oxygen-free copper in 3.5% NaCl solution," *Journal of Materials Science*, vol. 41, no. 13, pp. 4293–4297, 2006.

[23] H. Miyamoto, T. Kishi, T. Mimaki, A. Vinogradov, and S. Hashimoto, "Stress corrosion cracks propagation in ultrafine grain copper fabricated by an equal-channel angular pressing," in *Ultrafine Grained Materials, IV*, Y. T. Zhu, T. G. Langdon, Z. Horita, M. J. Zehetbauer, S. L. Semiatin, and Lowe T. C., Eds., p. 337, John Wiley & Sons, Inc., Hoboken, NJ, USA, 2002.

[24] H. Miyamoto, T. Mimaki, A. Vinogradov, and S. Hashimoto, "Mechanical, thermal and stress-corrosion properties of ultra-fine grain copper," *Annales de Chimie Science des Matériaux*, vol. 27, pp. S197–S206, 2002.

[25] H. Miyamoto, A. Vinogradov, and S. Hashimoto, "Susceptibility to stress corrosion cracking in ammonia of nanostructured Cu-10wt%Zn alloy produced by severe plastic deformation," *Materials Science Forum*, vol. 584-586, pp. 887–892, 2008.

[26] H. Nakano, S. Oue, S. Taguchi, S. Kobayashi, and Z. Horita, "Stress-corrosion cracking property of aluminum-magnesium alloy processed by equal-channel angular pressing," *International Journal of Corrosion*, vol. 2012, Article ID 543212, 2012.

[27] S. A. Nikulin, S. O. Rogachev, A. B. Rozhnov, M. V. Gorshenkov, V. I. Kopylov, and S. V. Dobatkin, "Resistance of alloy Zr-2.5% Nb with ultrafine-grain structure to stress corrosion cracking," *Metal Science and Heat Treatment*, vol. 54, no. 7-8, pp. 407–413, 2012.

[28] M. M. Sharma and C. W. Ziemian, "Pitting and stress corrosion cracking susceptibility of nanostructured Al-Mg alloys in natural and artificial environments," *Journal of Materials Engineering and Performance*, vol. 17, no. 6, pp. 870–878, 2008.

[29] A. Turnbull, K. Mingard, J. D. Lord et al., "Sensitivity of stress corrosion cracking of stainless steel to surface machining and grinding procedure," *Corrosion Science*, vol. 53, no. 10, pp. 3398–3415, 2011.

[30] A. Vinogradov, H. Miyamoto, T. Mimaki, and Hashimoto S., "Corrosion, stresss corrosion cracking and fatigue of ultra-fine grain copper fabricated by severe plastic deformation," *Annales de Chimie Science des Matériaux*, vol. 27, pp. 65–75, 2002.

[31] T. Yamasaki, H. Miyamoto, T. Mimaki, A. Vinogradov, and S. Hashimoto, "Stress corrosion cracking susceptibility of ultra-fine grain copper produced by equal-channel angular pressing," *Materials Science and Engineering A*, vol. 318, no. 1-2, pp. 122–128, 2001.

[32] B. N. Mordyuk, G. I. Prokopenko, M. A. Vasylyev, and M. O. Iefimov, "Effect of structure evolution induced by ultrasonic peening on the corrosion behavior of AISI-321 stainless steel," *Materials Science and Engineering A*, vol. 458, no. 1-2, pp. 253–261, 2007.

[33] T. Balusamy, S. Kumar, and T. S. N. S. Narayanan, "Effect of surface nanocrystallization on the corrosion behaviour of AISI 409 stainless steel," *Corrosion Science*, vol. 52, no. 11, pp. 3826–3834, 2010.

[34] Y. H. Zhao, Y. T. Zhu, X. Z. Liao, Z. Horita, and T. G. Langdon, "Tailoring stacking fault energy for high ductility and high strength in ultrafine grained Cu and its alloy," *Applied Physics Letters*, vol. 89, no. 12, Article ID 121906, 2006.

[35] H. Miyamoto, K. Ota, and T. Mimaki, "Viscous nature of deformation of ultra-fine grain aluminum processed by equal-channel angular pressing," *Scripta Materialia*, vol. 54, no. 10, pp. 1721–1725, 2006.

[36] K. Edalati and Z. Horita, "High-pressure torsion of pure metals: influence of atomic bond parameters and stacking fault energy on grain size and correlation with hardness," *Acta Materialia*, vol. 59, no. 17, pp. 6831–6836, 2011.

[37] C. X. Huang, W. Hu, G. Yang et al., "The effect of stacking fault energy on equilibrium grain size and tensile properties of nanostructured copper and copper-aluminum alloys processed by equal channel angular pressing," *Materials Science and Engineering A*, vol. 556, pp. 638–647, 2012.

[38] S. Komura, Z. Horita, M. Nemoto, and T. G. Langdon, "Influence of stacking fault energy on microstructural development in equal-channel angular pressing," *Journal of Materials Research*, vol. 14, no. 10, pp. 4044–4050, 1999.

[39] J. Wan, K. Zhou, S. Lu, X. Xu, and Y. Jian, "Influence of stacking fault energy on grain-refining during severe plastic deformation," *Chinese Journal of Mechanical Engineering*, vol. 44, pp. 126–131, 2008.

[40] S. S. Hazra, E. V. Pereloma, and A. A. Gazder, "Microstructure and mechanical properties after annealing of equal-channel angular pressed interstitial-free steel," *Acta Materialia*, vol. 59, no. 10, pp. 4015–4029, 2011.

[41] J. Lian, R. Z. Valiev, and B. Baudelet, "On the enhanced grain growth in ultrafine grained metals," *Acta Metallurgica Et Materialia*, vol. 43, no. 11, pp. 4165–4170, 1995.

[42] A. Takayama, X. Yang, H. Miura, and T. Sakai, "Continuous static recrystallization in ultrafine-grained copper processed by multi-directional forging," *Materials Science and Engineering A*, vol. 478, no. 1-2, pp. 221–228, 2008.

[43] W. Q. Cao, A. Godfrey, W. Liu, and Q. Liu, "Annealing behavior of aluminium deformed by equal channel angular pressing," *Materials Letters*, vol. 57, no. 24-25, pp. 3767–3774, 2003.

[44] W. Q. Cao, A. Godfrey, W. Liu, and Q. Liu, "EBSP study of the annealing behavior of aluminum deformed by equal channel angular processing," *Materials Science and Engineering A*, vol. 360, no. 1-2, pp. 420–425, 2003.

[45] R. Armstrong, I. Codd, R. M. Douthwaite, and N. J. Petch, "The plastic deformation of polycrystalline aggregates," *Philosophical Magazine*, vol. 7, no. 73, pp. 45–58, 1962.

[46] D. Terada, S. Inoue, and N. Tsuji, "Microstructure and mechanical properties of commercial purity titanium severely deformed by ARB process," *Journal of Materials Science*, vol. 42, no. 5, pp. 1673–1681, 2007.

[47] X. Huang, N. Hansen, and N. Tsuji, "Hardening by annealing and softening by deformation in nanostructured metals," *Science*, vol. 312, no. 5771, pp. 249–251, 2006.

[48] T. R. Pinchback, S. P. Clough, and L. A. Heldt, "Stress corrosion cracking of alpha brass in a tarnishing ammoniacal environment: fractography and chemical analysis," *Metallurgical Transactions A*, vol. 6, no. 8, pp. 1479–1483, 1975.

[49] E. Mattsson, "Stress corrosion in brass considered against the background of potential/pH diagrams," *Electrochimica Acta*, vol. 3, no. 4, pp. 279–291, 1961.

[50] R. Nishimura and T. Yoshida, "Stress corrosion cracking of Cu-30% Zn alloy in Mattsson's solutions at pH 7.0 and 10.0 using constant load method—a proposal of SCC mechanism," *Corrosion Science*, vol. 50, no. 4, pp. 1205–1213, 2008.

[51] P. R. Swann, "Dislocation substructure vs transgranular stress corrosion susceptibility of single phase alloys," *Corrosion*, vol. 19, no. 3, pp. 102t–114t, 1963.

[52] D. Tromans and J. Nutting, "Stress Corrosion cracking of face-centered-cubic alloys," *Corrosion*, vol. 21, no. 5, pp. 143–160, 1965.

[53] H. Vehoff, H. Stenzel, and P. Neumann, "Experiments of bicrystals concerning the influence of localized slip on the nucleation and growth of intergranular stress corrosion cracks," *Zeitschrift fuer Metallkunde/Materials Research and Advanced Techniques*, vol. 78, no. 8, pp. 550–556, 1987.

[54] M. W. Grabski and R. Korski, "Grain boundaries as sinks for dislocations," *Philosophical Magazine*, vol. 22, no. 178, pp. 707–715, 2006.

[55] A. Vinogradov, T. Mimaki, S. Hashimoto, and R. Valiev, "On corrosion of ultra-fine grained copper produced by equi-channel angular pressing," *Materials Science Forum*, vol. 312–314, pp. 641–646, 1999.

[56] V. Y. Gertsman, R. Birringer, R. Z. Valiev, and H. Gleiter, "On the structure and strength of ultrafine-grained copper produced by severe plastic deformation," *Scripta Metallurgica et Materiala*, vol. 30, no. 2, pp. 229–234, 1994.

[57] S. Lartigue and L. Priester, "Stability of extrinsic grain boundary dislocations in relation with intergranular segregation and precipitation," *Acta Metallurgica*, vol. 31, no. 11, pp. 1809–1819, 1983.

[58] W. Łojkowski and M. W. Grasbaski, "On material purity influence on the spreading temperature of grain boundary dislocations," *Scripta Metallurgica*, vol. 13, no. 6, pp. 511–514, 1979.

[59] W. Lojkowski, J. W. Wyrzykowski, and J. Kwiecinski, "Interpretation of grain boundary nonequilibrium in terms of geometrically necessary and statistically stored dislocations," *Journal de Physique Colloque*, vol. C1-51, no. 1, p. C1-239, 1990.

[60] J. May, H. W. Höppel, and M. Göken, "Strain rate sensitivity of ultrafine-grained aluminium processed by severe plastic deformation," *Scripta Materialia*, vol. 53, no. 2, pp. 189–194, 2005.

[61] N. Q. Chinh, P. Szommer, Z. Horita, and T. G. Langdon, "Experimental evidence for grain-boundary sliding in ultrafine-grained aluminum processed by severe plastic deformation," *Advanced Materials*, vol. 18, no. 1, pp. 34–39, 2006.

[62] M. Yamashita, T. Mimaki, S. Hashimoto, and S. Miura, "Misorientation dependence of susceptibility to intergranular Stress-Corrosion-Cracking in symmetrical fbffk110fbfft-tilt Cu-9at.%Al alloy bicrystals," *Scripta Metallurgica*, vol. 22, no. 7, pp. 1087–1091, 1988.

[63] M. Yamashita, T. Mimaki, S. Hashimoto, and S. Miura, "Stress corrosion cracking of [110] and [100] tilt boundaries of α-Cu-Al alloy," *Philosophical Magazine A*, vol. 63, no. 4, pp. 707–726, 1991.

[64] V. Y. Gertsman and S. M. Bruemmer, "Study of grain boundary character along intergranular stress corrosion crack paths in austenitic alloys," *Acta Materialia*, vol. 49, no. 9, pp. 1589–1598, 2001.

[65] K. T. Aust, U. Erb, and G. Palumbo, "Interface control for resistance to intergranular cracking," *Materials Science and Engineering A*, vol. 176, no. 1-2, pp. 329–334, 1994.

[66] C. Cheung, U. Erb, and G. Palumbo, "Application of grain boundary engineering concepts to alleviate intergranular cracking in Alloys 600 and 690," *Materials Science and Engineering A*, vol. 185, no. 1-2, pp. 39–43, 1994.

[67] Y. Pan, B. L. Adams, T. Olson, and N. Panayotou, "Grain-boundary structure effects on intergranular stress corrosion cracking of Alloy X-750," *Acta Materialia*, vol. 44, no. 12, pp. 4685–4695, 1996.

[68] D. C. Crawford and G. S. Was, "The Role of grain boundary misorientation in intergranular cracking of Ni-16Cr-9Fe in 360 °C argon and high-Purity water," *Metallurgical Transactions A*, vol. 23, no. 4, pp. 1195–1206, 1992.

15

Developing Field Test Procedures for Chloride Stress Corrosion Cracking in the Arabian Gulf

Hanan Farhat (iD)

College of the North Atlantic-Qatar, Doha 24449, Qatar

Correspondence should be addressed to Hanan Farhat; hanan.sharef@gmail.com

Academic Editor: Francisco Javier Perez Trujillo

Oil and gas production and petrochemical plants in the Arabian Gulf are exposed to severe environmental conditions of high temperature and humidity. This makes these plants susceptible to chloride-induced stress corrosion cracking (CSCC). The laboratory testing fails to provide the exact field environmental conditions. A cost efficient field test setup for CSCC was designed and developed for the Arabian Gulf. The setup included designing self-sustained loading devices, samples, and sample racks. The samples were exposed to a stress equivalent to 80% and 100% of their yield strength. This paper describes the developed test procedures to establish testing with high level of accuracy and repeatability. It also discusses the design aspects and the challenges that were met.

1. Introduction

Chloride stress corrosion cracking (CSCC) takes place when specific alloys are exposed to tensile stress, temperature, and a chloride containing environment [1–3]. During CSCC, the alloy is attacked over most of its surface, while fine cracks progress through it [4]. Chloride stress corrosion cracking can occur fast when evaporation exists even at room temperature [5–9]. A number of catastrophic CSCC failures of roof construction in swimming pool environments has resulted in human causalities over the past decades [5, 6].

Several studies were performed in simulated field conditions for CSCC testing [10–21] as conventional laboratory tests did not represent the real field conditions. In one such a study, Turnbull et al. [13] managed to set new threshold temperature for CSCC of 22Cr and 25Cr duplex stainless steels under evaporation seawater conditions [13]. Yet, there is a continuous argument that laboratory test conditions can never represent the real field condition. The susceptibility to SCC depends on many environmental factors, including temperature variation during the day, concentration of chloride ions in the environment, which is a factor of the distance from the sea, and altitude above sea level [8, 9, 22]. The environment in the Arabian Gulf's oil and gas production sites is very corrosive and is characterized by high temperature and high humidity, representing continuous evaporation conditions in most of the year. Corrosion protection in these sites is very challenging. Most of the oil and gas installations are located near the sea, which in turn, enhances the build-up of chloride species on the surfaces of the alloys. Cases of severe CSCC in the region were reported [23]. These environmental conditions were difficult to replicate in the laboratory. This led to the need to develop a high accuracy field test setup for CSCC. One of the main objectives is to test the resistance of different corrosion resistance alloys to this cracking in the Arabian Gulf and to determine the actual time-to-CSCC failure of these alloys. Moreover, the impact of load and/or stress and temperature on the susceptibility to CSCC is to be investigated. To the author's knowledge CSCC field testing has never been applied in the Arabian Gulf before.

2. Materials and Methods

2.1. Material. Specimens from seven types of alloys were machined following NACE TM0177-2005 standard test method A [24]. The types of alloy, their yield strength, and elongation are given in Table 1.

The test was conducted at different stress levels to determine the threshold stress for CSCC. The load was applied using sustainable load devices (proof rings). The proof rings

TABLE 1: The alloys properties that were recorded from the tensile tests.

Specimen Type (UNS)	Yield Strength (0.2% offset) (MPa)	Ultimate Tensile Strength (MPa)	Strain(mm/mm)	Reduction of Area (%)
S30400	448	724	0.49	74
S31603	500	689	0.53	76
S32100	362	590	0.48	76
S32003	507	710	0.44	62
S32750	662	841	0.53	83
N08904	545	689	0.50	83
N08825	507	703	0.42	75

were manufactured from UNS 32205 duplex stainless steel, while the sample racks were made from aluminum 6061 alloy (UNS A96061). To apply the load on the samples, 316L stainless steel nuts were used. Detailed design and manufacturing procedures are provided in the results and discussion in Section 3.

The selection of test method A that includes the use of proof rings was made because the method is simple, compact, and easy. It provides flexibility in the choices of the size of the test specimen and range of stress levels. In addition, it can be used to determine the threshold stress for cracking, which is one of the objectives of this research study. NACE TM0177-2005 standard test method B for bent-beam test that follows ASTM G30 standards [25] is simple and economical. This method was ruled out because the actual stress distribution in the test specimen is not known. Therefore, the effect of different applied stresses cannot be studied using this method. Likewise, NACE TM0177-2005 standard test method C-Ring that follows ASTM G38 standards [26] was not used because of the difficulty of machining the test specimens in this method and the decreased precision in stressing, which is essential in this study [24].

3. Results and Discussion

3.1. Design Stress Calculations. Design stress calculation and analysis for the combination of specimen and proof ring under a load of 100% of the specimen's yield strength were performed using Solidworks software at room temperature. The analysis was conducted to determine the stress distribution on the samples and the proof rings. The samples and proof rings' material were identified in each stress analysis. The analysis was performed for two main tasks:

(i) The stresses have to be concentrated mainly on the samples' gauge length. Therefore the size of the sample has to be chosen carefully. The stress analysis was conducted on two sizes of specimens selected based on NACE TM0177-2005 standard [24]: (1) standard test specimen (gauge diameter is 6.35 ± 0.13 mm, and length is 25.4 mm) and (2) subsize test specimen (gauge diameter is 3.81 ± 0.05 mm, and length is 25.4 mm).

(ii) The proof rings should not be subject to high stresses that could cause their distortion and affect the amount of applied stress on the specimens.

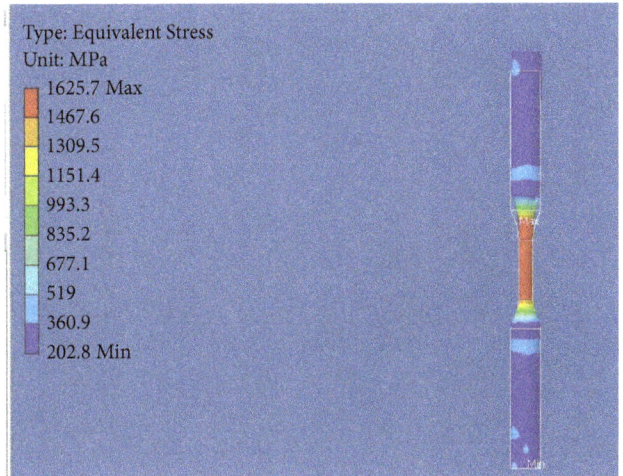

FIGURE 1: Design stress analysis for a subsize specimen showing the stress concentrating in the gauge length.

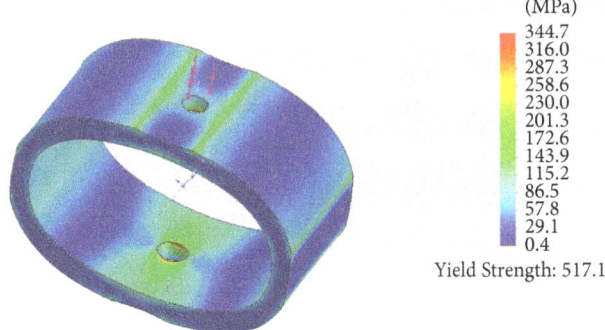

FIGURE 2: Design stress analysis for a proof ring 3" Schedule 80 pipe with the subsize specimen showing very small stresses around the hole.

Two sizes of proof rings were compared with a combination of two different specimens (subsize and standard specimen selected based on NACE TM0177-2005 standard [24]). The results of the stress analysis are presented in Figures 1–4. The stresses were found to concentrate in the specimen's gauge length as can be seen in Figure 1. The stresses increased whenever the colour changed from dark blue to red, as indicated by the load bar on the left of the figure. The increase

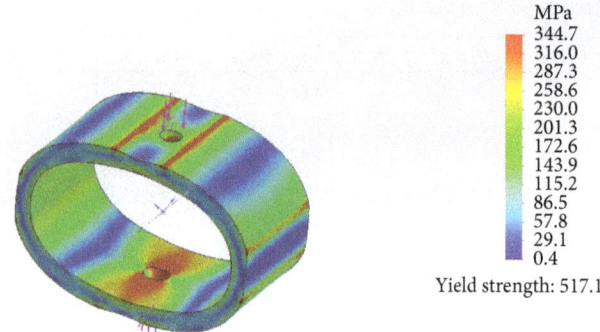

MPa
344.7
316.0
287.3
258.6
230.0
201.3
172.6
143.9
115.2
86.5
57.8
29.1
0.4

Yield strength: 517.1

FIGURE 3: Design stress analysis for a proof ring 3" Schedule 80 pipe with the standard specimen showing more stresses covering a larger area around the holes.

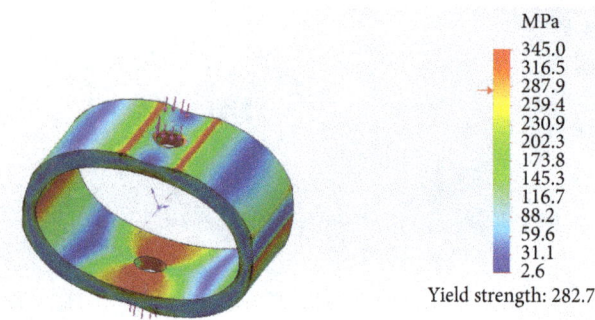

MPa
345.0
316.5
287.9
259.4
230.9
202.3
173.8
145.3
116.7
88.2
59.6
31.1
2.6

Yield strength: 282.7

FIGURE 4: Design stress analysis for a proof ring 3" Schedule 40 pipe with the standard specimen showing large amount of stresses covering a larger area around the holes, and on the ring sides.

of stresses in the gauge length is to be expected, due to its smaller size compared to the rest of the specimen. This is the area where cracking and failure is expected to take place. The stress analysis on the proof rings, on the other hand, showed that when using the subsize samples, all the stresses in the ring were below yielding, except for the area around the hole, where localized higher stresses existed, as can be seen in red colour inside the hole in Figure 2. Still these stresses were in the acceptable limits and are below the proof ring's yield strength. In contrast, when using the standard specimen, the analysis showed higher stresses on wider areas around the holes as illustrated in Figure 3. These stresses are high stresses and are covering much wider region, compared to the stresses in the subsize specimen. Furthermore, the stress analysis on proof rings made from UNS 32205 duplex stainless steel – 3" Schedule 80 pipe (Figure 2) was compared with analysis of proof rings made from UNS 32205 duplex stainless steel – 3" Schedule 40 pipe (Figure 4). The subsize specimens were used in the two combination. The analysis results revealed that the rings that were made from 3" Schedule 80 pipe showed very minimum amount of stress, compared to very large stress that was noticed around the hole and on the sides of the rings in the specimens that were made from 3" Schedule 40 pipe. The stress analysis results made the base for the decision of choosing to use the subsize specimen instead of standard

specimen and manufacturing the rings from 3" Schedule 80 pipe.

3.2. Design of Proof Rings. Sustained-load devices (proof rings) were used to apply load on the samples. The option of using coated or galvanized carbon steel pipe to make the rings was compared to the option of using duplex stainless steel pipe to make the rings. It was decided that applying coating on each induvial ring would be an expensive process and any defect or scratch in the coating could lead to galvanic corrosion of the carbon steel ring, and results in failing it, which may affect the load on the samples. Therefore, a UNS 32205 duplex stainless steel – 3" Schedule 80 pipe was used to make proof rings based on the stress analysis finding. The engineering drawing of the ring is illustrated in Figure 5.

3.3. Design of Test Specimens. The samples are designed following NACE TM0177-2005 standard [24]. The subsize tensile test specimen was chosen to be used based on the stress analysis results. One advantage this specimen has is that its small size enables shorter failure times compared to the standard size specimen, which will provide test results in a shorter time. Figure 6 illustrates the subsize specimen dimensions. Machining the samples was performed using Computer Numerical Control (CNC) machining following NACE TM0177-2005 standard [24], where the surface roughness was obtained by mechanical polishing and was ~ 0.21 μm. One hundred specimens from each alloy were machined to be placed in the field. After machining, the specimens were degreased with solvent and were cleaned in ethanol using ultrasonic cleaner. Gloves were used to handle the specimens to prevent the contamination of the gauge length. the specimens were stored in a desiccator until exposure.

3.4. Design of Sample Racks. The sample rack design took into consideration the test site location and environment. Wind speed was expected to reach a maximum of 32 m/s. The maximum deflection of each shelf was calculated to be 2.45 mm across the total length of 0.75 m. The racks were made from aluminum 6061 alloy (UNS A96061) to resist the corrosion in the sites. The shelves were tilted 30° with horizontal access to provide enough ambient exposure of all the specimens in the rack. Since different alloys will corrode/fail at different timing, each rack was designed so that it does not contribute to the applied load on the samples. Each rack accommodates 35 samples. Four racks were placed at each site to accommodate a total of 140 samples per site. In the onshore sites, each rack was fixed in a concrete stand to ensure that it will not get moved by the high wind. The racks that were placed offshore were bolted to the platforms. Figure 7 provides images of racks placed in onshore and offshore site.

3.5. Quality Control Testing. Several tests were conducted to control the quality of the samples, rings, and the sample racks. The tests included the following.

3.5.1. Visual Inspection. Using a 10X magnifier equipped with light, each sample was checked visually for any scratches or

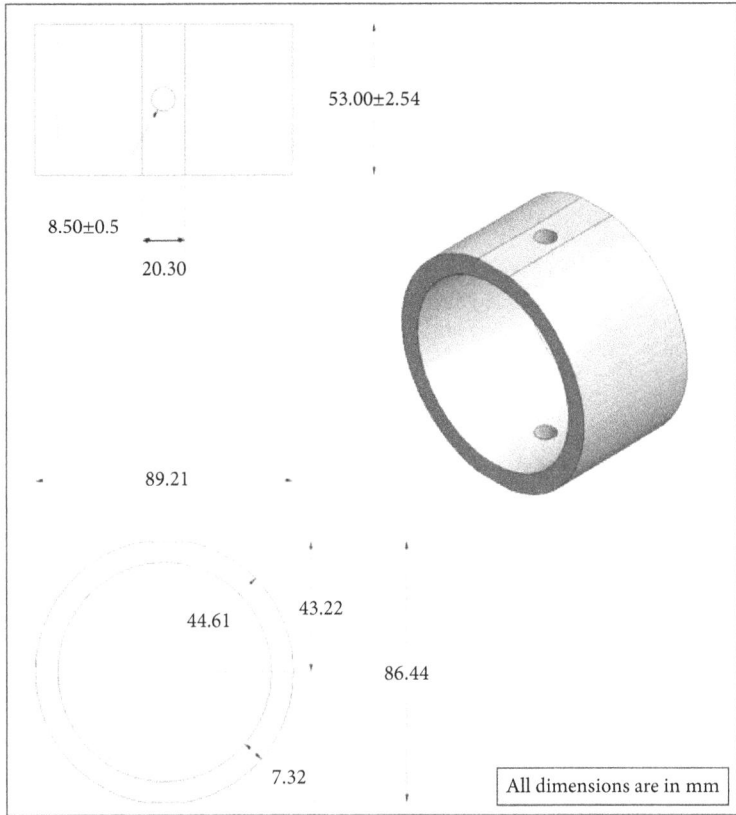

FIGURE 5: The design and dimensions of the proof rings.

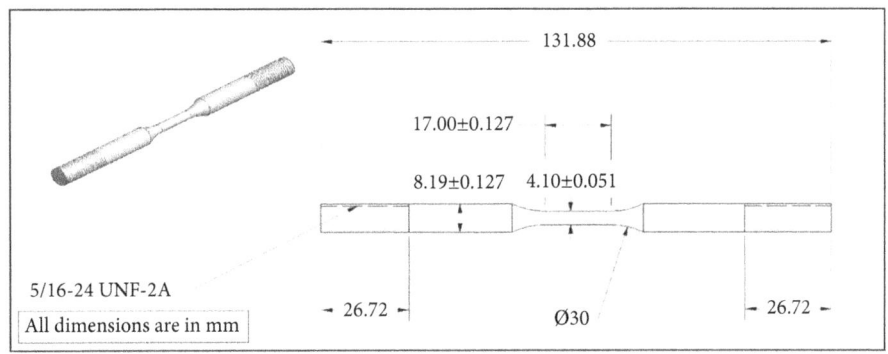

FIGURE 6: The dimensions of the subsize samples that were used in the CSCC field testing.

indentations in the gauge length. Few samples were noticed to have minor scratches in the gauge length area. These samples were excluded from field exposure, as scratches could act as preferred sites for pitting and cracking.

3.5.2. Tensile Testing of the Samples.
Tensile testing was performed at room temperature. Three samples of each alloy were tested. A total of twenty-one stress-strain curves were generated. The results of this test are provided in Table 1. The yield strength of each alloy was identified as 0.2% offset. The average yield strength was calculated using the values that were measured in each sample. Values of 80% and 100% of the yield strength were recorded in each curve. The

corresponding elongation was identified and used later, when applying loads to the samples.

3.6. Load Application.
Seven hundred samples and rings were prepared to be placed in the field. Before applying the load, each sample and ring were given a code and the codes were engraved on the samples (away from the gauge length) and on the ring. The code indicates the type of alloy, applied load, exposure location, and the location in the sample rack (shelf number). Before applying the load the sample length was measured as follows.

3.6.1. Sample Length Measurement.
The length of each sample was measured using two different methods. Overall length

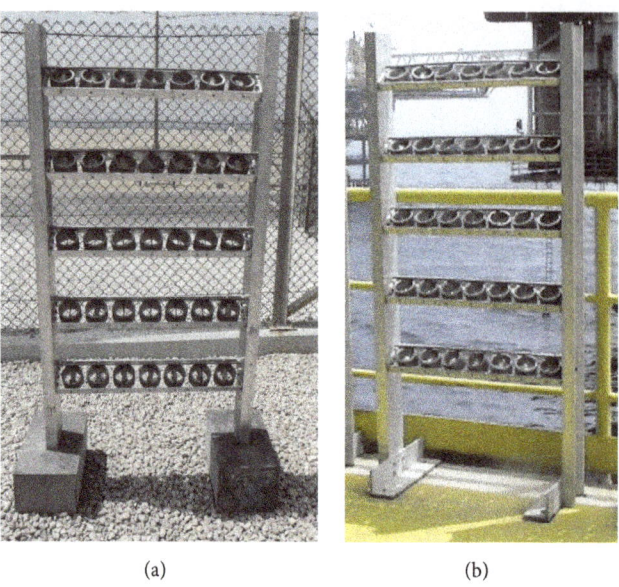

FIGURE 7: Sample racks (a) in onshore site and (b) in offshore site.

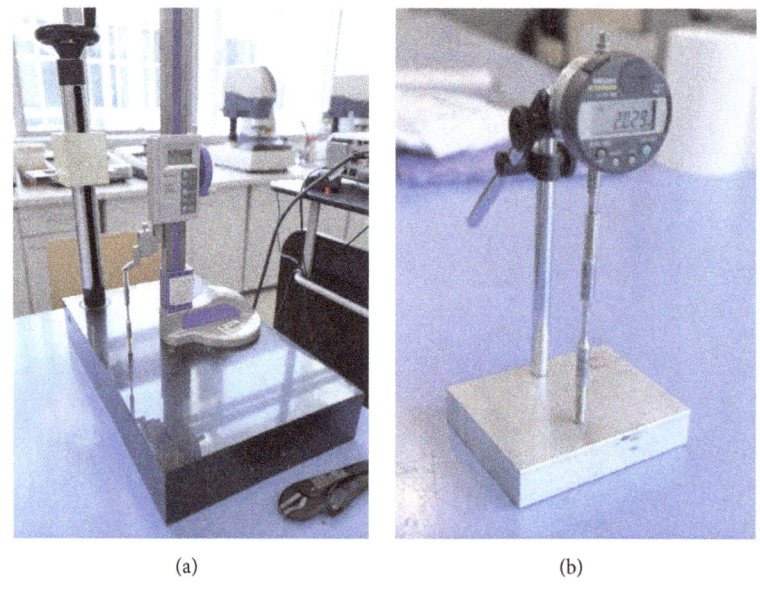

FIGURE 8: (a) Digital height gauge used for length measurement noted to be operator dependent. (b) Lab built device for elongation measurement proved to be not dependent on operator with high degree of repeatability.

was measured using a rectangular surface plate and a digital height gauge (Figure 8(a)). The measurement was taken up to two millimeter decimal digits. This method resulted in some variation, operator dependence, and repeatability problems.

The second method was done using a lab built device that includes a digital indicator and two steel balls (Figure 8(b)). Each of the samples had been machined with a center drill taper in both ends. The process took advantage of the center drill taper. Measuring using the taper and the steel balls provided a means of measuring the change in specimen length relative to a base point. The device was calibrated before each measurement using a standard sample. Measurement technique was found to be not sensitive to the

condition of the ends of each specimen and was not operator dependent and had a high degree of repeatability.

3.6.2. Load Application on the Samples. Once the samples were engraved and cleaned and their length was measured, they were assembled to their respective rings and hand tightened in place using two nuts on both sides. The nuts were AISI 316L stainless steel and were selected to fit a sample thread of 5/16" UNF.

Nuts from each type of alloy were difficult to find; therefore it was decided to use 316L stainless steel nuts. The use of 316L stainless steel nuts could cause galvanic corrosion between the specimens and nuts or the ring and nuts.

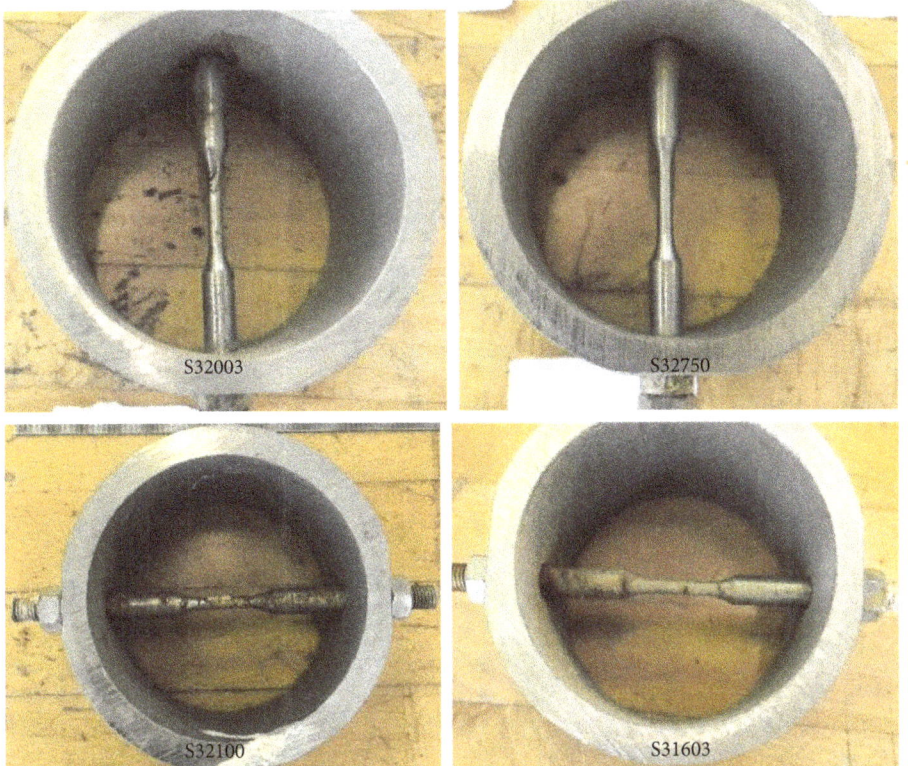

FIGURE 9: Retrieved samples after two years of exposure showing galvanic corrosion between some of the samples and the rings.

Attempts to use Teflon insulators did not work with the threaded nuts and samples. The nuts were greased before using and were observed during exposure for any signs of cracking or galvanic corrosion.

Each sample was torqued to a specific load using digital torque wrenches. The torque wrenches were calibrated before the application of the loads. Two loads were applied: 80% and 100% of the yield strength of the samples. The sample code indicates which load should be applied. Each load represented an elongation that was predetermined from the stress-strain curve of each alloy. The load application included the following steps:

(i) The 2 nuts at the bottom of the sample were tightened using a wrench.

(ii) A predetermined torque was applied to the top nut on the specimen. The torque produces an elongation that was determined from the stress-strain curve to reflect a specific load (80% or 100% yield strength).

(iii) The specimen was then placed back in the measuring fixture and the difference in the length was observed.

(iv) This procedure was repeated until the required change in specimen length was achieved.

(v) Once the required load and elongation has been attained, then the torqueing procedure is complete. All the measurement data is recorded and saved.

(vi) The sample is then kept aside for 7 days to allow enough time for cold creep, if any.

(vii) The sample's length was measured after 7 days and the difference in elongation between the last measured value and the present value was observed.

(viii) If the elongation was decreased, then the sample was torqued again using the same steps that were described above.

(ix) The final elongation was measured and recorded and the 4th nut was tightened onto the sample using a wrench. This step completes the retorqueing procedure.

In order to overcome the possible torsion of the samples while applying the load, two aluminum sleeves were machined to fit around the samples (not the gauge length area). A vise grip was used to hold the sleeves in place. The sleeves were used so that the vise grip does not damage the sample surface. The vise grip is used to keep the sample from rotating and to prevent torsion.

The samples were placed in the sample racks and were exposed in the onshore and offshore environment. The samples were inspected once every month. The test was terminated when the samples failed. The failure was identified as either

(a) visual observation of cracks on the gauge length or

(b) complete separation of the test specimen.

During inspection, the nuts were also inspected for any signs of cracking. Galvanic corrosion was observed between some of the samples and rings as can be seen in Figure 9.

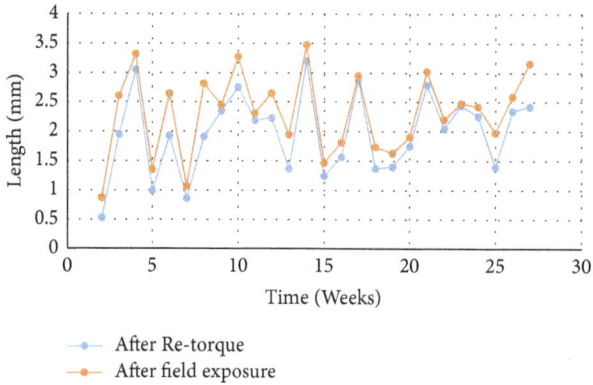

FIGURE 10: Comparison of samples' length measurements before and after exposure in one of the onshore sites showing no evidence of stress relaxation.

(a)   (b)

FIGURE 11: Samples exposed in (a) offshore site and (b).

Alloy N08825 and alloy N08904 did not show any signs of galvanic corrosion. The galvanic corrosion was observed only outside the gauge length area and did not affect the CSCC testing. Galvanic corrosion was also observed between some of the samples and the nuts. The length of each sample was measured after retrieval. No reduction in length was observed, indicating that there was no stress relaxation and that the galvanic corrosion did not reduce the applied load on the samples. A slight increase in the length was observed. It could be attributed to expansion due to the heat that the samples were exposed to in the Middle East environment. The length measurement of samples that were exposed for two years in one of the onshore sites is provided in Figure 10. No cracking was observed in any of the nuts.

The test setup was used for over five years and proved to work efficiently in the Arabian Gulf environment, where the sample racks and proof rings survived the environmental conditions as can be seen in Figure 11. When comparing the cost of the test setup with the cost of buying individual proof ring devices, it is unmistakable that this test setup is cost effective.

4. Conclusions

A field exposure test setup was built for CSCC. The following conclusions were made:

(i) Sustained-load devices (Proof Rings) were efficient in applying loads of 80% and 100% yield strength on the samples.

(ii) The use of subsize specimens resulted in low stress levels on the proof rings, which in turn increased their durability and reduced their deformation during the testing.

(iii) Standard size specimens proved to produce a high stress level around the hole in the proof rings.

(iv) Rings made from UNS 32205 duplex stainless steel – 3" Schedule 80 pipe are subject to minor stresses compared to rings made from UNS 32205 duplex stainless steel – 3" Schedule 40 pipe, when the same load and specimen size are applied.

(v) The sample racks were found to withstand the environmental conditions in the Arabian Gulf. Their design allows for the removal and installation of samples without disturbing the loads on the tested specimens.

(vi) A lab built device that includes a digital indicator was successful in measuring the samples length and elongation and was not operator dependent and had a high degree of repeatability.

(vii) The field test set up proved to work well in the Arabian Gulf onshore and offshore sites and proved to be cost effective.

Conflicts of Interest

The author declares that they have no conflicts of interest.

Acknowledgments

The author would like to thank Mr. Jim Fox, Ms. Vesna Covic-Palikuca, and Mr. Luay Hussein from the College of North Atlantic-Qatar for their contribution in the design and stress analysis. Thanks also are due to Dr. Roy Johnsen of the Norwegian University of Science and Technology, for his contribution to this research study, and to Qatar Petroleum, who made use of the field test set up and supported this study.

References

[1] D. A. Jones, *Principles and Prevention of Corrosion*, Prentice-Hall, NJ, USA, 1996, pp. 236-238.

[2] M. G. Fontana, *Corrosion Engineering*, McGraw-Hill, Singapore, 1987, pp. 109-112.

[3] S. A. Bradford, *Corrosion Control*, CASTI Publishing Inc., Bradford, England, 2004, pp. 171-174.

[4] Keneedy Space Center Corrosion Technology Laboratory, "Stress Corrosion Cracking," http://corrosion.ksc.nasa.gov/stresscor.htm, 2014.

[5] J. W. Oldfield and B. Todd, "Room temperature stress corrosion cracking of stainless steels in indoor swimming pool atmospheres," *British Corrosion Journal*, vol. 26, no. 3, pp. 173–182, 1991.

[6] J. W. Fielder, B. V. Lee, D. Dulieu, and J. Wilkinson, "The Corrosion of Stainless Steels in Swimming Pools," in *Applications of Stainless Steel '92*, vol. 2, pp. 762–772, Stockholm, Sweden, 9–11 June, 1992.

[7] R. B. Griffin, "ASM Handbook: Corrosion in Marine Atmospheres," 13C, pp. 42-57, 2005.

[8] M. E. R. Gustafsson and L. G. Franzén, "Dry deposition and concentration of marine aerosols in a coastal area, SW Sweden," *Atmospheric Environment*, vol. 30, no. 6, pp. 977–989, 1996.

[9] G. R. Meira, M. C. Andrade, I. J. Padaratz, M. C. Alonso, and J. C. Borba Jr., "Measurements and modelling of marine salt transportation and deposition in a tropical region in Brazil," *Atmospheric Environment*, vol. 40, no. 29, pp. 5596–5607, 2006.

[10] S. Huizinga, J. G. De Jong, W. E. Like, B. McLoughlin, and S. J. Paterson, "Offshore 22Cr Duplex Styainless Steel Cracking-Failure and Prevention," in *Corrosion 2005*, NACE International, Houston, TX, USA, 2005.

[11] U. Steinsmo, T. Rogne, and J. Drugli, "Aspects of testing and selecting stainless steels for seawater applications," *Corrosion*, vol. 53, pp. 955–964, 1997.

[12] H. Andersen, P. Arnvig, W. Wasielewska, L. Wegrelius, and C. Wolfe, "SCC of Stainless Steel under Evaporative Conditions," *Corrosion*, vol. 251, pp. 251/1–251/17, 1998.

[13] G. Hinds and A. Turnbull, "Threshold temperature for stress corrosion cracking of duplex stainless steel under evaporative seawater conditions," *Corrosion*, vol. 64, no. 2, pp. 101–106, 2008.

[14] A. Turnbull, S. Zhou, P. Nicholson, and G. Hinds, "Chemistry of concentrated salts formed by evaporation of seawater on duplex stainless steel," *Corrosion*, vol. 64, no. 4, pp. 325–333, 2008.

[15] A. Turnbull, P. Nicholson, and S. Zhou, "Chemistry of concentrated salts formed by evaporation of formation water and the impact on stress corrosion cracking of duplex stainless steel," *Corrosion*, vol. 63, no. 6, pp. 555–560, 2007.

[16] A. Almubarak, M. Belkharchouche, and A. Hussain, "Stress corrosion cracking of sensitized austenitic stainless steels in Kuwait petroleum refineries," *Anti-Corrosion Methods and Materials*, vol. 57, no. 2, pp. 58–64, 2010.

[17] L. Caseres and T. S. Mintz, "Atmospheric Stress Corrosion Cracking Susceptibility of Welded and Unwelded 304, 304L, and 316L Austenitic Stainless Steels Commonly Used for Dry Cask Storage Containers Exposed to Marine Environments," U.S.NRC Report, 2010.

[18] C. Ornek, X. Zhong, and D. L. Engelberg, "Low-Temperature environmentally assisted cracking of grade 2205 duplex stainless steel beneath a MgCl2:FeCl3 salt droplet," *Corrosion*, vol. 72, no. 3, pp. 384–399, 2016.

[19] L. Miller, T. S. Mintz, X. He et al., "Effect of Stress Level on the Stress Corrosion Cracking Initiation of Type 304L Stainless Steel Exposed to Simulated Sea Salt," https://www.nrc.gov/docs/ML1322/ML13220A332.pdf, 2013.

[20] B. M. Gordon, "Outside Diameter Stress Corrosion Cracking of stainless Steel in Light Water Reactors," in *Corrosion 2013*, NACE Int., Houston, TX, USA, 2013.

[21] T. S. Mintz, X. He, L. Miller et al., "Coastal Salt Effects on the Stress Corrosion Cracking of Type 304 Stainless Steel," in *Corrosion 2013*, NACE Int., Houston, Texas, 2013.

[22] Y. Toshima, Y. Ikeno, Y. Fujiwara, and Y. Nakao, "Long-Term Exposure Test for External Stress Corrosion Cracking on Austenitic Stainless Steels in Coastal Areas," in *Corrosion 2000*, NACE Int., Houston, TX, USA, 2000.

[23] P. A. Barker and H. H. Bech, "Material Selection for Threaded Instrument Fittings in Topside Offshore Service in the Arabian Gulf," in *Proceedings of the SPE International Production and Operations Conference & Exhibition*, Doha, Qatar, 2013.

[24] NACE, *Standard Test Method TM0177-2005: Laboratory Testing of Metals for Resistance to Sulfide Stress Cracking and Stress Corrosion Cracking in H_2S Environments*, NACE Int., Houston, TX, USA, 2005.

[25] ASTM, *G30-97: Standard Practice for Making and Using U-Bend Stress-Corrosion Test Specimens*, ASTM Int., West Conshohocken, 2009, pp. 1-7.

[26] ASTM, *G38-01: Standard Practice for Making and Using C-Ring Stress-Corrosion Test Specimens*, ASTM Int., West Conshohocken, 2013, pp. 1-8.

16

Seismic Behavior of Corroded RC Bridges: Review and Research Gaps

Kaveh Andisheh, Allan Scott, and Alessandro Palermo

Department of Civil Engineering, University of Canterbury, Private Bag 4800, Christchurch 8140, New Zealand

Correspondence should be addressed to Kaveh Andisheh; kavehandisheh@gmail.com

Academic Editor: Flavio Deflorian

Chloride-induced corrosion and its effect on structural and seismic performance of reinforced concrete (RC) structures have been the topic of several research projects in past decades. This literature review summarizes the state of the art by presenting a brief description of chloride-induced corrosion, its main characteristics and influencing factors, a summary of experimental published data, and existing corrosion-induced deterioration models together with numerical and experimental methods used to evaluate corroded RC bridge pier. This literature review highlights the need for reliable deterioration models for RC structures and appropriate analysis methods are needed for design of new structures or assessment of existing civil engineering structures especially in seismic areas.

1. Introduction

In recent years, growing attention has been given to the effects of corrosion on the structural performance of reinforced concrete (RC) structures. According to National Association of Corrosion Engineers (NACE), the direct annual cost of corrosion of infrastructure was more than $22 billion in the USA in 2002. American Society of Civil Engineers (ASCE) has reported that the USA should invest $2.2 trillion over the next five years to repair and upgrade more than 300,000 bridges in the USA that are approaching the end of their design life [1]. While RC structures in pristine condition can be expected to satisfy the code requirements of a given era, corrosion of reinforcing steel affects the seismic performance of the structure over time. Therefore, old corroded RC structures become vulnerable to probable future earthquakes. It should be noted that there are two well-known forms of corrosion: carbonation-induced and chloride-induced corrosion. Carbonation-induced corrosion is defined as a chemical reaction between atmospheric carbon dioxide and the product of cement hydration, mainly calcium hydroxide [2]. Chloride-induced corrosion is defined as an electrochemical reaction between chloride product (such as

iron(II)-chloride) and water. In this paper, chloride-induced corrosion has been studied. The vast majority of deterioration in RC structures is a result of corrosion of reinforcing steel due to ingress of chloride ions from either deicing salts or marine environment. Corrosion changes effective characteristics and mechanical properties of materials, leading to possible degraded seismic performance of corroded RC structures. This problem is very critical for bridges and more importantly for bridge piers since they dissipate earthquake energy through the formation of plastic hinges. Corrosion is a time-dependent process. Therefore, lifetime analysis is needed to evaluate seismic and structural performance of corroded structures. Long-term seismic performance of RC structures subject to corrosion includes three main parts that are shown in Figure 1: (1) chloride-induced corrosion, (2) deterioration of RC structures or elements due to corrosion, and (3) lifetime (time-dependent) seismic analysis and performance of corroded RC bridge pier.

Figure 2 shows an overview of the aforementioned three main parts. Two critical phenomena are the reduction in cross section area of reinforcing steel and the formation of corrosion by-product, leading to cracking and spalling of concrete in RC structures. Hence, corrosion-induced

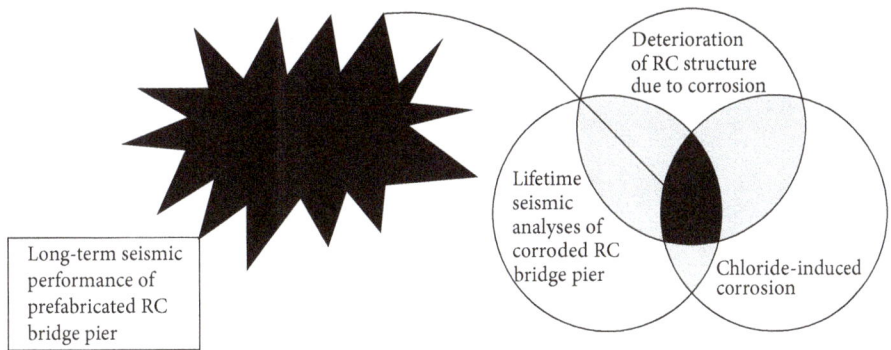

FIGURE 1: Outline of long-term seismic performance of corroded structures.

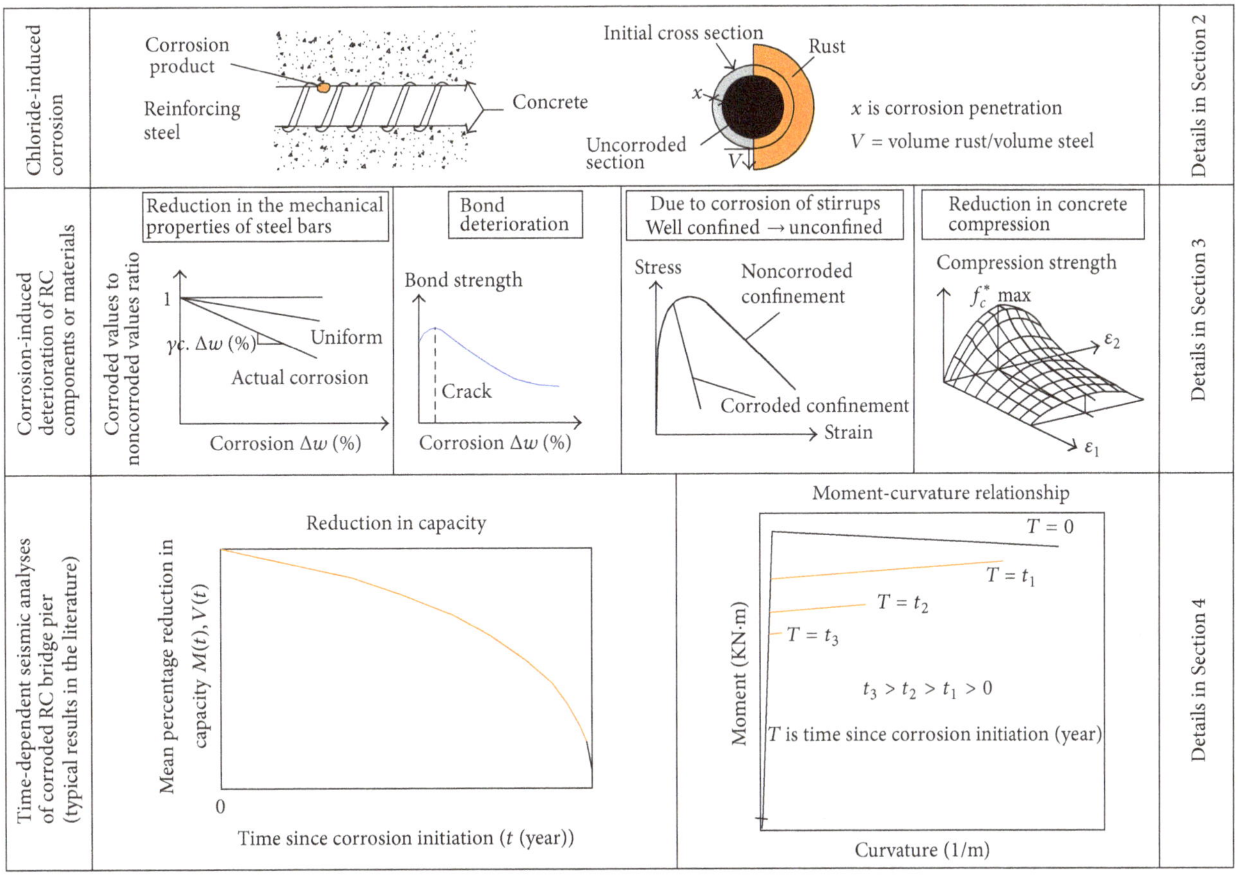

FIGURE 2: Overview of main parts of long-term seismic performance of corroded RC bridge pier.

deterioration of RC structures can be classified into four groups as follows:

(1) Reduction in mechanical properties of steel reinforcements.

(2) Deterioration of bond between steel and concrete.

(3) Degradation of confinement (decreasing shear strength).

(4) Damage to concrete material.

Traditional seismic analysis cannot be used for RC structures subjected to corrosion hazard for the following reasons. The first reason is that corrosion depends on time, so mechanical properties of structural elements are a function of time. The second reason is lack of robust analytical/numerical cyclic models to predict behavior of corroded RC structures subjected to earthquakes. Hence, lifetime analysis of corroded RC structures is needed, taking into consideration the corresponding deterioration models for corroded RC structures, amount of corrosion, and important factors influencing corrosion process such as corrosion initiation time. Figure 3 illustrated force-displacement response of a corroded bridge pier over time. The outcomes of lifetime seismic analysis of corroded RC structures can be represented in terms of

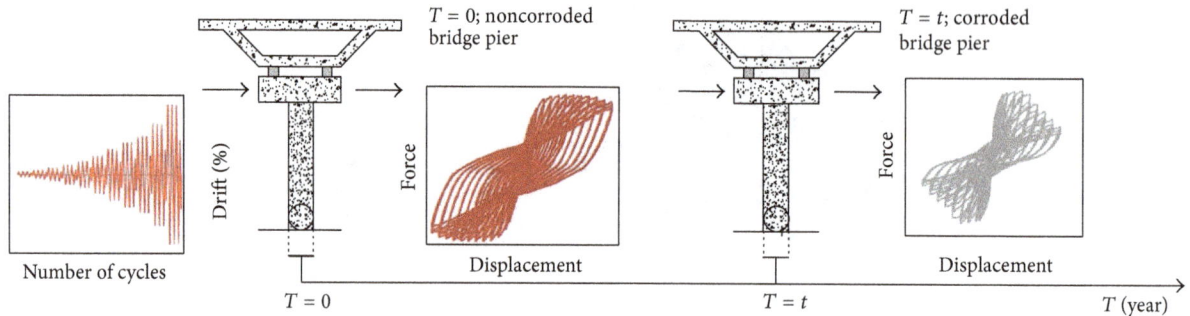

FIGURE 3: Force-displacement response of a corroded bridge pier over time.

reduction in structural capacity or increase in probability of failure over time.

This paper summarizes the state of the art by presenting a brief description of chloride-induced corrosion, its main characteristics and influencing factors, a summary of experimental published data, and existing corrosion-induced deterioration models together with numerical and experimental methods used to evaluate corroded RC bridge pier. The main objective of this paper is highlighting research gaps and critical need to further studies in this field.

2. Chloride-Induced Corrosion

Corrosion of reinforcing steel embedded in concrete is an electrochemical process. The process is initiated as soon as aggressive ions such as chloride penetrate the concrete cover and reach the steel reinforcement. Once the corrosion process commences, not only does the cross-sectional area of the corroding reinforcing steel decrease but also corrosion by-products such as rust are formed. The irregular loss of cross section leads to alterations in mechanical properties of reinforcing steels. The average volume of rust is approximately 2–4 times greater than that of the steel resulting in the development of tensile stresses in concrete, which ultimately lead to cracking and spalling of the cover concrete [3, 4]. Moreover, bond between steel and concrete decreases. It should be noted that a low level of corrosion can result in a slight increase in bond strength, but increasing corrosion level leads to reduction in bond between concrete and steel reinforcement [5–10].

2.1. Chloride Content in Reinforced Concrete Structures: Initial Stage and Threshold Value. Chloride content is the amount of chloride ion at the surface of steel reinforcement. To initiate corrosion, it should reach a certain level called critical chloride content (C_{crit}). C_{crit} is a threshold value needed to propagate chloride ion. However, there is difference between the scientific and practical definitions of C_{crit}. In scientific definition, C_{crit} is the threshold required to propagate on the surface of the steel, while in practical definition it is associated with the acceptable deterioration of reinforcing steel.

Angst et al. [11] have summarized the values of C_{crit} experimentally measured from steel embedded in cement based material in laboratory condition, from real structures

and from steel directly immersed in solution, reported by 32 published articles. The maximum and minimum values of C_{crit} based on the review of aforementioned experimental results together with maximum allowable total Cl⁻ % cement weight proposed by various ACI documents are presented in Table 1.

Moreover, Hussain et al. [12] estimated critical chloride of steel embedded in cement based material and showed that threshold of free Cl⁻, independent from C_3A content, varies from 0.22 to 0.29% cw (cement weight), while threshold of total chloride, dependent on C_3A content, varies from 0.48 to 1.2% cw for various amounts of C_3A content. The results agree with the associated range represented in Table 1. Ann and Song [13] stated that measurement accuracy of C_{crit} in terms of free Cl⁻ and Cl⁻/OH⁻ ratio is relatively low. Expressing C_{crit} in terms of total Cl⁻ (% cement weight) takes into consideration inhibiting effect of cement and the aggressive nature of chloride. Angst et al. [11] also have reported important factors influencing C_{crit} based on reviewing 24 articles. The important factors influencing C_{crit} have been categorized into three groups: steel type and condition, concrete and binder properties, and external factors. In addition to this, Alonso et al. [14] concluded that the type of steel does not significantly affect the critical chloride value, but after depassivation, the average rate of corrosion is slightly higher in ribbed bar. Glass and Buenfeld [15] showed that chloride binding reduced free chloride due to the removal of chloride ions from the pore solution of concrete. It also reduced total chloride content at depth. Maruya et al. [16] concluded that, because of condensation and ion absorption due to pore wall in wetting and drying cycles, increasing the cycle raises total chloride in RC structures. Polder [17] stated that, from a theoretical point of view, the effect of concrete resistivity on critical chloride value still remains unclear. Based on information above, the table presented by Angst et al. [11] has been updated. The update also includes additional new factors marked as "∗". Finally, Table 2 presents important factors influencing critical chloride content in terms of total Cl⁻, Cl⁻/OH⁻ ratio, and free Cl⁻.

It is worth noting that Angst et al. [18] have developed a probabilistic model to investigate the effect of specimen size on C_{crit} measured in laboratories. They have concluded that increasing sample's geometrical dimension decreases C_{crit}, but to apply this result to steel embedded in concrete it has to be verified through experimental studies.

TABLE 1: Maximum and minimum critical chloride content based on 32 experimental works reviewed by [11] compared with C_{crit} proposed by ACI codes.

| | C_{crit} from steel embedded in cement based material | | | C_{crit} real structures | C_{crit} from the steel directly immersed in solution | |
| | | | | Maximum-minimum | | |
	Total Cl⁻ % binder weight	Cl⁻/OH⁻ ratio	Free Cl⁻ (mol/L); 0.07–1.16 % bw	Total Cl⁻ (% bw)	Cl⁻/OH⁻ ratio	Free Cl⁻ (mol/L)
Maximum-minimum	0.04–8.34	0.09–45	0.045–4; 0.07–1.16	0.1–1.96	0.01–4.9	0.0056–0.42

C_{crit} expressed as total chloride content relative to cement weight proposed by various ACI documents for prestressed RC structures in service

	ACI 357	ACI 222	ACI 201
ACI documents	0.06	0.08	Not stated
C_{crit} expressed as Cl⁻ % binder weight	0.1	0.2	0.1

TABLE 2: Important factors and their effects on critical chloride content.

Factor	Effect on critical chloride content		
	Total Cl⁻ % cement weight	Cl⁻/OH⁻ ratio	Free Cl⁻
Steel type and condition			
Defect at steel-concrete interface	↓	↓	↓
Polishing and sandblasting	↑	↑	↑
Steel potential (>−200 mV SCE)	O	O	O
Steel potential (<−200 mV SCE)	↓	↓	↓
*Steel type**	O	O	O
Concrete and binder properties			
w/c ratio	↓	↓	↓
*Chloride binding**	↓↑	O	↓ O
pH	↑	↑	↑
*Electrical resistivity**	↑ NC	↑	↑
Silica fume	↓	↓	↓
Fly ash	↓↑ O[b]	↓ O[b]	↓ O[b]
GGBS (ground granulated blast furnace slag)	↓↑ O[b]	O	O
SRPC (low C_3A + C_4AF content)	↓	NS	NS
External factors			
Moisture in rather dry concrete	↓	↓	↓
Moisture in nearly saturated concrete	↑	↑	↑
Moisture variation	↓	↓	↓
Oxygen availability	↓	↓	↓
Temperature	↓	↓	↓
*Condensation in wet and drying cycle**	↑	NS	NS

↓↑ indicates decreasing and increasing critical value with an increase of concerning factor; NS: result not stated; NC: the effect of the concerning factor is unclear; O: the concerning factor has no effect; b: contradicting results reported in the literature. SRPC: sulphate resistant Portland cement.
∗ indicates additional new factors if compared with the table presented by Angst et al. [11].

A number of methods for determining the total chloride content and the free chloride content are applied in practical applications. For measuring the total chloride content, drilled cores from hardened concrete are analyzed. The total chloride content in concrete powder-nitric acid solution can be measured by a number of methods such as titration, use of ion selective electrodes, or spectrophotometric methods. However, a more expensive but very accurate method is to determine the total chloride content in concrete powder using X-ray fluorescence spectrometry (XRF). To determine the free chloride content, pore solution expression, leaching techniques, and ion selective electrodes are used in the literature [11]. Further information can be found in [11] and its corresponding references.

2.2. Pitting Corrosion: Limit Step to Start Pitting Corrosion.
It has been shown that the formation of macrocell, that is, small anodic area in comparison with large cathodic area, is observed in pitting (localized) corrosion [19–23]. It has been confirmed that three transitions occur in pitting corrosion: the first is transition from initiation stage to propagation stage, called depassivation [24]; the second is transition from depassivation to repassivation (the repassivation phase called metastable); and the third is transition from metastable to pit growth. As mentioned before, the first transition is related to existing critical chloride content. In metastable, nucleation

occurs, also called repassivation, depending on chemistry or metallurgy condition. Then in case of maintaining aggressive chemical composition in pit cavity, the transition from nucleation to stable pit growth occurs. This transition is due to the simultaneous ingress of H^+ and Cl^- and other anions into pit cavity [25–27]. Bertolini et al. [28] stated that pitting corrosion for reinforcement of steel in concrete is due to the acidification of pit cavity and ingress of Cl^- into the pit. Broomfield [29] found that steel reinforcement corrosion starts with the formation of a number of pits. Increasing the number of pits causes them to join up and form a general corrosion. Angst et al. [30] concluded that a transition from anodic to cathodic control occurs in pitting corrosion. However, it is not clear in which chloride content this transition takes place, so further investigation in this area should be carried out. It is clear that pitting corrosion occurs due to existence of high amount of chloride ion in a certain location. This means that corrosion potential is greater than pitting potential in that location [11]. Since pitting corrosion causes significant cross section loss in reinforcing steels, in structural analysis the amount of cross section loss due to pitting corrosion is a very important parameter. Therefore, researchers estimated a factor called "pitting factor" that is used to calculate pit depth and related loss of steel cross section. Pitting factor is the ratio of maximum pit depth on average corrosion penetration. Pitting factors reported

TABLE 3: Pitting factors obtained from literature.

Authors and date	$D_{(mm)}$	$L_{(mm)}$	Time (day)	Number of samples	i_{corr} (μA/cm^2)	Pitting factor: (R) Mean; COV
González et al. [73]	8	500	6 years	NS	0.1–7.0	4.4–5.9; NS
	16	400	30		10–100	5.9–16.1; NS
Rodriguez et al. [74]	6	650	100–200	18	100	4–11.7; 0.05–0.22
Rodriguez et al. [75]	12	2300	100–200	16	100	4.0; 0.15
Torres-Acosta and Martínez-Madrid [31]	13	310	700	35	NS	5.5; 0.59
	8					7.00; 0.18
	16					7.68; 0.16
	24	800	50 years		1	8.08; 0.16
Stewart [76]	28					8.23; 0.15
	32					8.36; 0.15
	36					8.48; 0.15
	10	100				1.65; 0.22
Cairns et al. [77]	16	NS	100–400	25	10–50	23.8; 0.56
Torres-Acosta et al. [78]	10	1500	40, 80, 200	8	80	11.7; NS
Stewart and Al-Harthy [79]	16	100	78	32	160–185	6.20; 0.18
	27				125–150	7.10; 0.17

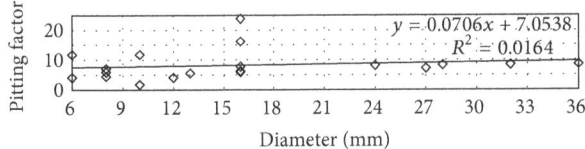

FIGURE 4: Pitting factor for different diameter sizes of bars regresses from results represented in Table 5.

in the literature have been collected from 8 experimental investigations and are summarized in Table 3.

Figure 4 shows the relationship between diameter size of reinforcing steel and pitting factor regressed from the results represented in Table 5. According to this figure, pitting factor increases with the growth of diameter sizes of reinforcing steels.

Average corrosion penetration can be calculated based on mass loss due to corrosion and estimation of associated equivalent diameter of corroded bar [31]. It has been stated that pitting factor rises with increase of reinforcing steel bars diameter [32].

Using Faraday's law, assuming hemispherical form for pits, the maximum pit depth is as follows [33]:

$$p(t) = 0.0116 \times R \times i_{corr}(l) \times t, \tag{1}$$

where $p(t)$ is the maximum pit depth (mm), R is pitting factor, $i_{corr}(l)$ is corrosion current density (μA/cm^2), and t is time (year).

2.3. Predicting the Rate of Corrosion.
Rate of corrosion and time of the commencement of corrosion are very important factors influencing the deterioration of RC structures as they are related to residual capacity of corroded structures. Many

factors affect the corrosion rate which are classified into three groups: named steel condition, concrete and binder properties, and external factors. Based on the empirical and mathematical models developed by past studies, Table 4 shows important factors affecting corrosion rate and initiation time of corrosion [34–52]. Among all factors, one may notice that increasing total chloride raises corrosion rate. This means that all factors affecting total chloride (see Table 2) influence corrosion rate. Increasing saturation degree of pore in empirical model causes both reduction and rising of corrosion rate. On the other hand, the mathematical model showed that increasing the degree of saturation pore from 30% to 50% causes increasing corrosion rate, while further increase from 60% to 100% causes reduction in corrosion rate [36].

While measuring corrosion accurately is difficult, there are some simple methods based on corrosion potential and corrosion rate that can be used by researchers and practical engineers to estimate active corrosion in RC structures. For example, according to ASTM C-876-91, if corrosion potential, V, is less than -0.35, probability of active corrosion is more than 95% [53]. Elsener et al. [54] have stated that corrosion potential ranging from -0.4 to -0.6 means that steel is corroding. Liang et al. [55] have reported that using corrosion current density to corrosion duration ratio the grade of corrosion can be evaluated according to Table 5.

2.4. Corrosion By-Products and Corrosion-Induced Cracking.
As discussed earlier, when corrosion initiates, corrosion by-products are formed. The volume of corrosion by-products is greater than that of steel. Therefore, volumetric expansion causes tensile stress leading to propagation of cracks into concrete cover. Zhao et al. [56] have suggested the expansion coefficient of 2.64, 2.85, and 3.02 for samples corroded in

TABLE 4: Important factors and their effects on corrosion rate and start time of corrosion.

Factor	Empirical corrosion rate (r_{corr}) model	Mathematical corrosion rate	
		Start time	r_{corr}
Steel condition			
Temperature at steel level	↑		
Galvanic coupling	↑		
Corroding area to exposure area			↓
Anodic and cathodic resistivity			↑
Concrete and binder properties			
w/c ratio	↑		↑
Total Cl⁻%	↑		
pH	↓		
Electrical resistivity	↓		↓
Permeability	↑		
Cracks		↓	NC
Size distribution of aggregates		↓	
Initiation crack width		↓	
Degree saturation of pore	↓↑		(30%–50) ↑; (60%–100%) ↓
External factors			
Slag concentration	↓		
Chloride conductivity			
RH	↑		
Time	↓		
Oxygen availability	↑		
Aging of oxides in dry concrete	↓		
Aging of oxides in wet concrete	O		
Cover	↓		↓

↓↑ indicates decreasing and increasing corrosion rate value with an increase of concerning factor; NC: the effect of the concerning factor is unclear; NS: not stated; O: the concerning factor has no effect.

TABLE 5: Simple method to evaluate grade of corrosion.

Corrosion grade	Noncorroded	Slight corrosion	Moderate corrosion	Severe corrosion
Corrosion current density [$\mu A/cm^2$]/corrosion duration [year]	<0.0066	0.0066–0.05	0.05–2.5	>2.5
Corrosion duration [year]	>15	10 to 15	2 to 10	<2

NaCl solution, near or on the coast and in splash zone, respectively. Table 6 presents the volume expansion of different components of corrosion by-product found by past studies.

Predicting the expansion volume of rusts is very important to improve the knowledge of service life of reinforced concrete structures. The variation in expansion coefficient reported by past studies clearly indicates that the need for further studies on rust compositions is demanded.

With respect to mechanical properties and characteristics of corrosion by-products, Caré et al. [57] have stated that Young's modulus of rust layers depends on the diameter of uncorroded steel bar and thickness of the rust layers. Zhao et al. [56] have shown that environmental parameters such as amount of humidity and oxygen availability vary the volume expansion coefficient. Increasing the amount of humidity and oxygen raises the expansion coefficient.

Past studies investigated corrosion-induced cracking and shared the influencing factors and measured crack width based on experimental data. A summary of 18 reviewed works has been presented in Table 7 showing crack width measured in addition to some details such as amount, current density, and/or type of corrosion and time of exposure. Table 7 can give an overall view on corrosion-induced cracking. It is worth noting that the maximum crack width reported in the literature so far has been 6 mm [31]. Few models have been developed for predicting width of crack in the literature. Andrade et al. [58], for example, presented a simple formula to predict average width of crack in elements exposed to natural corrosion [59]:

$$w = k \left[\frac{P_x}{C/\phi} \right], \qquad (2)$$

TABLE 6: Volume expansion of corrosion by-product components.

References	Volume expansion coefficient of different components of corrosion by-product										
	FeO	Fe_3O_4	Fe_2O_3	$\alpha FeOH$	$\gamma FeOH$	$Fe_2O_3 \cdot H_2O$	$\beta FeOH$	$Fe(OH)_2$	$Fe(OH)_3$	$Fe(OH)_3 \cdot 3H_2O$	$Fe_2O_3 \cdot 3H_2O$
[62]	1.82	2.08	2.17	—	—	—	—	3.76	4.24	6.46	—
[80]	1.74	2.1	2.11, 2.26	2.9	3.12	—	3.5	3.7	4		6.24
[65]	1.8	2	2.2	—	—	—	—	3.75	4.2	6.4	—
[57]	—	2.08	2.12	2.91	3.03	—	3.48	—	—	—	—
[81]	1.77	2.1	2.14	2.92	3.06	3.12	—	3.71	4.82	—	6.5
[56]	—	2.1	—	2.95	3.07	—	3.53	—	—	—	—

TABLE 7: Relationship between crack widths with corrosion obtained from literature.

References	Type of specimen	Time (day)	i_{corr} ($\mu A/cm^2$)	A_{corr} (%)	Crack width (mm)
[82]	Beam	NS	2000	<4.5	1.3
[83]	Beam	3–15	0.5	2.5–12	0.1–0.75
	Bond-pull		500	NS	NS
[58]	Prism	1–100	10, 100	0.5–2.5	0.05–0.5
[84]	Beam	28		0.8–9.2	NS
	Bond-pull	1–28		3.6–19.2	0.06–0.46
[74]	Column	106–204	100	9.1–17.8	0.8–4.0
[75]	Beam	100–200	100	10.1–26.3	0.2–0.6
[85]	Slabs	1–2.5	3	1–75	NS
[86]	Beam	126 hours	5×10^8	<1	NS
[43]	Prism	68–221	10, 100	NS	0.06–1
[87]	Prism	NS	5 V current	0, 12	0.35, 0.8
[88]	Beam	15–18	1–4	2.5–10	NS
[89]	Beam	16–64 h	3	1.25–5	NS
[31]	Prism	700		Up to 51	0–6
[90]	Prism	815, 766, 380, 306	100, 200, 350, 500	4.38, 7.3, 6.5, 7.26	0.25–1
[91]	Beam	17 years	Saline Environment	26	1.6
	Beam	14 years		12	1.8
[92]	Slab	2–9 months	100	NS	0.05–1.5
[93]	T-beam	2–15 years	Average: 0.128	NS	0.31–3.94
	Column	2–15 years		NS	0.2–0.51
[94]	Wall spec.	20–56 weeks	$R_p = \pm 20$ mV	Up to 6.5%	0.01–1

where w is the crack width (mm), k is a nondimensional factor, C/ϕ is concrete cover/diameter of the bar ratio, and P_x is penetration of the corrosion in time t (year) and equal to

$$P_x = 0.0115 r_{corr} t, \qquad (3)$$

where r_{corr} (mm/year) is the corrosion rate.

Andrade et al. [59] validated their formula with 15-year-old RC specimens. They proposed $k = 9.5$ for their formula. Du et al. [60] showed that both decrease of w/c (water to cement) ratio and increase of cover are important factors to resist cracking due to ingress of chloride. It is clear that rate of corrosion is always an important factor affecting crack-induced corrosion.

The relationship between corrosion current density, i_{corr} (mA/m^2), and corrosion rate, r_{corr} ($\mu m/year$), can be expressed in the following equation [61]:

$$r_{corr} = \frac{0.327 \left(M i_{corr} \right)}{n\rho}, \qquad (4)$$

where M (g/mol) is the atomic weight, n is the ion valence, and ρ is the density (g/cm^3).

Predicting time of corrosion cracking is another important factor in corrosion-induced cracking topic and is used for predicting the service life of corroded RC structures. Predicting the time of corrosion cracking has been addressed by a number of researchers including [55]. In this regard, a few mathematical models have been developed by [62–67]. Figure 5 shows the relationship between percentage of cross

TABLE 8: Corrosion-induced reduction factors of mechanical properties of steel reinforcing.

References	Sample (D); test	Corrosion method and condition	A_{corr}%	α_y	α_u	α_E	$\alpha_{\varepsilon u}$
[95]	Ribbed (8–32); t Plain (8–32); t	Open air (environment)	0–0.5	0.0	0.0		
[96]	Ribbed (12); t	Acc. 0.5–2.0 mA/cm^2	0–11	0.45	0.33		
[97]	Bare; t	Open air (environment); Arabian coast	0–1	0.0	0.0		
[98]	Ribbed (8); tensile Plain (8); tensile	Acc. 0.5 mA/cm^2	0–28	0.16 0.28	0.44 0.68		
[99]	RC-ribbed (10–25); t RC-plain (8–14); t	Open air (environment); carbonation	0–67	0.04	0.05		
[100]	RC	Open air (environment); chloride	0–25	0.6	0.63		
[101]	Ribbed (10); tensile	Acc. 13.0 mA/cm^2	0–25	0.21	NS		
[102]	RC beam (6, 12); B	Chlorides; saline environment	0–20*	NS	NS		3.5
[103]	Rib-RC (8, 16, 32); t Plain-RC (8, 16); t	Acc. 0.5–2.0 mA/cm^2; 3.5% NaCl; tensile	0–25 0–25	0.12 0.49	0.15 0.65		
[104]	6 mm 12 mm	Acc. 2.0 mA/cm^2	0–75 0–80		1.98** 0.74**	NS NS	4.6 NS
[105]	NS	Open air (Environment); chloride	0–>30	0.00	0.00		
[60]	Ribbed (8, 16, 32) and RC; tensile	Acc. 0.5–2.0 mA/cm^2; 3.5% NaCl; yield force	0–25	1.14–1.28*			
		Acc. 0.5–2.0 mA/cm^2; 3.5% NaCl; ultimate force	0–25	1.22–1.39**			
		Acc. 0.5–2.0 mA/cm^2; 3.5% NaCl; yield strength	0–25	0.16–0.36			
		Acc. 0.5–2.0 mA/cm^2; 3.5% NaCl; ultimate strength	0–25	0.26–0.48			
	Plain (8,16) and RC; tensile	Acc. 0.5–2.0 mA/cm^2; 3.5% NaCl; yield force	0–25	1.60, 1.48*			
		Acc. 0.5–2.0 mA/cm^2; 3.5% NaCl; ultimate force	0–25	1.71, 1.26**			
		Acc. 0.5–2.0 mA/cm^2; 3.5% NaCl; yield strength	0–25	0.79, 0.58			
		Acc. 0.5–2.0 mA/cm^2; 3.5% NaCl; ultimate strength	0–25	0.94, 0.44			
[77]	RC cube; tensile RC cylinders; t	Acc. 0.01–0.05 mA/cm^2; cyclic wet-dry Acc. (electric); 5% NaCl	0–3	1.2	1.1	NS	3
[69]	S400, 10 mm	Salt spray corrosion	1.5–8.5	1.47	1.31	NS	6.97
[68]	RC ribbed (10, 13); t RC ribbed (10, 13); t	Accelerated uniform (electric); 3% NaCl Acc. pitting (chloride induced); cyclic wet-dry	0–35 0–35	1.24 1.98	1.07 1.57	1.75 1.15	1.95 2.95
[106]	RC beam; bending	Open air (environment); 1–12 years	0–50	NS	1.97*	NS	NS
	RC beam; bending	Accelerated (electric); 3% NaCl	0–50	NS	1.59*	NS	NS
	Ribbed; tensile	Open air (environment); 1–12 years	0–50	NS	1.41	NS	NS
	Ribbed; tensile	Accelerated (electric); 3% NaCl	0–50	NS	1.34	NS	NS
[71]	BS B500B; 10 mm; t	NS	9.5–19.6	1.21	1.38	NS	1.91
[72]	RC plain (6.5 mm); t RC ribbed (12); t	Naturally carbonation-induced corroded Acc. 0.1 mA/cm^2; 5% NaCl	14–38 4–28	1.12 1.1	1.36 1.22	NS NS	NS NS

$*$ indicates factor associated with yield force.

$**$ indicates reduction factor associated with ultimate force.

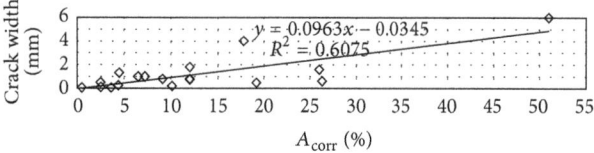

FIGURE 5: Crack width versus percentage loss of cross section regresses from results represented in Table 8.

The reason for rising crack width with corrosion percentage is that more corrosion by-products in higher level of corrosion lead to increase of crack width.

2.5. Research Gaps. The main research gaps in chloride-induced corrosion can be summarized as follows:

(i) More accurate value of critical content of chloride concentration for real RC structures.

(ii) Factors and their effects on the critical content of chloride.

(iii) Robust pitting factors for real corroded structures.

(iv) Factors and their effects on corrosion rate and time of initiation.

section loss in steel bars and crack width (mm) based on the results represented in Table 8. Figure 5 shows increasing percentage loss of cross section due to corrosion raising crack width.

(v) Robust corrosion rate and time of initiation prediction.

(vi) More accurate values volume expansion for corrosion by-product and rust components, especially for real corroded RC structures.

Research studies aiming to fill the above research gaps will lead to decrease of uncertainties in estimation of chloride-induced corrosion.

3. Corrosion-Induced Deterioration of RC Structures

As discussed earlier, the two main outcomes of corrosion are decreasing cross section area of steel reinforcement and volumetric expansion caused by corrosion by-products. As a result, mechanical properties of steel reinforcement such as modules of elasticity, force, stress and strain at yield, and ultimate points alter with corrosion. Regarding cyclic behavior of steel reinforcement, in particular, corrosion changes energy dissipating characteristic and number of cycles needed for failure. Bond between concrete and steel varies in corroded reinforced concrete members. The stress-strain model of confined concrete in compression region is affected by corrosion and maximum compression stress of concrete decreases because of cracks propagated into concrete cover due to corrosion. Up to date, there is no experimental study showing the effects of corrosion on stress-strain relationship on RC columns. It is worth noting that the similar experimental study is in progress by authors. Therefore, materials characteristics of corroded reinforced concrete members have to be applied for analyses and simulations of corroded structures.

3.1. Effect of Corrosion on Mechanical Properties of Steel Reinforcing. Irregular decreases in cross-sectional area of steel reinforcing cause changes in mechanical properties of reinforcements. A number of monotonic tensile tests on bare bars and RC elements and bending tests on RC beams and slabs have been carried out to estimate the reduction factors corresponding to the mechanical properties. Reduction factors indicate the percentage of reductions in mechanical properties that will happen for 1% reduction in cross section, and they have been estimated from experimental results and reported by past studies. In this paper, a survey on 18 experimental works has been done and the results and references have been presented in Table 8. The following investigated mechanical properties were included: yield and ultimate (stress or force) strength, elongation, and module of elasticity. Equations (5)–(10) are typical models regressed from experimental data used by past studies that can be used to calculate mechanical properties of corroded steel reinforcements:

$$\sigma_y^c = \left[100 - \alpha_y \times A_{corr}\%\right]\sigma_y, \tag{5}$$

$$\sigma_u^c = \left[100 - \alpha_u \times A_{corr}\%\right]\sigma_u, \tag{6}$$

$$E_s^c = \left[100 - \alpha_E \times A_{corr}\%\right]E_s, \tag{7}$$

$$\delta_s^c = \left[100 - \alpha_{\varepsilon u} \times A_{corr}\%\right]\delta_s, \tag{8}$$

$$F_y^c = \left[100 - \alpha_y^* \times A_{corr}\%\right]F_y, \tag{9}$$

$$F_u^c = \left[100 - \alpha_u^{**} \times A_{corr}\%\right]F_u, \tag{10}$$

where σ_y^c, σ_u^c, E_s^c, δ_s^c, F_y^c, and F_u^c are yield stress, ultimate stress, module of elasticity, elongation, yield force, and ultimate force of corroded bars, respectively, α_y, α_u, α_E, $\alpha_{\varepsilon u}$, α_y^*, and α_u^{**} are their associated reduction factors, $A_{corr}\%$ is the percentage loss of cross section, and σ_y, σ_u, E_s, δ_s, F_y, and F_u are yield stress, ultimate stress, module of elasticity, elongation, yield force, and ultimate force of noncorroded bars.

While the results presented in the literature have a wide variation, some conclusions reported by the above reviewed references are as follows:

(i) Very low corrosion may not affect the mechanical properties of the steel reinforcing.

(ii) Usually, reduction factors for environment corrosion and plain steel reinforcement are higher than accelerated corrosion and deformed steel reinforcement.

(iii) The greatest reduction factor is related to elongation. This is very important for seismic behavior of RC structures.

(iv) Usually, pitting corrosion and irregularities in corrosion increase the reduction factors. On the other hand, reduction factors for pitting corrosion are greater than those for general corrosion [68].

(v) The reduction factors for corroded bare steel reinforcement and those corroded while embedded in concrete are similar [60].

(vi) The effects of the type (plain or deformed type) and diameter of reinforcing steels on reduction factors can be neglected [60].

To illustrate the variation of the published reduction factors, the minimum and maximum reduction factors of four mechanical properties of steel reinforcement based on the data represented in Table 8 are shown in Figure 6. The mechanical properties include elongation, modulus of elasticity (*E*), yield stress, and ultimate stress. Since linear regression has been employed by all past studies to estimate reduction factors, the minimum and maximum reduction factors shown in Figure 6 are represented based on linear regression.

Figure 6 shows that corrosion deteriorates the mechanical properties of reinforcing steel. However, there are big variations in results published in the literature based on monotonic tests. The results also show that the maximum reduction factors and the greatest difference between minimum and maximum reduction factors have been reported for elongation.

A few number of studies identified cyclic behavior of corroded steel reinforcements. Apostolopoulos and Papadopoulos [69] have shown that a mass loss less than 2% and 3% causes 22% and 47% reduction to the number of maximum

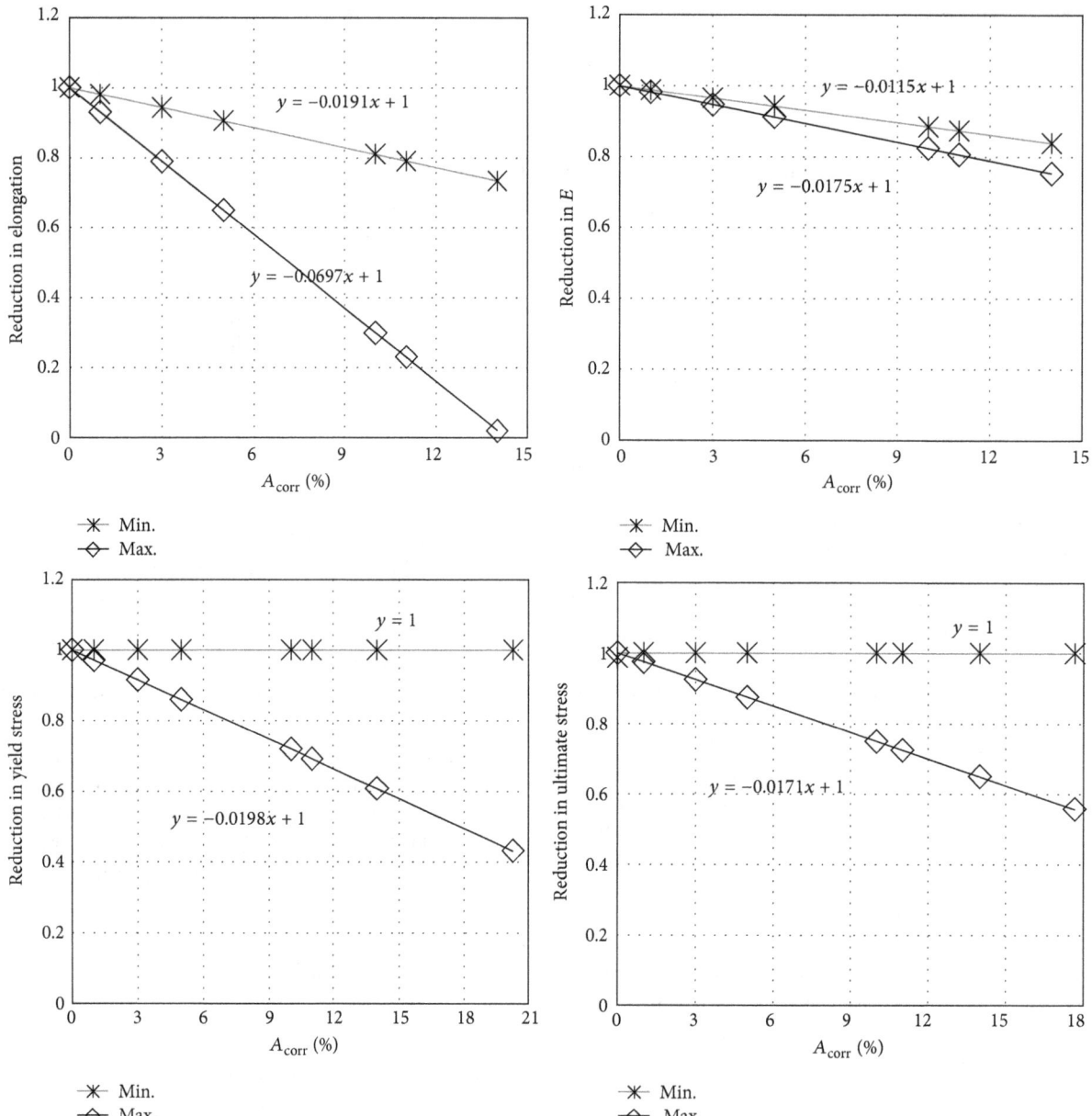

FIGURE 6: Maximum and minimum reduction factors for elongation, modulus of elasticity, yield stress, and ultimate stress.

cycles required for rupture, respectively. Apostolopoulos and Pasialis [70] have studied the low cycle fatigue behavior of smooth and ribbed steel reinforcement for different degrees of corrosion. They have reported that smooth bars showed a better cyclic behavior than that of ribbed bars for low strain amplitude and up to 8% loss of mass due to corrosion. On the other hand, smooth bars can dissipate more energy and need higher number of cycles to fail in low strain magnitude (±1%) than those of ribbed bars. These advantages disappear as strain amplitude increases. Hawileh et al. [71] studied the effect of corrosion on cyclic behavior of BS B500B bars. They have demonstrated that corrosion decreases low cycle fatigue life of the bars. They have pointed out that lower strain amplitude (±4%) causes more reduction in dissipating energy

and more cycles are needed for failure than those of higher strain amplitude (±6%). Zhang et al. [72] have found that increasing the degree of corrosion causes reduction in fatigue life of corroded bars. They also have claimed that the impact of corrosion on fatigue behavior and naturally corroded steel bar is more than that on monotonic behavior and artificially corroded steel bar, respectively.

3.2. The Effects of Corrosion on Bond Strength between Steel and Concrete. Table 9 shows the effect of the percentage of corrosion on bond strength based on 16 experimental works reviewed by the authors. As an overall trend, low corrosion percentage increases the bond strength, while high percentage of corrosion always decreases the bond strength.

TABLE 9: Effect of corrosion on bond strength between concrete and steel reinforcement.

Authors and date	Sample (D); test	Corrosion method	Time exposure	A_{corr}%	Bond strength; % of change
[82]	14 mm bending	2000 $\mu A/cm^2$	NS	0.55	↑; 40
	10 mm pullout			0.55	↑; 42
	10 mm pullout			3.5–7.4	↓; 0–66
	14 mm pullout			0.65	↑; 28
	14 mm pullout			2.6–5.7	↓; 0–65
	20 mm pullout			0.43	↑; 25
	20 mm pullout			1.6–4	↓; 0–47
[107]	12 mm, pullout	0.4 A current electric	NS	<4	↑; 17
				5–7	↓; 30–69
				8–12	↓; 70–78
				12–80	↓; 78–86
[84]	12 mm, pullout	2–64 A current electric	<28 days	0.7	↑; 18
				2.42, 12	↓; 5, 60
[108]	19 mm, pullout	Saturated in $Ca(OH)_2$	<5 weeks	NS	↑; NS
			>5 weeks	NS	↓; NS
[5]	10 mm, bending	0.1 A current electric	21–63 days	20	↓; 74
[6]	19 mm, pullout	12 A current electric	>3 days	0–5.2	↓; 78
[7]	13 mm, pullout	1 A current electric	NS	3	↓; 35
			NS	16.8	↓; 77
			NS	13*	↓; 13
			NS	24*	↓; 38
[109]	10 mm, pullout, 15 mm cover specimen	140 $\mu A/cm^2$	NS	5	↑; ≅15
				10	↓; ≅35
[110]	10 mm, bending	12 A current electric	NS	2	↑; 30
				2.8, 15	↓; 0, 77
[111]	Deformed pullout	0–2 A current electric	10–12 days	4, 9	↓; 45, 68
	Deformed pullout			3.8*, 6*	↓; 4, 12
	Smooth, pullout			3.3	↑; 21
[8]	20 mm, pullout	0–2 A current electric	NS	4	↓; 45
				3.8*	↓; 5
[112]	20 mm, pullout	500 $\mu A/cm^2$	8–48 h	0.2, 0.36	↑; 50, 20
			56–96 h	0.4, 0.76	↓; 20, 84
Berto et al. [9]	10 mm, pullout	NS	NS	4.27	↓; 12
				7.8	↓; 75
Chung et al. [113]	13 mm, pullout	12 A current electric	>3 days	<3	↑; 40
				3–7	↓; 27
Kivell [10]	Deformed, 20 mm pullout	0.1 A current electric	10–50 days	0.6*, 11*	↓; 6, 50
				20*	↓; 76
	Def. 20 mm cyclic			18.6*	↓; 59

↓↑ indicates decreasing and increasing bond strength of corroded bar in comparison with sound bar; * indicates confined reinforced concrete. It should be noted that increasing cycles also cause a reduction in bond strength.

Type of steel bars and confinement are the important factors that influence the change of bond strength due to corrosion. In spite of the above general trends, the variation is very high indicating the significance of further investigation in this area. Since chloride-induced corrosion is a function of concrete cover and stirrups always have less cover than longitudinal bars, further research should be performed to consider this problem that to the best of the authors' knowledge there is no report on in the literature. Moreover, high level of corrosion causes a critical reduction in bond under cyclic loading, while corrosion under 5% increases bond capacity [114]. It has been reported that confinement efficiently decreases bond degradation under cyclic loading [114].

The information collected in Table 9 has been graphically presented in Figure 7. Figure 7, therefore, shows bond strength of corroded to noncorroded steel reinforcement ratio over corrosion percentage based on past published

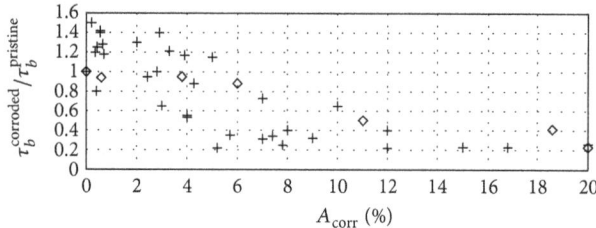

+ Unconfined
◇ Confined

FIGURE 7: The relationship between bond strength of corroded to noncorroded reinforcing steels and corrosion for confined and unconfined RC samples based on collected experimental results.

experimental studies. The data have been classified into two groups including confined and unconfined RC samples.

3.3. The Effects of Corrosion on Stress-Strain Model of Confined Concrete.

As far as confinement is concerned, corrosion of lateral steel bars alters confinement properties of reinforced concrete members. However, there is no evidence indicating how the stress-strain model of a confined concrete changes due to corrosion. Mander et al. [115] stated that "confinement is defined as sufficient lateral reinforcement in the form of the circular or rectangular arrangement of steel." They also stated that "the aim is to confine reinforced concrete members under compression to avoid the buckling of longitudinal bars, and to prevent shear failure." Confinement is a critical factor in plastic hinge region, because it ensures the ductility capacity demanded in seismic events. Transversal steel reinforcements are the closest steel bars to the surface of RC members. Therefore, they are corroded more severely than longitudinal bars. The effect of corrosion on confinement is very rare and only one report [116] was found on this subject in the literature. Ou et al. [116] analytically calculated confining strength ratio based on corroded steel reinforcing and ultimate strain of confined concrete based on reduction in mechanical properties of reinforcing steels. It is clear that further investigation is needed in this area. A study in this area is in progress by authors at the University of Canterbury.

3.4. The Effects of Corrosion on Concrete Strength of Reinforced Concrete Structures.

As discussed earlier, corrosion causes propagating cracks into concrete core, influencing compression and tensile strength of concrete material. A few studies have identified the effects of cracks on tensile and compression strength of concrete materials. Vecchio and Collins [117], for example, presented the following equation that addresses the effect of cracks on the compressive strength of concrete:

$$f_c^* = \frac{f_c}{0.8 - 0.34\varepsilon_1/\varepsilon_c}, \tag{11}$$

where f_c^* is compression stress of cracked concrete, f_c is maximum compression stress of noncracked concrete, and $\varepsilon_1/\varepsilon_c$ is the ratio of principle tensile strain to maximum strain corresponding to maximum compression stress (f_c). It is clear that $\varepsilon_1/\varepsilon_c$ is negative.

The above equation has been improved by the following research studies [118, 119].

Further investigation is critically needed to identify the effects of corrosion on concrete cracks and consequent concrete strength.

3.5. Time-Dependent Deterioration Models for Reinforced Concrete Members.

Corrosion and consequent degradation are time-dependent. Therefore, deterioration models are time variant mathematical equations showing relationships between the deteriorated mechanical properties and time. For example, with replacement of A_{corr} with an equation showing relationship between A_{corr} and time, (4) to (9) will be time-dependent deterioration models for mechanical properties of reinforcing steels.

There are two different types of deterioration models for RC structures called macro model and micro model. The macro deterioration model has been developed based on growing micro cracks. Growing micro cracks lead to macro cracks and also to accelerated ingress of aggressive ions [120, 121].

The micro deterioration model frequently used in the literature has been developed based on three models including transport model of aggressive ions, electrochemical model of corrosion, and structural model. The structural model can be developed corresponding to decrease of dimension, reduction in strength, or increase of cracks [121]. Fick's second law of diffusion is used for ion transport model [24]. There are a few studies that have been developed on the degradation of reinforced concrete structures due to corrosion [3, 122–124].

3.6. Main Research Gaps.

The main research gaps in corrosion-induced deterioration of RC structures can be summarized as follows:

(i) Robust deterioration model for corroded steel reinforcement considering cyclic behavior of steel reinforcement in seismic events.

(ii) Robust deterioration model to predict bond between steel and concrete in corroded RC structures.

(iii) Corrosion-induced stress-strain model of confined concrete.

(iv) Corrosion influencing compression strength of concrete.

4. Evaluation of Seismic Performance of Corroded Reinforced Concrete Bridge Piers

According to what has been discussed so far, corrosion degrades mechanical and structural characteristics leading to negative impact on seismic performance of RC structures. Bridge piers are earthquake-resistant elements of bridges.

Therefore, assessing seismic performance of bridge piers exposed to corrosion is very important. Numerical simulation of corroded bridge piers is very complicated, and a number of uncertainties have limited utilizing the numerical simulation. On the other hand, both numerical and experimental investigations are needed for a comprehensive study on long-term seismic performance of corroded bridge piers. A number of studies have developed methods and formulations to predict initiation and propagation time of corrosion, corrosion cracking time, time of breaking of bond between steel and concrete, minimum load carry capacity, maximum deformation, maximum permeability, or failure probability [55, 125–130]. Then, developing methodologies for performance-based earthquake engineering and developing seismic fragility of bridges made a basis to evaluate seismic performance of RC bridges and other structures using fragility curves based on probability of failure [131–137]. The seismic performance assessment using fragility function can be applied for either a member or the whole bridge structure. Recently, Akiyama and Frangopol [138] have presented a procedure to estimate life-cycle seismic reliability of corroded bridge piers based on integration of probabilistic assessment of seismic and airborne chloride hazard. Ou et al. [116] have developed a simple seismic evaluation of corroded RC bridges based on nonlinear static pushover analysis. They have presented seismic capacity and demand of the RC bridges in terms of peak ground acceleration (PGA). Actually, the number of years that the seismic demand (collapse PGA) becomes greater than the seismic capacity (design PGA) has been calculated for real RC bridges.

However, as mentioned earlier, difficulties and uncertainties in numerical simulation of RC bridge piers subjected to corrosion and seismic hazards indicate critical needs for further investigation, and advanced numerical methods are needed in this content. The next generations of numerical methods to evaluate seismic performance of corroded RC bridge piers are possibly as follows:

(i) Developing a new formulation of finite element methods based on fiber element or fiber beam method [139].

(ii) Real-time signal processing and finite element model updating of existing bridges based on vibration and corrosion potential measurements.

(iii) Artificial intelligence methods such as genetic algorithm [140].

4.1. Numerical Methods to Simulate Degradation of Reinforced Concrete Bridge Piers Exposed to Corrosion.

Numerical methods to simulate seismic behavior of RC bridge piers exposed to corrosion can be classified into three groups including cross section, member, and system level analysis. This classification is similar to the one applied for noncorroded RC bridge piers. The integration of nonlinear analysis and finite element method is a popular numerical method that has been used for corroded and noncorroded RC bridge piers [140–143]. According to the literature, remarkable results obtained from numerical simulations of seismic behavior

of bridges with corroded RC piers can be summarized as follows:

(i) Corrosion alters mechanism of collapse [143].

(ii) Corrosion decreases load carrying capacity due to increasing seismic demand and decreasing seismic capacity leading to increase of probability of failure [144–146].

(iii) Corrosion increases uncertainty in probabilistic model-based analysis models [147].

Cross section level analysis is probably the oldest numerical method among the three methods used to simulate deterioration of RC bridge piers. Corrosion causes damage to concrete material and bond between steel and concrete leading to loss in section ductility. The loss of section ductility can be calculated using moment-curvature analysis [148, 149]. There are some studies where degradation of RC bridge piers caused by corrosion has been investigated using moment-curvature analysis of cross section. Seismic capacity of corroded cross section can also be achieved using the cross section level analysis [143, 145, 150, 151].

Member level analysis is a numerical method providing an opportunity to evaluate the seismic performance of whole corroded bridge piers. Finite element formulation is used to simulate seismic behavior of a corroded bridge pier. To this aim, a relationship, for example, between lateral force and displacement is developed [152]. The simulation should take into consideration deterioration models to simulate degradation due to corrosion. Failure modes and seismic response of corroded RC bridge piers can be obtained from the member level analysis.

System level analysis is aiming to assess dynamic response of a corroded bridge exposed to ground motions. There are a number of studies in the literature where the system level analysis of corroded bridges has been done using fragility estimation [124, 144–146, 153]. However, Lv et al. [142], for example, have evaluated the effects of corrosion on seismic performance of curved beam with height piers using time-history analysis of the bridge finite element model. They found that corrosion deteriorates the seismic performance of the bridge, and two main factors including pier-height and pier-corrosion are responsible for increasing plastic strain.

4.2. Large Scale Experimental Tests to Evaluate Seismic Performance of Corroded Bridge Piers.

As mentioned earlier, numerical simulations of corroded bridge piers are very complicated and they probably cannot capture all the effects of corrosion on seismic performance of bridge piers. Therefore, large scale experimental tests need to assess seismic performance of corroded bridge piers. Some studies have been reported on the effects of corrosion on cyclic behavior of RC columns [154, 155]. However, according to the best knowledge of the authors, large scale seismic experimental test on corroded bridge piers is rare and only one case [139] was found in the literature. Dietz et al. [139] designed a reinforced concrete bridge pier to EC2. They corroded the bridge pier using accelerated corrosion technique by ponding a part of the pier in NaCl solution for 6 months. The corroded bridge pier was then subjected to lateral cyclic loading up to 50 KN using a hydraulic actuator. They measured deflection at the top, rotation at the base, strains in the concrete and steel bars, and width of cracks. From a structural point of view, the bridge pier rigidly connected to the foundation is high damage system because formation of plastic hinge at the end(s) of the pier is the mechanism of dissipating energy in seismic events. The high damage system is the traditional seismic resistant system that has been criticized by past studies because of high repair time and cost and problems arising from traffic interruption [156].

4.3. Seismic, Structural, and Durability Behavior of Repaired Bridge Piers Exposed to Corrosion-Induced Damage. The need for retrofit of corroded bridge piers has been addressed by past studies. There is a traditional method to rehabilitate corroded bridge pier that includes two stages: first, all critically corroded areas should be removed; then, an overlay of materials with low-permeability has been used [157, 158]. Gergely et al. [159] rehabilitated corroded bridge pier samples using fiber-reinforced plastic composites. They compared seismic performance of the piers through numerical and large scale experimental tests and found that shear capacity and ductility have been improved significantly. However, to simulate the effect of corrosion on steel reinforcement, they cut three stirrup loops of column and three stirrup loops of each side of the cap-beam near the joint. They noted that an advantage of using FRP composite is that it does not increase weight of column. Toutanji [160] has found that the confinement provided by FRP warps improves compression strength and ductility of RC columns. However, durability is affected by the type of epoxy used. Demers and Neale [161] have showed that the type of FRP influences ductility and strength of RC columns. Pantazopoulou et al. [162] compared alternative methods to assess seismic performance of repaired corroded bridge piers using external fiber-reinforced polymer (FRP) wraps. They stated that the best repair strategy in terms of postrepair corrosion, strength recovery, and ductility was cleaning the damaged surface (without removal of materials) and then jacketing using layers of FRP. However, further experiments are needed to confirm the efficiency of the proposed strategy in practice. Baiyasi and Harichandran [163] have concluded that a greater amount of glass fiber than carbon fiber was needed to achieve the equivalent structural performance of postcorrosion repair. Teng et al. [164] have reported that while FRP increases durability characteristics of RC columns, it possibly has some negative impacts on mechanical properties of the RC columns. A number of studies have revealed that FRP does not fully stop chloride-induced corrosion but decreases the rate of corrosion [162, 163, 165–173]. To answer an important question of what the best strategy is to repair corroded bridge pier using FRP, Sen [169] stated that the best strategy to protect concrete columns against chloride-induced corrosion is applying FRP jackets and filling gaps between the column and jackets using epoxy so that the following conditions are satisfied:

(i) Applying FRP jacket over the full length when no visible corrosion-induced sign can be found.

(ii) Utilizing appropriate epoxy as a surface corrosion barrier and filling gap between the column and jacket.

(iii) Utilizing at least two layers of FRP.

Sen [169] has argued that confined concrete by FRP warps changes corrosion diffusion. He also has recommended not using FRP warps before visible corrosion sign to minimize repair cost. Wootton et al. [170] have concluded that FRP warps protect the RC columns better than epoxy alone. It has been shown that low amount of corrosion (up to 4.2%) does not influence eccentric load carrying capacity of RC columns, and the strength of damaged columns fully warped with carbon FRP was higher than undamaged columns. The performance of full length covered by CFRP was better than that partially covered by CFRP [174]. Li et al. [175] have analyzed seismic performance of corroded RC columns confined by FRP and steel jacket. They showed that FRP and steel jacket enhance the seismic performance of RC columns, and applying both jackets improves seismic performance better than applying one alone. A recent survey carried out on corroded RC bridges in New York State emphasized the demand for retrofitting corroded RC bridge columns, in particular, corroded lap-splice, to decrease possible damage in future seismic events [176].

4.4. Development in Time-Dependent Seismic Evaluation of Corroded RC Bridge Piers Reported in the Literature. Traditional structural analysis is not able to analyze systems and structures under multiple time-invariant hazards. A time-dependent analysis during lifetime, therefore, is needed to take into consideration all hazards. In case of RC structures (in particular RC bridge pier) subjected to corrosion and earthquake, two hazards are corrosion-induced deterioration and seismic events. There are some studies in the literature, mainly published in recent years, showing that corrosion influences the seismic performance of bridge piers over time. However, different criteria have been used by past studies. Biondini et al. [151], for example, have presented time-dependent bending moment resistance of a bridge pier exposed to corrosion. They have shown that the bending moment resistance of the corroded bridge pier decreased over time. Time-dependent deformation capacity, drift ratio demand, shear capacity, and demand of a corroded bridge pier have been studied [93, 124, 144]. Time-dependent probability of failure (time-dependent fragility analysis) of corroded bridge piers has been developed by past studies that can be directly used for seismic analysis purposes [93, 124, 144, 145, 177, 178]. Moreover, a fragility increment function has been developed that is a function of time and given deformation or shear demand and can be used to predict fragility of corroded bridge piers in life-cycle analysis and risk assessment [147, 178].

4.5. Research Gaps. The main research gaps in evaluation of seismic performance of corroded reinforced concrete bridge piers can be summarized as follows:

(i) Robust numerical modelling of corroded bridge piers.

(ii) Large scale experimental tests on seismic behavior of corroded bridge piers and bridge structures (half scale and full scale tests).

(iii) Experimental tests on efficiency of repair methods used for corroded bridge piers.

(iv) Robust numerical model to evaluate time-dependent seismic performance of RC bridges exposed to corrosion.

5. Conclusions

In this paper, chloride-induced corrosion, the effects of corrosion on structural and mechanical properties of RC structures or elements, and seismic performance of corroded RC bridge pier have been reviewed. To meet this aim, a large number of published papers with all their experimental and numerical details have been collected and reviewed. From the present literature review, the following main conclusions are drawn:

(1) The results of published papers represented in this review paper have been obtained from samples or structures mainly utilizing ordinary Portland cement material, and the behavior of upcoming and more recent cement materials need further investigations.

(2) Damage prediction of RC structures due to chloride-induced corrosion significantly depends on estimation of important input parameters such as corrosion rate, critical content, limit step to start pitting corrosion, and corrosion-induced cracking. However, results reported by past studies exhibit critical problems including contradictory results and uncertainty in experimental techniques and reported results. Moreover, the obtained results cannot be transferred to real structures. Therefore, further investigations are needed in these areas.

(3) More reliable deterioration models are highly demanded for seismic evaluation and analysis of corroded RC structures. On the other side, the deterioration models have been mainly developed for artificial corroded samples, while the relationship between natural and artificial corrosion in many aspects is unknown. Hence, this area is an important direction of research.

(4) Seismic analysis of corroded RC structures (in particular bridge piers) is very complicated, and many uncertainties limit utilizing of numerical methods. Hence, developing more recent numerical methods and large scale experimental tests is needed for the analysis of RC structures under multiple hazards (corrosion and earthquake).

(5) While few researchers based on their modelling have already started to build fragility functions to be integrated in a lifetime seismic performance framework, still many gaps need to be covered in testing and modelling.

(6) Long-term seismic performance of RC bridge pier exposed to chloride incorporates the three main subareas reviewed in Sections 2 to 4 in this paper aiming to develop the following steps that are in common with LCA of corroded RC bridge piers:

(7) Time-dependent deterioration models.

(8) Time-dependent seismic performance of corroded structure.

Competing Interests

The authors declare that they have no competing interests.

Acknowledgments

The research program was supported by the Natural Hazards Research Platform (NHRP) project named "Advanced Bridge Construction and Design for New Zealand (ABCD – NZ Bridges)," 2011–2015. The authors gratefully acknowledge the support from NHRP.

References

[1] B. Hansen, "ASCE's infrastructure report card gives nation a D, estimate cost at #.2 trillion," *ASCE News*, vol. 34, no. 2, pp. 1–4, 2009.

[2] R. R. Hussain and T. Ishida, "Critical carbonation depth for initiation of steel corrosion in fully carbonated concrete and development of electrochemical carbonation induced corrosion model," *International Journal of Electrochemical Science*, vol. 4, no. 8, pp. 1178–1195, 2009.

[3] K. A. T. Vu and M. G. Stewart, "Structural reliability of concrete bridges including improved chloride-induced corrosion models," *Structural Safety*, vol. 22, no. 4, pp. 313–333, 2000.

[4] D. Coronelli and P. Gambarova, "Structural assessment of corroded reinforced concrete beams: modeling guidelines," *Journal of Structural Engineering*, vol. 130, no. 8, pp. 1214–1224, 2004.

[5] K. Stanish, R. D. Hooton, and S. J. Pantazopoulou, "Corrosion effects on bond strength in reinforced concrete," *ACI Structural Journal*, vol. 96, no. 6, pp. 915–921, 1999.

[6] Y. Auyeung, P. Balaguru, and L. Chung, "Bond behavior of corroded reinforcement bars," *ACI Materials Journal*, vol. 97, no. 2, pp. 214–220, 2000.

[7] H.-S. Lee, T. Noguchi, and F. Tomosawa, "Evaluation of the bond properties between concrete and reinforcement as a function of the degree of reinforcement corrosion," *Cement and Concrete Research*, vol. 32, no. 8, pp. 1313–1318, 2002.

[8] C. Fang, K. Lundgren, M. Plos, and K. Gylltoft, "Bond behaviour of corroded reinforcing steel bars in concrete," *Cement and Concrete Research*, vol. 36, no. 10, pp. 1931–1938, 2006.

[9] L. Berto, P. Simioni, and A. Saetta, "Numerical modelling of bond behaviour in RC structures affected by reinforcement corrosion," *Engineering Structures*, vol. 30, no. 5, pp. 1375–1385, 2008.

[10] A. R. L. Kivell, "Effects of bond deterioration due to corrosion on seismic performance of reinforced concrete structures," 2012.

[11] U. Angst, B. Elsener, C. K. Larsen, and Ø. Vennesland, "Critical chloride content in reinforced concrete—a review," *Cement and Concrete Research*, vol. 39, no. 12, pp. 1122–1138, 2009.

[12] S. E. Hussain, A. S. Al-Gahtani, and Rasheeduzzafar, "Chloride threshold for corrosion of reinforcement in concrete," *ACI Materials Journal*, vol. 93, no. 6, pp. 534–538, 1996.

[13] K. Y. Ann and H.-W. Song, "Chloride threshold level for corrosion of steel in concrete," *Corrosion Science*, vol. 49, no. 11, pp. 4113–4133, 2007.

[14] C. Alonso, C. Andrade, M. Castellote, and P. Castro, "Chloride threshold values to depassivate reinforcing bars embedded in a standardized OPC mortar," *Cement and Concrete Research*, vol. 30, no. 7, pp. 1047–1055, 2000.

[15] G. K. Glass and N. R. Buenfeld, "The influence of chloride binding on the chloride induced corrosion risk in reinforced concrete," *Corrosion Science*, vol. 42, no. 2, pp. 329–344, 2000.

[16] T. Maruya, K. Hsu, H. Takeda, and S. Tangtermsirikul, "Numerical modeling of steel corrosion in concrete structures due to chloride ion, oxygen and water movement," *Journal of Advanced Concrete Technology*, vol. 1, no. 2, pp. 147–160, 2003.

[17] R. B. Polder, "Critical chloride content for reinforced concrete and its relationship to concrete resistivity," *Materials and Corrosion*, vol. 60, no. 8, pp. 623–630, 2009.

[18] U. Angst, A. Rønnquist, B. Elsener, C. K. Larsen, and Ø. Vennesland, "Probabilistic considerations on the effect of specimen size on the critical chloride content in reinforced concrete," *Corrosion Science*, vol. 53, no. 1, pp. 177–187, 2011.

[19] C. Andrade, I. R. Maribona, S. Feliu, J. A. González, and S. Feliu Jr., "The effect of macrocells between active and passive areas of steel reinforcements," *Corrosion Science*, vol. 33, no. 2, pp. 237–249, 1992.

[20] M. Raupach, "Chloride-induced macrocell corrosion of steel in concrete—theoretical background and practical consequences," *Construction and Building Materials*, vol. 10, no. 5, pp. 329–338, 1996.

[21] M. Raupach, *Corrosion of Reinforcement in Concrete: Mechanisms, Monitoring, Inhibitors and Rehabilitation Techniques*, Woodhead, Cambridge, UK, 2007.

[22] J. Warkus and M. Raupach, "Numerical modelling of macrocells occurring during corrosion of steel in concrete," *Materials and Corrosion*, vol. 59, no. 2, pp. 122–130, 2008.

[23] J. Warkus and M. Raupach, "Modelling of reinforcement corrosion—geometrical effects on macrocell corrosion," *Materials and Corrosion*, vol. 61, no. 6, pp. 494–504, 2010.

[24] K. Tuutti, *Corrosion of Steel in Concrete*, Swedish Cement and Concrete Research Institute, Stockholm, Sweden, 1982.

[25] P. C. Pistorius and G. T. Burstein, "Metastable pitting corrosion of stainless steel and the transition to stability," *Philosophical Transactions of the Royal Society A: Mathematical, Physical and Engineering Sciences*, vol. 341, no. 1662, pp. 531–559, 1992.

[26] G. T. Burstein, P. C. Pistorius, and S. P. Mattin, "The nucleation and growth of corrosion pits on stainless steel," *Corrosion Science*, vol. 35, no. 1–4, pp. 57–62, 1993.

[27] N. J. Laycock and R. C. Newman, "Localised dissolution kinetics, salt films and pitting potentials," *Corrosion Science*, vol. 39, no. 10-11, pp. 1771–1790, 1997.

[28] L. Bertolini, B. Elsener, P. Pedeferri, and R. Polder, *Corrosion of Steel in Concrete: Prevention Diagnosis Repair*, Wiley-VCH, Weinheim, Germany, 2004.

[29] J. P. Broomfield, *Corrosion of Steel in Concrete: Understanding, Investigation and Repair*, Taylor & Francis, 2002.

[30] U. Angst, B. Elsener, C. K. Larsen, and Ø. Vennesland, "Chloride induced reinforcement corrosion: rate limiting step of early pitting corrosion," *Electrochimica Acta*, vol. 56, no. 17, pp. 5877–5889, 2011.

[31] A. A. Torres-Acosta and M. Martínez-Madrid, "Residual life of corroding reinforced concrete structures in marine environment," *Journal of Materials in Civil Engineering*, vol. 15, no. 4, pp. 344–353, 2003.

[32] M. G. Stewart, "Mechanical behaviour of pitting corrosion of flexural and shear reinforcement and its effect on structural reliability of corroding RC beams," *Structural Safety*, vol. 31, no. 1, pp. 19–30, 2009.

[33] D. V. Val and R. E. Melchers, "Reliability of deteriorating RC slab bridges," *Journal of Structural Engineering*, vol. 123, no. 12, pp. 1638–1644, 1997.

[34] O. B. Isgor and A. G. Razaqpur, "Predicting the initiation and propagation of corrosion in reinforced concrete structures," in *Proceedings of the 4th Structural Specialty Conference of Canadian Society for Civil Engineering*, Montreal, Canada, June 2002.

[35] C. Alonso, C. Andrade, and J. A. González, "Relation between resistivity and corrosion rate of reinforcements in carbonated mortar made with several cement types," *Cement and Concrete Research*, vol. 18, no. 5, pp. 687–698, 1988.

[36] W. López and J. A. González, "Influence of the degree of pore saturation on the resistivity of concrete and the corrosion rate of steel reinforcement," *Cement and Concrete Research*, vol. 23, no. 2, pp. 368–376, 1993.

[37] G. Balabanić, N. Bićanić, and A. Đureković, "Mathematical modeling of electrochemical steel corrosion in concrete," *Journal of Engineering Mechanics*, vol. 122, no. 12, pp. 1113–1122, 1996.

[38] G. Balabanić, N. Bićanić, and A. Đureković, "The influence of w/c ratio, concrete cover thickness and degree of water saturation on the corrosion rate of reinforcing steel in concrete," *Cement and Concrete Research*, vol. 26, no. 5, pp. 761–769, 1996.

[39] M. J. Katwan, T. Hodgkiess, and P. D. Arthur, "Electrochemical noise technique for the prediction of corrosion rate of steel in concrete," *Materials and Structures*, vol. 29, no. 5, pp. 286–294, 1996.

[40] H. Yalçyn and M. Ergun, "The prediction of corrosion rates of reinforcing steels in concrete," *Cement and Concrete Research*, vol. 26, no. 10, pp. 1593–1599, 1996.

[41] S. C. Kranc and A. A. Sagüés, "Modeling the time-dependent response to external polarization of a corrosion macrocell on steel in concrete," *Journal of the Electrochemical Society*, vol. 144, no. 8, pp. 2643–2652, 1997.

[42] C. Alonso, C. Andrade, X. R. Nóvoa, M. Izquierdo, and M. C. Pérez, "Effect of protective oxide scales in the macrogalvanic behaviour of concrete reinforcements," *Corrosion Science*, vol. 40, no. 8, pp. 1379–1389, 1998.

[43] C. Alonso, C. Andrade, J. Rodriguez, and J. M. Diez, "Factors controlling cracking of concrete affected by reinforcement corrosion," *Materials and Structures*, vol. 31, no. 211, pp. 435–441, 1996.

[44] T. Liu and R. W. Weyers, "Modeling the dynamic corrosion process in chloride contaminated concrete structures," *Cement and Concrete Research*, vol. 28, no. 3, pp. 365–379, 1998.

[45] M. Raupach and J. Gulikers, "A simplified method to estimate corrosion rates—a new approach based on investigations of macrocells," in *Durability of Building Materials and Components 8: Service Life and Durability of Materials and Components*, vol. 1, chapter 36, p. 376, 1999.

[46] K. Takewaka, T. Yamaguchi, and S. Maeda, "Simulation model for deterioration of concrete structures due to chloride attack," *Journal of Advanced Concrete Technology*, vol. 1, no. 2, pp. 139–146, 2003.

[47] J. Gulikers, "Theoretical considerations on the supposed linear relationship between concrete resistivity and corrosion rate of steel reinforcement," *Materials and Corrosion*, vol. 56, no. 6, pp. 393–403, 2005.

[48] B. Huet, V. L'hostis, G. Santarini, D. Feron, and H. Idrissi, "Steel corrosion in concrete: determinist modeling of cathodic reaction as a function of water saturation degree," *Corrosion Science*, vol. 49, no. 4, pp. 1918–1932, 2007.

[49] A. Scott and M. G. Alexander, "The influence of binder type, cracking and cover on corrosion rates of steel in chloride-contaminated concrete," *Magazine of Concrete Research*, vol. 59, no. 7, pp. 495–505, 2007.

[50] I. Martínez and C. Andrade, "Examples of reinforcement corrosion monitoring by embedded sensors in concrete structures," *Cement and Concrete Composites*, vol. 31, no. 8, pp. 545–554, 2009.

[51] M. Otieno, H. Beushausen, and M. Alexander, "Prediction of corrosion rate in RC structures—a critical review," in *Modelling of Corroding Concrete Structures*, C. Andrade and G. Mancini, Eds., vol. 5 of *RILEM Bookseries*, Springer, Berlin, Germany, 2011.

[52] M. Otieno, H. Beushausen, and M. Alexander, "Prediction of corrosion rate in reinforced concrete structures—a critical review and preliminary results," *Materials and Corrosion*, vol. 63, no. 9, pp. 777–790, 2012.

[53] ASTM, *Standard Test Method for Half-Cell Potentials of Uncoated Reinforcing Steel in Concrete*, ASTM C876-91, ASTM, West Conshohocken, Pa, USA, 1999.

[54] B. Elsener, C. Andrade, J. Gulikers, R. Polder, and M. Raupach, "Hall-cell potential measurements—potential mapping on reinforced concrete structures," *Materials and Structures*, vol. 36, no. 261, pp. 461–471, 2003.

[55] M.-T. Liang, L.-H. Lin, and C.-H. Liang, "Service life prediction of existing reinforced concrete bridges exposed to chloride environment," *Journal of Infrastructure Systems*, vol. 8, no. 3, pp. 76–85, 2002.

[56] Y. Zhao, H. Ren, H. Dai, and W. Jin, "Composition and expansion coefficient of rust based on X-ray diffraction and thermal analysis," *Corrosion Science*, vol. 53, no. 5, pp. 1646–1658, 2011.

[57] S. Caré, Q. T. Nguyen, V. L'Hostis, and Y. Berthaud, "Mechanical properties of the rust layer induced by impressed current method in reinforced mortar," *Cement and Concrete Research*, vol. 38, no. 8-9, pp. 1079–1091, 2008.

[58] C. Andrade, C. Alonso, and F. J. Molina, "Cover cracking as a function of bar corrosion: part I-experimental test," *Materials and Structures*, vol. 26, no. 8, pp. 453–464, 1993.

[59] C. Andrade, A. Muñoz, and A. Torres-Acosta, "Relation between crack width and corrosion degree in corroding elements exposed to the natural atmosphere," in *Proceedings of 7th International Conference on Fracture Mechanics of Concrete and Concrete Structures (FraMCoS '07)*, May 2010.

[60] Y. G. Du, L. A. Clark, and A. H. C. Chan, "Residual capacity of corroded reinforcing bars," *Magazine of Concrete Research*, vol. 57, no. 3, pp. 135–147, 2005.

[61] W. H. Peter, R. A. Buchanan, C. T. Liu et al., "Localized corrosion behavior of a zirconium-based bulk metallic glass relative to its crystalline state," *Intermetallics*, vol. 10, no. 11-12, pp. 1157–1162, 2002.

[62] Y. Liu, *Modeling the Time-to-Corrosion Cracking of the Cover Concrete in Chloride Contaminated Reinforced Concrete Structures*, Virginia Polytechnic Institute and State University, Blacksburg, Va, USA, 1996.

[63] Y. Liu and R. E. Weyers, "Modeling the time-to-corrosion cracking in chloride contaminated reinforced concrete structures," *ACI Materials Journal*, vol. 95, no. 6, pp. 675–681, 1998.

[64] S. J. Pantazopoulou and K. D. Papoulia, "Modeling cover-cracking due to reinforcement corrosion in RC structures," *Journal of Engineering Mechanics*, vol. 127, no. 4, pp. 342–351, 2001.

[65] K. Bhargava, A. K. Ghosh, Y. Mori, and S. Ramanujam, "Modeling of time to corrosion-induced cover cracking in reinforced concrete structures," *Cement and Concrete Research*, vol. 35, no. 11, pp. 2203–2218, 2005.

[66] K. Bhargava, A. K. Ghosh, Y. Mori, and S. Ramanujam, "Analytical model for time to cover cracking in RC structures due to rebar corrosion," *Nuclear Engineering and Design*, vol. 236, no. 11, pp. 1123–1139, 2006.

[67] T. El Maaddawy and K. Soudki, "A model for prediction of time from corrosion initiation to corrosion cracking," *Cement and Concrete Composites*, vol. 29, no. 3, pp. 168–175, 2007.

[68] H.-S. Lee and Y.-S. Cho, "Evaluation of the mechanical properties of steel reinforcement embedded in concrete specimen as a function of the degree of reinforcement corrosion," *International Journal of Fracture*, vol. 157, no. 1-2, pp. 81–88, 2009.

[69] C. A. Apostolopoulos and M. P. Papadopoulos, "Tensile and low cycle fatigue behavior of corroded reinforcing steel bars S400," *Construction and Building Materials*, vol. 21, no. 4, pp. 855–864, 2007.

[70] C. A. Apostolopoulos and V. P. Pasialis, "Effects of corrosion and ribs on low cycle fatigue behavior of reinforcing steel bars S400," *Journal of Materials Engineering and Performance*, vol. 19, no. 3, pp. 385–394, 2010.

[71] R. A. Hawileh, J. A. Abdalla, A. Al Tamimi, K. Abdelrahman, and F. Oudah, "Behavior of corroded steel reinforcing bars under monotonic and cyclic loadings," *Mechanics of Advanced Materials and Structures*, vol. 18, no. 3, pp. 218–224, 2011.

[72] W. Zhang, X. Song, X. Gu, and S. Li, "Tensile and fatigue behavior of corroded rebars," *Construction and Building Materials*, vol. 34, pp. 409–417, 2012.

[73] J. A. González, C. Andrade, C. Alonso, and S. Feliu, "Comparison of rates of general corrosion and maximum pitting penetration on concrete embedded steel reinforcement," *Cement and Concrete Research*, vol. 25, no. 2, pp. 257–264, 1995.

[74] J. Rodriguez, L. Ortega, and J. Casal, "Load bearing capacity of concrete columns with corroded reinforcement," in *Proceedings of 4th International Symposium*, Special Publication No 183, Corrosion of Reinforcement in Concrete Construction, Cambridge, UK, July 1996.

[75] J. Rodriguez, L. M. Ortega, and J. Casal, "Load carrying capacity of concrete structures with corroded reinforcement," *Construction and Building Materials*, vol. 11, no. 4, pp. 239–248, 1997.

[76] M. G. Stewart, "Spatial variability of pitting corrosion and its influence on structural fragility and reliability of RC beams in flexure," *Structural Safety*, vol. 26, no. 4, pp. 453–470, 2004.

[77] J. Cairns, G. A. Plizzari, Y. Du, D. W. Law, and C. Franzoni, "Mechanical properties of corrosion-damaged reinforcement," *ACI Materials Journal*, vol. 102, no. 4, Article ID 102-M29, pp. 256–264, 2005.

[78] A. A. Torres-Acosta, S. Navarro-Gutierrez, and J. Terán-Guillén, "Residual flexure capacity of corroded reinforced concrete beams," *Engineering Structures*, vol. 29, no. 6, pp. 1145–1152, 2007.

[79] M. G. Stewart and A. Al-Harthy, "Pitting corrosion and structural reliability of corroding RC structures: experimental data and probabilistic analysis," *Reliability Engineering and System Safety*, vol. 93, no. 3, pp. 373–382, 2008.

[80] T. D. Marcotte, *Characterization of chloride-induced corrosion products that form in steel-reinforced cementitious materials [Ph.D. thesis]*, University of Waterloo, Waterloo, Canada, 2001.

[81] M. Raupach, G. Weizhong, and J. Wei-Liang, "Korrosionsprodukte und deren Volumenfaktor bei der Korrosion von Stahl in Beton," *Beton- und Stahlbetonbau*, vol. 105, no. 9, pp. 572–578, 2010.

[82] G. J. Al-Sulaimani, M. Kaleemullah, I. A. Basunbul, and Rasheeduzzafar, "Influence of corrosion and cracking on bond behavior and strength of reinforced concrete members," *ACI Structural Journal*, vol. 87, no. 2, pp. 220–231, 1990.

[83] Y. Tachibana, K.-I. Maeda, Y. Kajikawa, and M. Kawamura, *Mechanical Behaviour of RC Beams Damaged by Corrosion of Reinforcement*, Elsevier Applied Science, 1990.

[84] J. G. Cabrera, "Deterioration of concrete due to reinforcement steel corrosion," *Cement and Concrete Composites*, vol. 18, no. 1, pp. 47–59, 1996.

[85] A. A. Almusallam, A. S. Al Gahtani, A. S. A. Gahtani, M. Maslehuddin, M. M. Khan, and A. R. Aziz, "Evaluation of repair materials for functional improvement of slabs and beams with corroded reinforcement," *Proceedings of the Institution of Civil Engineers—Structures and Buildings*, vol. 122, no. 1, pp. 27–34, 1997.

[86] R. Huang and C. C. Yang, "Condition assessment of reinforced concrete beams relative to reinforcement corrosion," *Cement and Concrete Composites*, vol. 19, no. 2, pp. 131–137, 1997.

[87] L. Amleh and S. Mirza, "Corrosion influence on bond between steel and concrete," *ACI Structural Journal*, vol. 96, no. 3, pp. 415–423, 1999.

[88] P. S. Mangat and M. S. Elgarf, "Flexural strength of concrete beams with corroding reinforcement," *ACI Structural Journal*, vol. 96, no. 1, pp. 149–158, 1999.

[89] P. S. Mangat and M. S. Elgarf, "Strength and serviceability of repaired reinforced concrete beams undergoing reinforcement corrosion," *Magazine of Concrete Research*, vol. 51, no. 2, pp. 97–112, 1999.

[90] T. A. El Maaddawy and K. A. Soudki, "Effectiveness of impressed current technique to simulate corrosion of steel reinforcement in concrete," *Journal of Materials in Civil Engineering*, vol. 15, no. 1, pp. 41–47, 2003.

[91] T. Vidal, A. Castel, and R. François, "Analyzing crack width to predict corrosion in reinforced concrete," *Cement and Concrete Research*, vol. 34, no. 1, pp. 165–174, 2004.

[92] K. Vu, M. G. Stewart, and J. Mullard, "Corrosion-induced cracking: experimental data and predictive models," *ACI Structural Journal*, vol. 102, no. 5, pp. 719–726, 2005.

[93] J. Zhong, P. Gardoni, and D. Rosowsky, "Seismic fragility estimates for corroding reinforced concrete bridges," *Structure and Infrastructure Engineering*, vol. 8, no. 1, pp. 55–69, 2012.

[94] R. Zhang, A. Castel, and R. François, "Concrete cracking due to chloride-induced reinforcement corrosion—influence of steel-concrete interface defects due to the 'top-bar effect'," *European Journal of Environmental and Civil Engineering*, vol. 16, no. 3-4, pp. 402–413, 2012.

[95] M. Maslehuddin, I. M. Allam, G. J. Al-Sulaimani, A. Al-Mana, and S. N. Abduljauwad, "Effect of rusting of reinforcing steel on its mechanical properties and bond with concrete," *ACI Materials Journal*, vol. 87, no. 5, pp. 496–502, 1990.

[96] C. Andrade, C. Alonso, D. Garcia, and J. Rodriguez, "Remaining lifetime of reinforced concrete structures: effect of corrosion in the mechanical properties of the steel, life predication of corrodible structures," in *Proceedings of the NACE*, pp. 12/1–12/11, Cambridge, UK, September 1991.

[97] I. M. Allam, M. Maslehuddin, H. Saricimen, and A. I. Al-Mana, "Influence of atmospheric corrosion on the mechanical properties of reinforcing steel," *Construction and Building Materials*, vol. 8, no. 1, pp. 35–41, 1994.

[98] M. Saifullah, *Effect of reinforcement corrosion on bond strength in reinforced concrete [Ph.D. thesis]*, The University of Birmingham, Birmingham, UK, 1994.

[99] P. S. Zhang, M. Lu, and X. Y. Li, "The mechanical behavior of corroded bar," *Journal of Industrial Buildings*, vol. 25, no. 257, pp. 41–44, 1995.

[100] S. Morinaga, "Remaining life of reinforced concrete structures after corrosion cracking," *Durability of Building Materials and Components*, vol. 71, pp. 127–136, 1996.

[101] H. Lee, F. Tomosawa, and T. Noguchi, "Effects of rebar corrosion on the structural performance of singly reinforced beams," *Durability of Building Materials and Components*, vol. 7, pp. 571–580, 1996.

[102] A. Castel, R. François, and G. Arliguie, "Mechanical behaviour of corroded reinforced concrete beams—part 1: experimental study of corroded beams," *Materials and Structures*, vol. 33, no. 9, pp. 539–544, 2000.

[103] Y. G. Du, *Effect of reinforcement corrosion on structural concrete ductility [Ph.D. thesis]*, The University of Birmingham, Birmingham, UK, 2001.

[104] A. A. Almusallam, "Effect of degree of corrosion on the properties of reinforcing steel bars," *Construction and Building Materials*, vol. 15, no. 8, pp. 361–368, 2001.

[105] R. Palsson and M. S. Mirza, "Mechanical response of corroded steel reinforcement of abandoned concrete bridge," *ACI Structural Journal*, vol. 99, no. 2, pp. 157–162, 2002.

[106] M. Oyado, T. Kanakubo, T. Sato, and Y. Yamamoto, "Bending performance of reinforced concrete member deteriorated by corrosion," *Structure and Infrastructure Engineering*, vol. 7, no. 1-2, pp. 121–130, 2011.

[107] A. A. Almusallam, A. S. Al-Gahtani, A. R. Aziz, and Rasheeduzzafar, "Effect of reinforcement corrosion on bond strength," *Construction and Building Materials*, vol. 10, no. 2, pp. 123–129, 1996.

[108] X. Fu and D. D. L. Chung, "Effect of corrosion on the bond between concrete and steel rebar," *Cement and Concrete Research*, vol. 27, no. 12, pp. 1811–1815, 1997.

[109] K. Soudki and T. Sherwood, "Bond behavior of corroded steel reinforcement in concrete wrapped with carbon fiber reinforced polymer sheets," *Journal of Materials in Civil Engineering*, vol. 15, no. 4, pp. 358–370, 2003.

[110] L. Chung, S.-H. Cho, J.-H. J. Kim, and S.-T. Yi, "Correction factor suggestion for ACI development length provisions based on flexural testing of RC slabs with various levels of corroded reinforcing bars," *Engineering Structures*, vol. 26, no. 8, pp. 1013–1026, 2004.

[111] C. Fang, K. Lundgren, L. Chen, and C. Zhu, "Corrosion influence on bond in reinforced concrete," *Cement and Concrete Research*, vol. 34, no. 11, pp. 2159–2167, 2004.

[112] A. Ouglova, Y. Berthaud, F. Foct, M. François, F. Ragueneau, and I. Petre-Lazar, "The influence of corrosion on bond properties between concrete and reinforcement in concrete structures," *Materials and Structures*, vol. 41, no. 5, pp. 969–980, 2008.

[113] L. Chung, J.-H. Jay Kim, and S.-T. Yi, "Bond strength prediction for reinforced concrete members with highly corroded reinforcing bars," *Cement and Concrete Composites*, vol. 30, no. 7, pp. 603–611, 2008.

[114] C. Fang, K. Gylltoft, K. Lundgren, and M. Plos, "Effect of corrosion on bond in reinforced concrete under cyclic loading," *Cement and Concrete Research*, vol. 36, no. 3, pp. 548–555, 2006.

[115] J. B. Mander, M. J. Priestley, and R. Park, "Theoretical stress-strain model for confined concrete," *Journal of Structural Engineering*, vol. 114, no. 8, pp. 1804–1826, 1988.

[116] Y.-C. Ou, H.-D. Fan, and N. D. Nguyen, "Long-term seismic performance of reinforced concrete bridges under steel reinforcement corrosion due to chloride attack," *Earthquake Engineering & Structural Dynamics*, vol. 42, no. 14, pp. 2113–2127, 2013.

[117] F. J. Vecchio and M. P. Collins, "The modified compression-field theory for reinforced concrete elements subjected to shear," *ACI Structural Journal*, vol. 83, no. 2, pp. 219–231, 1986.

[118] F. J. Vecchio and M. P. Collins, "Compression response of cracked reinforced concrete," *Journal of Structural Engineering*, vol. 119, no. 12, pp. 3590–3610, 1993.

[119] M. Capé, *Residual service-life assessment of existing R/C structures [M.S. thesis]*, Chalmers University of Technology, Gothenburg, Sweden; Milan University of Technology, Milan, Italy, 1999.

[120] P. Mehta, "Concrete Technology at the Crossroad—Problem and Opportunities," in *Conerete Technology: Past, Present and Future*, ACI SP144-1, American Concrete Institute, 1994.

[121] P. A. M. Basheer, S. E. Chidiac, and A. E. Long, "Predictive models for deterioration of concrete structures," *Construction and Building Materials*, vol. 10, no. 1, pp. 27–37, 1996.

[122] P. Thoft-Christensen, F. Jensen, C. Middleton, and A. Blackmore, *Assessment of the Reliability of Concrete Slab Bridges*, Department of Building Technology and Structural Engineering, 1996.

[123] D. V. Val, M. G. Stewart, and R. E. Melchers, "Effect of reinforcement corrosion on reliability of highway bridges," *Engineering Structures*, vol. 20, no. 11, pp. 1010–1019, 1998.

[124] D.-E. Choe, P. Gardoni, D. Rosowsky, and T. Haukaas, "Probabilistic capacity models and seismic fragility estimates for RC columns subject to corrosion," *Reliability Engineering & System Safety*, vol. 93, no. 3, pp. 383–393, 2008.

[125] Z. P. Bazant, "Physical model for steel corrosion in concrete sea structures—theory," *Journal of the Structural Division*, vol. 105, no. 6, pp. 1137–1153, 1979.

[126] Z. P. Bazant, "Physical model for steel corrosion in concrete sea structures—application," *Journal of the Structural Division*, vol. 105, pp. 1155–1166, 1979, ASCE 14652 Proceeding.

[127] J. De Brito and F. Branco, "Whole life costing in road bridges applied to service life prediction," in *Proceedings of the 3rd International Conference on Bridge Management, Bridge Management 3. Inspection, Maintenance and Repair*, pp. 603–612, University of Surrey, Guildford, UK, April 1996.

[128] S. Ng and F. Moses, "Prediction of bridge service life using time-dependent reliability analysis," *Bridge Management*, vol. 3, pp. 26–32, 1996.

[129] M. P. Enright and D. M. Frangopol, "Service-life prediction of deteriorating concrete bridges," *Journal of Structural Engineering*, vol. 124, no. 3, pp. 309–317, 1998.

[130] F. Biondini and D. M. Frangopol, "Probabilistic limit analysis and lifetime prediction of concrete structures," *Structure and Infrastructure Engineering*, vol. 4, no. 5, pp. 399–412, 2008.

[131] J. B. Mander and N. Basöz, "Seismic fragility curve theory for highway bridges," in *Proceedings of the 5th US Conference on Lifeline Earthquake Engineering*, Optimizing Post-Earthquake Lifeline System Reliability, ASCE, Seattle, Wash, USA, August 1999.

[132] J. Moehle, A. Lynn, K. Elwood, and H. Sezen, "Gravity load collapse of reinforced concrete frames during earthquakes," in *Proceedings of the 1st US-Japan Workshop on Performance-Based Design Methodology for Reinforced Concrete Building Structures*, Pacific Earthquake Engineering Research Center, University of California, 1999.

[133] M. Shinozuka, M. Q. Feng, J. Lee, and T. Naganuma, "Statistical analysis of fragility curves," *Journal of Engineering Mechanics*, vol. 126, no. 12, pp. 1224–1231, 2000.

[134] P. Gardoni, A. Der Kiureghian, and K. M. Mosalam, "Probabilistic capacity models and fragility estimates for reinforced concrete columns based on experimental observations," *Journal of Engineering Mechanics*, vol. 128, no. 10, pp. 1024–1038, 2002.

[135] K. A. Porter, "An overview of PEER's performance-based earthquake engineering methodology," in *Proceedings of the on Applications of Statistics and Probability in Civil Engineering (ICASP '9), Civil Engineering Risk and Reliability Association (CERRA '03)*, San Francisco, Calif, USA, 2003.

[136] J. Moehle and G. G. Deierlein, "A framework methodology for performance-based earthquake engineering," in *Proceedings of the 13th World Conference on Earthquake Engineering*, 2004.

[137] S. Matsuki, S. Billington, and J. Baker, "Impact of long-term material degradation on seismic performance of a reinforced concrete bridge," in *Proceedings of the 8th US National Conference on Earthquake Engineering*, San Francisco, Calif, USA, April 2006.

[138] M. Akiyama and D. M. Frangopol, "Long-term seismic performance of RC structures in an aggressive environment: emphasis on bridge piers," *Structure and Infrastructure Engineering*, vol. 10, no. 7, pp. 865–879, 2014.

[139] M. S. Dietz, L. Dihoru, O. Oddbjornsson et al., "Earthquake and large structures testing at the Bristol laboratory for advanced dynamics engineering," in *Role of Seismic Testing Facilities in Performance-Based Earthquake*, pp. 21–41, Springer, 2012.

[140] T. Less and H. Adeli, "Computational earthquake engineering of bridges," *Scientia Iranica*, vol. 17, no. 5, pp. 325–338, 2010.

[141] A. Palermo and S. Pampanin, "Enhanced seismic performance of hybrid bridge systems: comparison with traditional monolithic solutions," *Journal of Earthquake Engineering*, vol. 12, no. 8, pp. 1267–1295, 2008.

[142] H. Lv, J. Teng, and D. Zou, "Seismic performance under environment corrosion for curved beam bridges with high piers," in *Proceedings of the 2nd International Conference on Multimedia Technology (ICMT '11)*, pp. 1693–1696, IEEE, Hangzhou, China, July 2011.

[143] F. Biondini, E. Camnasio, and A. Palermo, "Lifetime seismic performance of concrete bridges exposed to corrosion," *Structure and Infrastructure Engineering*, vol. 10, no. 7, pp. 880–900, 2014.

[144] D.-E. Choe, P. Gardoni, D. Rosowsky, and T. Haukaas, "Seismic fragility estimates for reinforced concrete bridges subject to corrosion," *Structural Safety*, vol. 31, no. 4, pp. 275–283, 2009.

[145] J. Ghosh and J. E. Padgett, "Aging considerations in the development of time-dependent seismic fragility curves," *Journal of Structural Engineering*, vol. 136, no. 12, pp. 1497–1511, 2010.

[146] J. Simon, J. M. Bracci, and P. Gardoni, "Seismic response and fragility of deteriorated reinforced concrete bridges," *Journal of Structural Engineering*, vol. 136, no. 10, pp. 1273–1281, 2010.

[147] P. Gardoni and D. Rosowsky, "Seismic fragility increment functions for deteriorating reinforced concrete bridges," *Structure and Infrastructure Engineering*, vol. 7, no. 11, pp. 869–879, 2011.

[148] H. J. Dagher and S. Kulendran, "Finite element modeling of corrosion damage in concrete structures," *ACI Structural Journal*, vol. 89, no. 6, pp. 699–708, 1992.

[149] R. Capozucca, "Damage to reinforced concrete due to reinforcement corrosion," *Construction and Building Materials*, vol. 9, no. 5, pp. 295–303, 1995.

[150] F. Biondini, F. Bontempi, D. M. Frangopol, and P. G. Malerba, "Cellular automata approach to durability analysis of concrete structures in aggressive environments," *Journal of Structural Engineering*, vol. 130, no. 11, pp. 1724–1737, 2004.

[151] F. Biondini, F. Bontempi, D. M. Frangopol, and P. G. Malerba, "Probabilistic service life assessment and maintenance planning of concrete structures," *Journal of Structural Engineering*, vol. 132, no. 5, pp. 810–825, 2006.

[152] M. Akiyama, D. M. Frangopol, and H. Matsuzaki, "Life-cycle reliability of RC bridge piers under seismic and airborne chloride hazards," *Earthquake Engineering & Structural Dynamics*, vol. 40, no. 15, pp. 1671–1687, 2011.

[153] R. Kumar and P. Gardoni, "Modeling structural degradation of RC bridge columns subjected to earthquakes and their fragility estimates," *Journal of Structural Engineering*, vol. 138, no. 1, pp. 42–51, 2012.

[154] Y. Ma, Y. Che, and J. Gong, "Behavior of corrosion damaged circular reinforced concrete columns under cyclic loading," *Construction and Building Materials*, vol. 29, pp. 548–556, 2012.

[155] A. Meda, S. Mostosi, Z. Rinaldi, and P. Riva, "Experimental evaluation of the corrosion influence on the cyclic behaviour of RC columns," *Engineering Structures*, vol. 76, pp. 112–123, 2014.

[156] A. Palermo and M. Mashal, "Accelerated bridge construction(ABC) and seismic damage resistant technology: a New Zealand challenge," *Bulletin of the New Zealand Society for Earthquake Engineering*, vol. 45, no. 3, pp. 123–134, 2012.

[157] R. E. Weyers, B. D. Prowell, M. M. Sprinkel, and M. Vorster, "Concrete bridge protection, repair, and rehabilitation relative to reinforcement corrosion: a methods application manual," *Contract*, vol. 100, article 103, 1993.

[158] A. M. Vaysburd and P. H. Emmons, "How to make today's repairs durable for tomorrow—corrosion protection in concrete repair," *Construction and Building Materials*, vol. 14, no. 4, pp. 189–197, 2000.

[159] L. Gergely, C. P. Pantelides, R. J. Nuismer, and L. D. Reaveley, "Bridge pier retrofit using fiber-reinforced plastic composites," *Journal of Composites for Construction*, vol. 2, no. 4, pp. 165–174, 1998.

[160] H. A. Toutanji, "Durability characteristics of concrete columns confined with advanced composite materials," *Composite Structures*, vol. 44, no. 2-3, pp. 155–161, 1999.

[161] M. Demers and K. W. Neale, "Confinement of reinforced concrete columns with fibre-reinforced composite sheets—an experimental study," *Canadian Journal of Civil Engineering*, vol. 26, no. 2, pp. 226–241, 1999.

[162] S. J. Pantazopoulou, J. F. Bonacci, S. Sheikh, M. D. A. Thomas, and N. Hearn, "Repair of corrosion-damaged columns with FRP wraps," *Journal of Composites for Construction*, vol. 5, no. 1, pp. 3–11, 2001.

[163] I. Baiyasi and R. Harichandran, "Corrosion and wrap strains in concrete bridge columns repaired with FRP wraps," in *Proceedings of the 80th Annual Meeting of the Transportation Research Board*, (CD-ROM), Washington, DC, USA, January 2001.

[164] M.-H. Teng, E. D. Sotelino, and W.-F. Chen, "Performance evaluation of reinforced concrete bridge columns wrapped with fiber reinforced polymers," *Journal of Composites for Construction*, vol. 7, no. 2, pp. 83–92, 2003.

[165] C. Lee, J. F. Bonacci, M. D. A. Thomas et al., "Accelerated corrosion and repair of reinforced concrete columns using carbon fibre reinforced polymer sheets," *Canadian Journal of Civil Engineering*, vol. 27, no. 5, pp. 941–948, 2000.

[166] E. Berver, D. Fowler, J. Jirsa, H. Wheat, and M. Ford, "Corrosion in FRP-wrapped concrete members," in *Proceedings of the International Conference on Structural Faults and Repair, Engineering Technics*, London, UK, 2001.

[167] G. Mullins, R. Sen, A. Torres-Acosta et al., "Lateral capacity of corroded pile bents," Final Report, University of South Florida, 2001.

[168] A. S. Debaiky, M. F. Green, and B. B. Hope, "Carbon fiber-reinforced polymer wraps for corrosion control and rehabilitation of reinforced concrete columns," *ACI Materials Journal*, vol. 99, no. 2, pp. 129–137, 2002.

[169] R. Sen, "Advances in the application of FRP for repairing corrosion damage," *Progress in Structural Engineering and Materials*, vol. 5, no. 2, pp. 99–113, 2003.

[170] I. A. Wootton, L. K. Spainhour, and N. Yazdani, "Corrosion of steel reinforcement in carbon fiber-reinforced polymer wrapped concrete cylinders," *Journal of Composites for Construction*, vol. 7, no. 4, pp. 339–347, 2003.

[171] T. El Maaddawy, A. Chahrour, and K. Soudki, "Effect of fiber-reinforced polymer wraps on corrosion activity and concrete cracking in chloride-contaminated concrete cylinders," *Journal of Composites for Construction*, vol. 10, no. 2, pp. 139–147, 2006.

[172] M. F. Green, L. A. Bisby, A. Z. Fam, and V. K. R. Kodur, "FRP confined concrete columns: behaviour under extreme conditions," *Cement and Concrete Composites*, vol. 28, no. 10, pp. 928–937, 2006.

[173] S. Gadve, A. Mukherjee, and S. N. Malhotra, "Corrosion of steel reinforcements embedded in FRP wrapped concrete," *Construction and Building Materials*, vol. 23, no. 1, pp. 153–161, 2009.

[174] T. E. Maaddawy, "Behavior of corrosion-damaged RC columns wrapped with FRP under combined flexural and axial loading," *Cement and Concrete Composites*, vol. 30, no. 6, pp. 524–534, 2008.

[175] J. Li, J. Gong, and L. Wang, "Seismic behavior of corrosion-damaged reinforced concrete columns strengthened using combined carbon fiber-reinforced polymer and steel jacket," *Construction and Building Materials*, vol. 23, no. 7, pp. 2653–2663, 2009.

[176] R. S. Aboutaha, F. Jnaid, S. Sotoud, and M. Tapan, *Seismic Evaluation and Retrofit of Deteriorated Concrete Bridge Components*, Project no: 49111-23-22, University of Syracuse, Syracuse, NY, USA, 2013.

[177] A. Alipour, B. Shafei, and M. Shinozuka, "Performance evaluation of deteriorating highway bridges located in high seismic areas," *Journal of Bridge Engineering*, vol. 16, no. 5, pp. 597–611, 2011.

[178] D.-E. Choe, P. Gardoni, and D. Rosowsky, "Fragility increment functions for deteriorating reinforced concrete bridge columns," *Journal of Engineering Mechanics*, vol. 136, no. 8, pp. 969–978, 2010.

17

The Inhibition Effect of Sodium Glutarate towards Carbon Steel Corrosion in Neutral Aqueous Solutions

G. Chan-Rosado and M. A. Pech-Canul

Departamento de Física Aplicada, Cinvestav-Mérida, Km. 6 Ant. Carr. a Progreso, AP73, Cordemex, 97310 Mérida, YUC, Mexico

Correspondence should be addressed to M. A. Pech-Canul; maximo.pech@cinvestav.mx

Academic Editor: Yu Zuo

The inhibition effect of sodium glutarate towards corrosion of carbon steel in neutral 0.02 M NaCl solution was investigated with potentiodynamic polarization and electrochemical impedance measurements. Results of electrochemical measurements revealed a poor inhibitive action for low concentrations (1 mM and 5 mM) and a significant improvement in efficiency for concentrations of 32 mM or higher. The protective film exhibited excellent stability in the temperature range 22°C–55°C. Full chemical passivation was accomplished and analysis of the impedance spectra for the high concentrations of glutarate was consistent with the inhibition mechanism which assumes that the carboxylates support the passivation of carbon steel in aerated solutions by plugging the defect sites and that the passivation process is enhanced by adsorption of the carboxylates on the oxide-covered surface. Such mechanism was confirmed by the XPS analysis.

1. Introduction

The corrosion of carbon steel in neutral or slightly alkaline oxygen-containing solutions is a common problem in cooling water systems and an effective way to control the corrosion of steel is the addition of corrosion inhibitors [1]. Although several compounds have been reported to act as good corrosion inhibitors for steel in neutral media (e.g., molybdates, tungstates, polyphosphate, and phosphonate) [1–5], the use of carboxylic acid derivatives has been attractive since they are environmentally friendly and can be derived from fatty acids extracted from vegetable oils [6, 7].

The sodium salt of benzoic acid (C_6H_5COONa) is the classic carboxylic acid inhibitor used since the 1960s for protection of steel in neutral solutions [8–10]. Straight chain aliphatic carboxylates have also been reported by various authors as effective inhibitors. Mrowczynski and Szklaeska-Smialowska [11] reported the electrochemical behavior of iron in sulphate solutions in the presence of monocarboxylic acids with 6–10 carbon atoms in the molecule, showing that efficiency increases with increasing length of the hydrocarbon chain, provided that the concentration exceeds a critical value; they attributed the reduction in corrosion rate of iron to a synergistic action of these acids and dissolved oxygen. Reinhard and Rammelt [12–14] also investigated the passivation effect of some carboxylates, including straight chain mono- and dicarboxylates; they reported that dissolved oxygen forms a passive oxide film on iron and steel in the presence of the carboxylate above a critical minimum concentration. Hefter et al. [15] made a systematic study of the abilities of straight chain monocarboxylic ($C_nH_{2n+1}COO^-$) and dicarboxylic ($^-OOC[CH_2]_nCOO^-$) acids to inhibit the corrosion of mild steel, copper, and aluminum in neutral solutions. For mild steel, short chain length ($n \leq 5$) monocarboxylates were found to behave as either corrosive ($n = 0$) or weak inhibitors; from $n = 6$ to $n = 10$, the efficiency increased rapidly but decreased abruptly at $n = 11$ and declined further up to $n = 17$. In the case of dicarboxylates, they also observed that short chain lengths ($n \leq 3$) were either corrosive ($n = 0$) or weak inhibitive; also, the inhibitor effectiveness increased from $n = 4$ up to $n = 12$ and decreased sharply at $n = 14$. More recently, Lahem et al. [16] used electrochemical methods to investigate the efficiency of some dicarboxylates ($n = 1, 2, 4,$ and 8) as inhibitors for mild steel in neutral chloride solutions. Their results showed that dicarboxylates with $n \geq 4$ have good efficiency (better efficiency with increasing carbon chain length) and behave predominantly as anodic inhibitors. The proposed mechanism of inhibition involves the formation of an insoluble ferric compound (iron dicarboxylate complex) on the metal surface after an initial oxidation of iron into ferrous ions. In recent works, Rajendran et al. have shown that the inhibitive effect of succinic and adipic acid ($n = 2$ and 4, resp.) can be enhanced by addition of Zn^{2+} ions, due to a synergistic action [17, 18]. With FTIR, they determined that the protective film consists of a complex of the dicarboxylic acid with Zn^{2+} ions and $Zn(OH)_2$.

In the study reported by Hefter et al. [15], modest inhibition efficiencies were observed for dicarboxylates with chain lengths of $n = 2$ and $n = 3$. However, such result corresponds to a fixed concentration of 5 mM. It is worth investigating how the protectivity can be improved by increasing the concentration.

In this work, a dicarboxylate with $n = 3$ was chosen with the aim of providing further information with respect to that reported in the previous literature. The inhibitive effect of sodium glutarate ($NaOOC(CH_2)_3COONa$) on the corrosion of carbon steel in a near-neutral 0.02 M NaCl solution was investigated at ambient temperature for a range of concentrations (1 mM–100 mM) by potentiodynamic polarization and electrochemical impedance spectroscopy (EIS) measurements. The effect of temperature was investigated for the highest concentration. In addition, X-ray photoelectron spectroscopy (XPS) was used to analyze the composition of the protective layer. From these results, the corrosion inhibition mechanism was discussed.

2. Experimental

2.1. Materials. All experiments were carried out in a three-electrode electrochemical cell with a platinum foil as counter electrode and an Ag/AgCl reference electrode. The working electrode was made from a 1018 carbon steel rod embedded in epoxy resin, leaving an exposed area of $1\,cm^2$. The chemical composition of this material was Fe-98.703%, C-0.177%, Mn-0.636%, Cu-0.197%, Si-0.0578%, P-0.041%, and traces of Ni, Cr, Mo, Al, S, and Nb. Prior to each measurement, the working surface was abraded with SiC papers up to 600 grits, rinsed with distilled water and dried with a stream of hot air. The sodium glutarate solutions were prepared through neutralization of glutaric acid (obtained from Sigma-Aldrich®) by sodium hydroxide in distilled water. For the first set of experiments, at ambient temperature (22°C), solutions were prepared in a concentration range of 0.001 M to 0.1 M. The effect of temperature on inhibition efficiency was investigated for the 0.1 M glutarate solution conducting experiments at 3 other temperatures (35°C, 45°C, and 55°C) using a water bath temperature-controlled. The corrosive medium was a 0.02 M NaCl solution, pH 7.4. Prior to each test, the solutions were preaerated by air bubbling during 15 min.

2.2. Electrochemical Measurements. The electrochemical measurements were performed with a Gamry series G300 potentiostat-galvanostat. The working electrode was allowed

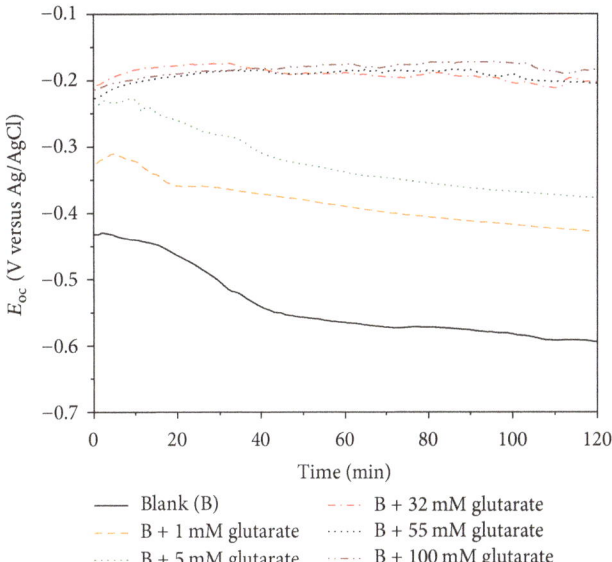

FIGURE 1: Variation with time of open circuit potential for carbon steel in air saturated 0.02 M NaCl solutions without and with additions of sodium glutarate.

to stabilize at open circuit conditions in the solution during 2 hours; then, EIS measurements were performed under potentiostatic conditions at the open circuit potential (E_{oc}) using a sine wave of 10 mV amplitude in the frequency range from 10 KHz to 10 mHz with 5 points/decade. At the end, a potentiodynamic curve was recorded with a sweep rate of $1\,mV\,s^{-1}$ in the range of ±300 mV versus E_{oc}.

2.3. XPS Analysis. X-ray photoelectron spectroscopy (XPS) was used to analyze the chemical composition of the protective layer formed on the surface of carbon steel after 2 h immersion in the NaCl solution containing 0.1 M of sodium glutarate. As a comparison, the analysis was also carried out for a reference sample of carbon steel not immersed in the solutions (covered with native oxide film). XPS measurements were carried out on a Thermo Scientific K-Alpha XPS spectrophotometer. A monochromatic Al Kα X-ray source was used, with a spot area of $400\,\mu m$. High resolution core level spectra for Fe 2p, C 1s, and O 1s were acquired in the angle-resolved mode. Results are presented for two takeoff angles, 0° and 50°, with respect to the normal.

3. Results and Discussion

3.1. Effect of Inhibitor Concentration at Ambient Temperature. Figure 1 shows the evolution with time of the open circuit potential for carbon steel in the 0.02 M NaCl solution without and with additions of sodium glutarate at different concentrations. In the blank solution, E_{oc} starts at about −430 mV and gradually shifts towards more negative values reaching a value of −590 mV (Ag/AgCl) after the 2 h immersion. This open circuit potential decay can be ascribed to breakdown of the air-formed oxide film. Such process was studied in

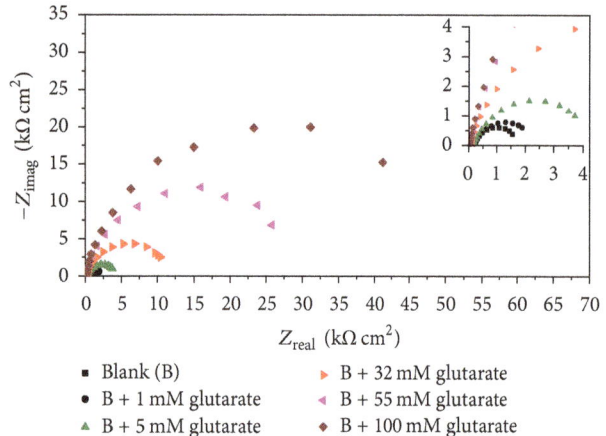

FIGURE 2: Typical Nyquist diagrams for carbon steel in the 0.02 M NaCl solution without (blank) and with additions of sodium glutarate.

Legend (Figure 2):
- ■ Blank (B)
- ● B + 1 mM glutarate
- ▲ B + 5 mM glutarate
- ▶ B + 32 mM glutarate
- ◀ B + 55 mM glutarate
- ◆ B + 100 mM glutarate

detail by Gilroy and Mayne [19] and Konno et al. [20] for air-formed oxide films on iron. Both research groups carried out open circuit potential measurements in corrosive solutions and observed that, due to dissolution of the oxide film, E_{oc} initially decreases gradually with immersion time and then it drops rapidly to −0.76 V (Ag/AgCl), the free corrosion potential of Fe. We did not observe the sudden decrease to −0.76 V; so it appears that during the 2 h immersion the air-formed oxide film is not completely removed yet. The addition of sodium glutarate leads to a positive shift in the initial value of E_{oc}. However, for the two lowest concentrations (1 mM and 5 mM), the potential stabilized in the region around −400 mV, suggesting that sodium glutarate assisted in repairing defects of the air-formed oxide film but not enough to ensure full protection. For glutarate concentrations ≥32 mM, the initial E_{oc} was ~−225 mV (Ag/AgCl) and remained around −200 mV throughout the immersion period. This ennoblement of potential in the presence of oxygen and of the nonoxidizing inhibitor anion has been ascribed in the early literature to a passivation process [21, 22]. As discussed by Brasher [22], the anion adsorbs on the metal surface at formation. Rammelt et al. [12, 14] observed this passivating effect for mild steel with carboxylates and suggested that the protective film consists mainly of Fe(III) oxide with inclusion of insoluble Fe(III) carboxylate. So, the results in Figure 1 suggest that solutions containing sodium glutarate with a concentration of 32 mM or higher provide much better conditions for a satisfactory inhibition compared to the solutions with lower concentration of glutarate (1 and 5 mM).

The typical impedance responses obtained for the carbon steel electrode after 2 h immersion in solutions with and without inhibitor are presented as complex plane plots in Figure 2. The inset shows a zoom in of the region 0–4 kΩ·cm² (in the x- and y-axis) to help visualize the spectra with lower impedance amplitude (i.e., for steel in the blank and solutions with 1 mM and 5 mM of sodium glutarate).

Although the impedance diagrams apparently look like a slightly depressed single capacitive semicircle, the Bode plot and the equivalent circuit analysis (see below) suggest that in all cases the Nyquist diagram actually consists of two, closely overlapped, time constants.

As indicated above, the carbon steel surface in the blank solution was initially covered with an air-formed oxide film which undergoes breakdown with increasing immersion time; when carboxylate is added to the chloride solution, it acts at local defects of the thin oxide layer sealing them according to the pore plugging concept [23]. Nevertheless, for low concentrations of glutarate (1 mM and 5 mM), full passivation could not be achieved. Thus, the impedance data obtained in the blank and the solutions with the two lowest additions of glutarate can be modeled with the equivalent circuit for an imperfectly covered electrode (Figure 3(a)) where R_s is the solution resistance, R_f and C_f correspond to the resistance and capacitance relative to the surface film, and C_{int} is the interfacial capacitance in parallel with R_p the polarization resistance, at the defect sites (where charge transfer takes place).

On the other hand, the E_{oc} values in Figure 1 give an indication that in solutions containing 32 mM or higher of sodium glutarate the carbon steel electrode is in a passive state. As discussed previously in related literature [13, 14, 22], the passivation process is enhanced by adsorption of the carboxylates on the oxide-covered surface. Accordingly, the impedance data obtained in solutions with sodium glutarate in concentrations ≥32 mM were modeled with the equivalent circuit (Figure 3(b)) for a two-layer system, with an oxide layer on the metal surface and an adsorption layer on it [13]; R_s is the solution resistance, R_{ads} and C_{ads} correspond to the resistance and capacitance of adsorption of an electrochemically indifferent species; C_{ox} and R_{ox} are the capacitance and resistance associated with the oxide layer.

Bode plots comparing experimental versus fitted data for carbon steel in 0.02 M NaCl solutions with different concentrations of sodium glutarate are presented in Figures 4 and 5. Very good fits to the experimental impedance spectra were obtained with the two equivalent circuit models in Figure 3. The optimum fit parameters are presented in Tables 1 and 2. As usual, for the fitting process, a constant phase element (CPE) is introduced to account for nonideal capacitive behavior due to surface inhomogeneity [24, 25]. Its admittance is given by $Y_{CPE} = Y_0(j\omega)^\alpha$, where ω is the sine wave modulation angular frequency, Y_0 is the base admittance (with dimensions $\Omega^{-1} s^\alpha cm^{-2}$), and α is and the empirical exponent ($0 \leq \alpha \leq 1$) which measures the deviation from the ideal capacitive behavior.

One characteristic of the CPE is that when $\alpha \approx 1$, then the base admittance becomes the capacitance ($Y_0 = C$) with the dimension F cm⁻². In Table 1, the α_f parameter associated with the capacitance of film C_f suggests a great departure from ideality. This can be attributed to surface roughness and porous nature of the layer with possible presence of corrosion products. However, it is interesting to observe that the increase trend of the film resistance R_f is consistent with the assumption that the carboxylate acts plugging the pores, making the film more resistive. On the other hand, the α and

TABLE 1: Parameters from equivalent circuit analysis of typical impedance data obtained in the blank and solutions containing glutarate with low concentrations (1 mM and 5 mM).

$\text{Conc}_{\text{inh}}/\text{mM}$	$R_s/(\Omega \text{ cm}^2)$	$Y_{0,f} \times 10^4/(\Omega^{-1} \text{ s}^\alpha \text{ cm}^{-2})$	α_f	$R_f/(\Omega \text{ cm}^2)$	$Y_{0,\text{int}} \times 10^4/(\Omega^{-1} \text{ s}^\alpha \text{ cm}^{-2})$	α_{int}	$R_p/(\text{k}\Omega \text{ cm}^2)$
0	168	4.84	0.75	44	12.01	0.81	1.64
1	131	7.08	0.67	160	5.52	0.80	2.27
5	120	3.66	0.74	209	1.66	0.77	4.27

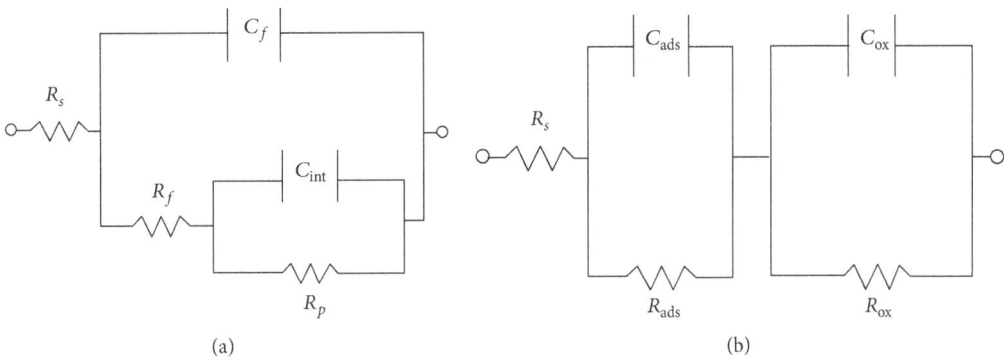

(a)   (b)

FIGURE 3: Equivalent circuit models used to fit the experimental results (a) in the blank and solutions with glutarate concentrations <32 mM and (b) for glutarate concentrations ≥32 mM.

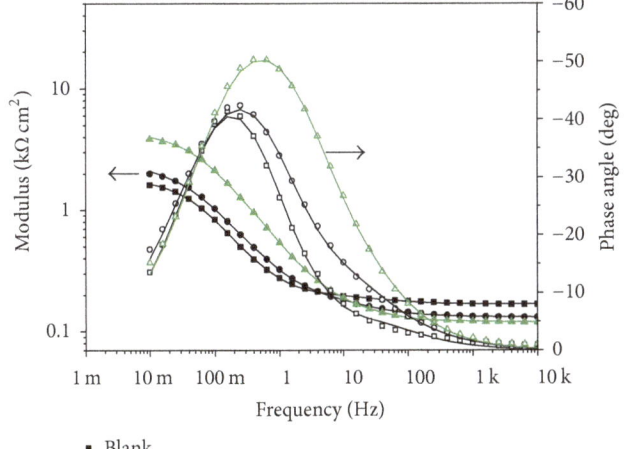

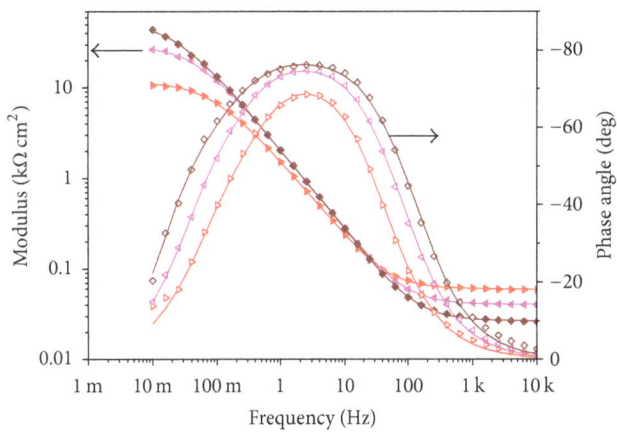

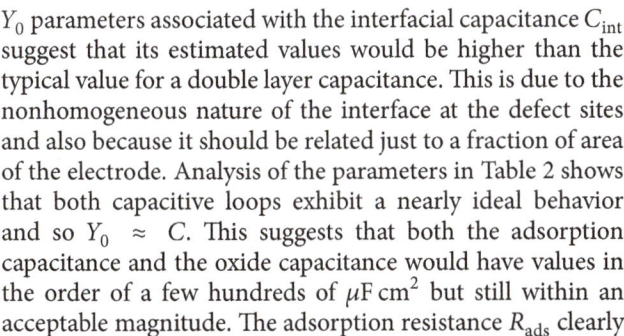

■ Blank
● Blank + 1 mM glutarate
▲ Blank + 5 mM glutarate

▶ Blank + 32 mM glutarate
◀ Blank + 55 mM glutarate
◆ Blank + 100 mM glutarate

FIGURE 4: Bode plots of carbon steel in aerated 0.02 M NaCl solution without and with low additions of sodium glutarate (1 mM and 5 mM). The solid lines correspond to fitted curves.

FIGURE 5: Bode plots of carbon steel in aerated 0.02 M NaCl solution with additions of sodium glutarate ≥32 mM. The solid lines correspond to fitted curves.

Y_0 parameters associated with the interfacial capacitance C_{int} suggest that its estimated values would be higher than the typical value for a double layer capacitance. This is due to the nonhomogeneous nature of the interface at the defect sites and also because it should be related just to a fraction of area of the electrode. Analysis of the parameters in Table 2 shows that both capacitive loops exhibit a nearly ideal behavior and so $Y_0 \approx C$. This suggests that both the adsorption capacitance and the oxide capacitance would have values in the order of a few hundreds of $\mu\text{F cm}^2$ but still within an acceptable magnitude. The adsorption resistance R_{ads} clearly increases with inhibitor concentration, thus giving indication that the adsorption layer becomes more densely packed [14]. Furthermore, the oxide resistance R_{ox} also increases with inhibitor concentration. Rammelt et al. [13] associated the better protective property of inhibitor mixtures with an increase in R_{ox}, arguing that pore plugging combined with strong adsorption results in an oxide film which becomes increasingly impermeable to the dissolution of ferrous ions. So these results show that although the carbon steel electrode has the same E_{oc} in solutions with 32 mM, 55 mM, or 100 mM of sodium glutarate, the protectivity increases with inhibitor concentration.

TABLE 2: Parameters from equivalent circuit analysis of typical impedance data obtained in solutions containing glutarate with concentrations ≥ 32 mM.

$Conc_{inh}/mM$	$R_s/(\Omega\ cm^2)$	$Y_{0,ads} \times 10^4/(\Omega^{-1}\ s^\alpha\ cm^{-2})$	α_{ads}	$R_{ads}/(k\Omega\ cm^2)$	$Y_{0,ox} \times 10^4/(\Omega^{-1}\ s^\alpha\ cm^{-2})$	α_{ox}	$R_{ox}/(k\Omega\ cm^2)$
32	60	4.38	0.95	0.97	1.80	0.83	10.75
55	41	5.48	0.95	1.24	1.16	0.86	28.50
100	28	1.63	0.85	11.01	2.07	0.97	37.15

TABLE 3: Average inhibition efficiency at 22°C estimated from average values of R_p and R_{ox}.

$Conc_{inh}/mM$	$R_p/(K\Omega\ cm^2)$	$R_{ox}/(k\Omega\ cm^2)$	$(\eta^{ac})_{avg}$
0	1.52 ± 0.21		
1	2.29 ± 0.31		33.4
5	4.24 ± 0.17		64.1
32		12.13 ± 2.3	87.4
55		24.60 ± 3.9	93.8
100		35.70 ± 2.1	95.7

Since the Stern-Geary equation is a relationship between the corrosion current density i_{corr} and the polarization resistance R_p, the inhibition efficiency η can be evaluated from the R_p values in Table 1 according to the formula $\eta^{ac} = (1 - R_p^B/R_p^I) \times 100$, where the superscript ac means that it is calculated from impedance parameters and the superscripts B and I refer to blank and inhibited solutions. Furthermore, for the impedance spectra fitted with the equivalent circuit in Figure 3(b), the parameter in Table 2 which is proportional to corrosion resistance is R_{ox}. So, in this case, the formula to calculate efficiency is $\eta^{ac} = (1 - R_p^B/R_{ox}^I) \times 100$. In this work, each experiment was repeated at least 3 times to ensure reproducibility. The parameters in Tables 1 and 2 correspond to the typical spectra presented in Figure 2; however, the fittings were carried out for all the experimental data. Therefore, average values of inhibition efficiency at ambient temperature were estimated from the average values of R_p and R_{ox} as presented in Table 3.

A modest efficiency of 64% is obtained with 5 mM addition of glutarate (a result in agreement with the work of Hefter at al. [15]). However, a significant improvement in protectivity is achieved by increasing the concentration from 5 mM to 32 mM or even better with 55 mM or 100 mM.

Figure 6 shows typical polarization curves obtained after 2 h immersion of carbon steel at 22°C in the 0.02 M NaCl solution without and with additions of sodium glutarate. Along with the shift in E_{corr} towards less negative values, there are some important features of the polarization curves: (a) the anodic current density (say 100 mV above E_{corr} for each polarization curve) decreases with increasing inhibitor concentration, (b) there is also a decrease trend in the corrosion current density (i_{corr}), (c) a passivation plateau starts to develop for concentrations ≥ 32 mM, and (d) it appears that there is also a modification in the kinetics of oxygen reduction for concentrations ≥ 32 mM.

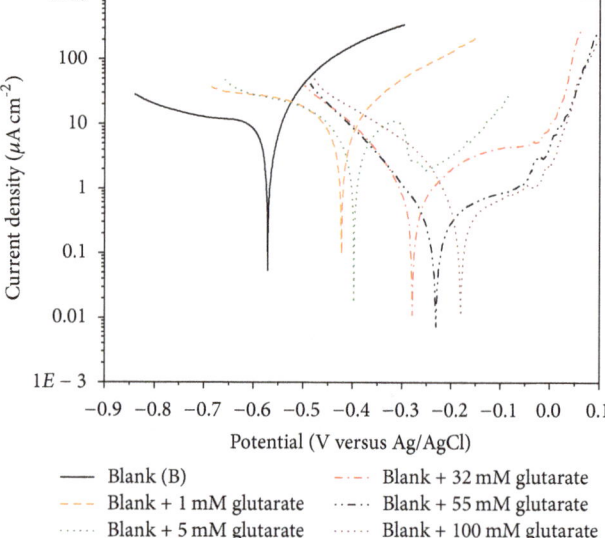

—— Blank (B)	-·-· Blank + 32 mM glutarate
- - - Blank + 1 mM glutarate	-··-·· Blank + 55 mM glutarate
······ Blank + 5 mM glutarate	······ Blank + 100 mM glutarate

FIGURE 6: Typical polarization curves of carbon steel in 0.02 M NaCl solution in the absence (blank) and in the presence of different concentrations of sodium glutarate.

TABLE 4: Corrosion current density for carbon steel in 0.02 M NaCl without and with additions of sodium glutarate. The average values of inhibition efficiency at 22°C are also included.

$Conc_{inh}/mM$	$i_{corr}/(\mu A\ cm^{-2})$	$(\eta^{dc})_{avg}$
0	6.61 ± 0.39	
1	5.0 ± 0.40	24.4
5	2.34 ± 0.50	64.6
32	0.75 ± 0.09	88.7
55	0.29 ± 0.07	95.6
100	0.26 ± 0.09	96.1

In Table 3, each experiment was repeated at least 3 times to ensure reproducibility. The same applies for the polarization curves. So, average values of the inhibition efficiency $\eta^{dc} = (1 - i_{corr}^I/i_{corr}^B) \times 100$ were calculated from the average values of i_{corr} obtained from the Tafel analysis of polarization curves. The results are presented in Table 4. A reasonable agreement with the values obtained from impedance measurements (Table 3) is observed.

3.2. *Effect of Temperature.* In order to evaluate the stability of protective films on the carbon steel surface as well as

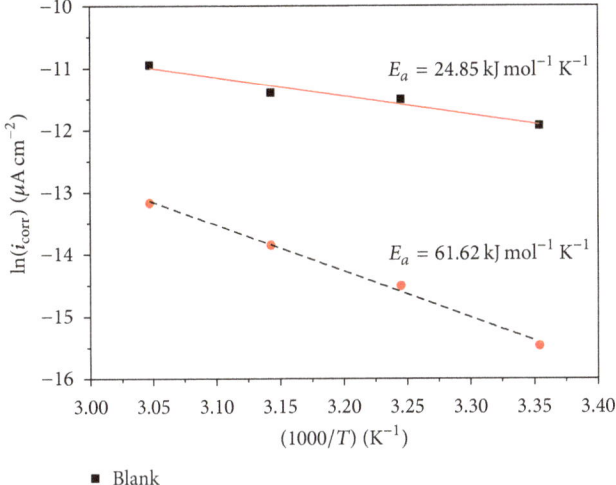

FIGURE 7: Arrhenius plot of the corrosion of carbon steel in 0.02 M NaCl solution in the absence and in the presence of 100 mM glutarate.

TABLE 5: Activation parameters for carbon steel in 0.02 M NaCl solution in the absence and in the presence of 100 mM sodium glutarate.

	$E_a/(\text{kJ mol}^{-1})$	$\Delta H_a/(\text{kJ mol}^{-1})$	$\Delta S_a/(\text{J mol}^{-1})$
Blank (B)	24.85	22.24	−269.34
B + 100 mM Glutarate	61.62	59.02	−175.09

activation parameters of the corrosion process, potentiodynamic polarization curves were obtained at three more temperatures (35°C, 45°C, and 55°C) for the blank and for solutions containing 100 mM of sodium glutarate. The results are presented as an Arrhenius type of plot in Figure 7.

It is evident that as temperature increases, so does the corrosion rate in both the blank and the inhibited solution. For instance, at 55°C, the values of i_{corr} are 17.62 and 1.91 µA cm^{-2}, respectively, in the blank and inhibited solution. Therefore, the inhibition efficiency decreased from 96.1% at 22°C to 89.1% at 55°C. It is expected that raising the temperature leads to higher anodic dissolution of carbon steel and partial desorption of the inhibitor from the metal surface. Nevertheless, these results show that sodium glutarate at a concentration of 100 mM forms a very stable and protective film even at 55°C.

The activation parameters (activation energy E_a, activation enthalpy ΔH_a, and activation entropy ΔS_a) were obtained from the Arrhenius equation (1) and the transition state equation (2) for corrosion current density:

$$i_{\text{corr}} = A \exp\left(-\frac{E_a}{RT}\right), \tag{1}$$

$$i_{\text{corr}} = \frac{RT}{Nh} \exp\left(\frac{\Delta S_a}{R}\right) \exp\left(-\frac{\Delta H_a}{RT}\right), \tag{2}$$

where R is the gas constant (8.314 J mol^{-1} K^{-1}), T is the temperature in K, A is the preexponential factor, N is Avogadro's number (6.02 × 10^{23} mol^{-1}), and h is Planck's constant (6.63 × 10^{-34} m^2 Kg s^{-1}). E_a is obtained from the slope of a ln(i_{corr}) versus $1/T$ plot; ΔH_a and ΔS_a are obtained from the slope and intercept, respectively, of a ln(i_{corr}/T) versus $1/T$ plot. The results are presented in Table 5.

Table 5 shows that E_a for the inhibited solution is higher than that for the blank. This suggests that the strong adsorption of glutarate serves to block active sites for metal dissolution on the carbon steel surface, thus creating an energy barrier for charge transfer, leading to a decrease in corrosion rate. The ΔH_a values are positive (i.e., the dissolution process is endothermic) and follow the same trend as the activation energy; in fact, the difference between E_a and ΔH_a is 2.6 kJ mol^{-1} which indicates that corrosion kinetics satisfy the following equation from activated complex theory:

$$E_a = \Delta H_a + RT. \tag{3}$$

The entropy of activation is negative and large. This implies that the activated complex represents association rather than dissociation step, indicating that a decrease in disorder takes place, going from reactants to the activated complex. Moreover, ΔS_a values are less negative in inhibited solution than that obtained in the blank. This behavior can be explained as a result of the replacement process of water molecules during adsorption of glutarate on the steel surface [26].

3.3. XPS Analysis of Protective Films. In Figure 8(a), the high resolution C 1s spectrum for native oxide film obtained at an angle of 0° (normal emission) is presented. Since the carbon steel sample with native oxide film was not immersed in the solution containing the carboxylate, the entire C 1s signal is attributed to carbonaceous contamination. The deconvolution of this and all the other spectra was carried out with the XPSPEAK software, using a Shirley background correction. In accordance with results reported in the literature for iron oxide layers [27, 28], the peak was resolved into three different contributions, characteristic of C-C/C-H, C-O, and O-C=O species. The hydrocarbon contribution (C-C/C-H) predominates over the two oxidized carbon species. Angle-resolved XPS measurements are useful for obtaining nondestructive information at different film depths. The depth of the film, from which the photoelectrons escape, decreases with an increase of the takeoff angle. Thus, the bulk of the film is probed at an angle of 0° (Figure 8(a)) whereas the outer surface is probed at 50°. As shown in Figure 8(b), the C 1s spectrum for native iron oxide film obtained at a takeoff angle of 50° is very similar to the one at 0°, with only a difference in the relative contribution of the C-O structure which increases in the outer surface.

When organic molecules are adsorbed on the oxidized iron or carbon steel surface, the carbon intensity in the XPS measurement is expected to increase, compared to that observed in a nontreated sample. This is not possible to observe in Figure 8; however, an indirect indication that the intensity of the C 1s signal increases in treated samples is the observation that the signal to noise ratio in Figures 8(c)

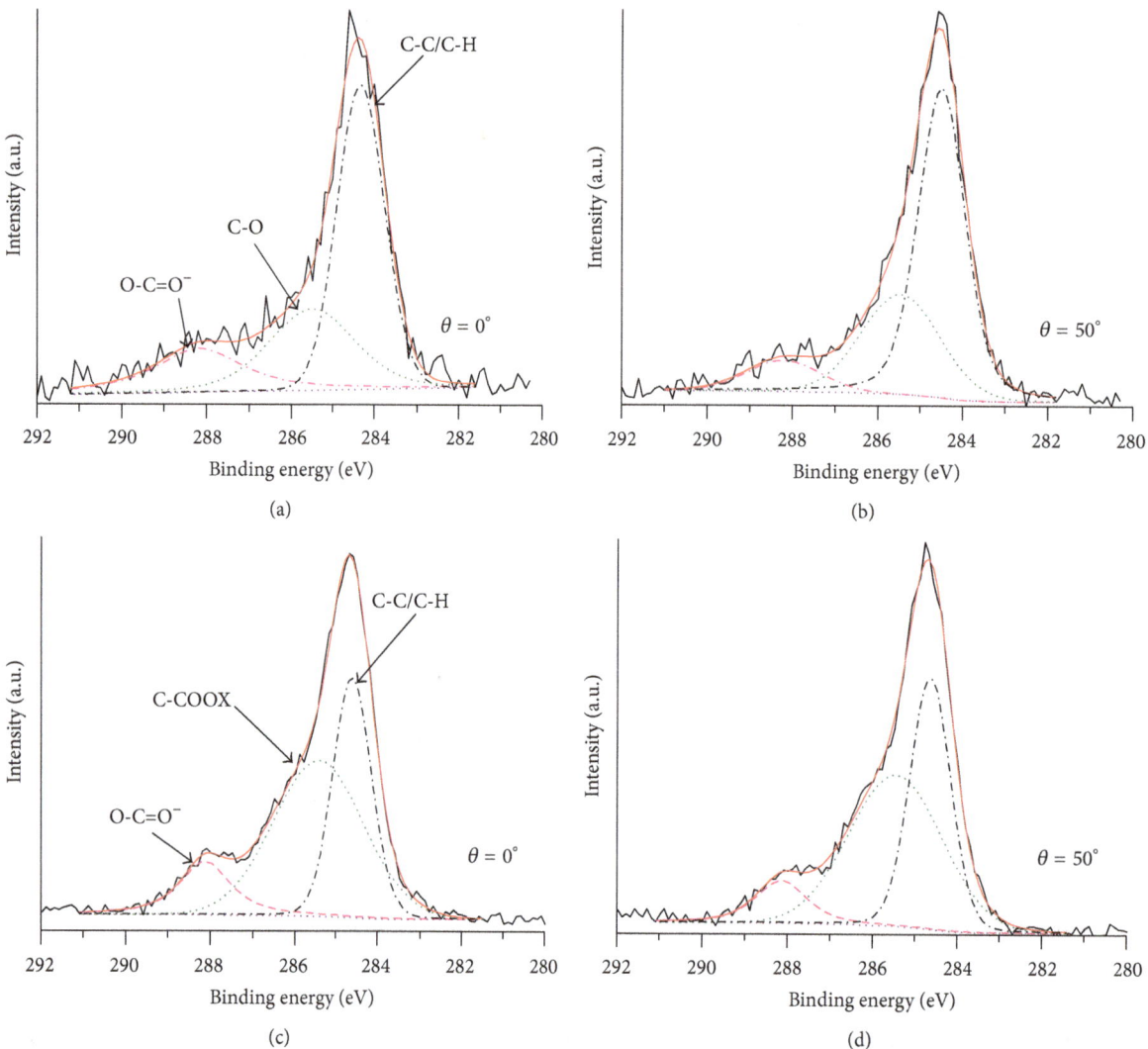

FIGURE 8: High resolution C 1s spectra for native oxide film on carbon steel (a, b) and for carbon steel sample after 2 h immersion in 0.02 M NaCl solution containing 0.1 M sodium glutarate (c, d). In each case, results for two different takeoff angles (0° and 50°) are presented.

and 8(d) (treated) is higher than that in Figures 8(a) and 8(b) (nontreated). A further evidence that the C 1s signals in Figures 8(c) and 8(d) are due to adsorption of the organic molecule and not to carbonaceous contamination is the fact that the relative contribution of oxidized carbon subpeaks is higher compared to that observed for the native oxide film. In fact, deconvolution of the C 1s peak in Figure 8(c) resulted in 3 subpeaks: the subpeak at a binding energy (BE) ~284.7 eV can be assigned to the carbon atoms in the saturated chains of the dicarboxylic acid, the one around 285.4 eV to the carbon atom attached to the carboxyl carbon (C-COOX), and the peak at BE ~288.3 eV to the carboxylate (O-C=O⁻), in agreement with results published in related literature [29–31].

A common feature of angle-resolved XPS is that as the takeoff angle increases the intensity of the high resolution spectrum decreases, because the analysis becomes more surface sensitive. So, although not shown, the intensity of the C 1s spectrum in Figure 8(d) is lower compared to that in

Figure 8(c). It is interesting, however, to observe that the C 1s signal obtained at 50° (Figure 8(d)) has the same shape compared with the one at 0° (Figure 8(c)) and also that the fitting process leads to the same subpeaks. According to Taheri et al. [31], if undissociated carboxylic acid group (C-COOH) was present in the surface, it would have a subpeak at a BE of ~289.1 eV. We did not detect such subpeak in the region more near to the surface (Figure 8(d)); so it appears that both carboxylic acid groups are deprotonated and the O-C=O⁻ subpeak gives evidence of the formation of coordinatively bonded carboxylate species at both ends.

As discussed above, the electrochemical measurements give evidence of chemical passivation of the carbon steel and the analysis of the impedance spectra in solutions with glutarate concentration higher than 32 mM suggests that carboxylates are adsorbed on the oxide-covered surface. In other words, the carboxylate ion is expected to be strongly chemisorbed on the passive oxide film. Several

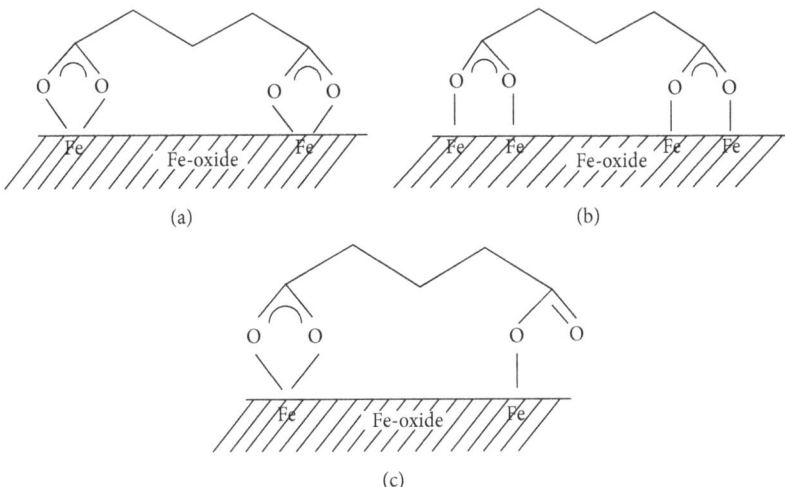

FIGURE 9: Schematic representation of possible modes of binding of glutarate with the passive film: (a) bidentate chelating at both ends, (b) bidentate bridging at both ends, and (c) bidentate chelating-monodentate.

experimental and theoretical studies on the adsorption of aliphatic mono- and dicarboxylic acids on metal oxide surfaces have been reported in the literature [32–36]. It is commonly acknowledged that chemisorption may occur in different modes, including monodentate, bridge bidentate, and chelating bidentate adsorbed structures. Rajendran et al. [37] reported that adipic acid adsorbs to iron oxide on carbon steel surface in a tetradentate manner. Figure 9 shows three possible modes of binding of glutarate to the passive film in this study. A detailed infrared study will be carried out in a future work to elucidate the actual binding mode.

Figure 10(a) shows the high resolution Fe $2p_{3/2}$ spectrum obtained at a takeoff angle of 0° for the native oxide film. Deconvolution of this spectrum suggests that it consists of 3 subpeaks, corresponding to metallic iron at BE ~706.3 eV, a broad subpeak at ~709.8 eV due to Fe^{2+} surface species, and a small subpeak at BE ~710.7 eV ascribed to ferric surface oxidation product Fe^{3+}, in good agreement with results published previously for iron oxides and oxyhydroxides [38–40]. The corresponding O 1s signal is presented in Figure 10(b). Deconvolution of this spectrum shows that the main contributions are from oxygen in the form of oxide (O^{2-}) and hydrous iron oxides (OH^-). A third subpeak appears at BE ~532.7 eV which can be ascribed to adsorbed water and oxidized carbon from carbonaceous contamination [27, 28, 31]. As pointed out by Temesghen and Sherwood [38], the use of XPS for the identification of different types of oxidized iron species is complicated because the core Fe2p region shows little difference between different iron compounds. Considering the results of deconvolution of spectra in Figures 10(a) and 10(b), it can be proposed that the native oxide film is composed of a hydrous iron oxide FeOOH (difficult to distinguish which one) and Fe $(OH)_2$, with possibly a small amount of Fe_3O_4. The Fe $2p_{3/2}$ and O 1s spectra obtained at a takeoff angle of 50° are presented in Figures 10(c) and 10(d). Deconvolution of these spectra

leads to the same subpeaks observed at a takeoff angle of 0°, but with differences in the relative contribution of each subpeak. For the Fe $2p_{3/2}$ signal, the intensity of the subpeak corresponding to Fe^0 decreases since at this angle the outer part of the film is being probed. Figure 10(d) shows that in comparison to Figure 10(b) the contribution of OH^- and H_2O/C-O increases. This behavior is consistent with the common observation that the outermost portion of the oxide film is more hydrated [41, 42].

Figure 11(a) shows the high resolution Fe $2p_{3/2}$ spectrum obtained at a takeoff angle of 0° for the carbon steel sample that was immersed in the solution containing 0.1 M Na-glutarate. It was also resolved into three subpeaks corresponding to Fe^0, Fe^{2+}, and Fe^{3+}. The presence of the metallic iron subpeak suggests that the passive film is very thin (with thickness in the order of the maximum escape depth of photoelectrons). Noteworthily, the subpeak corresponding to ferric species exhibits a greater contribution compared to the one observed in the native oxide film (Figure 10(a)). Deconvolution of the O 1s peak (also obtained at a takeoff angle of 0°) shows (Figure 11(b)) that it consists of 3 subpeaks. Apart from that, corresponding to oxygen in the form of oxide (O^{2-}) and of hydrous iron oxides (OH^-), a third subpeak at BE ~532.7 eV was observed, due to oxygen in the carboxylate ion and to adsorbed water. The dominant chemical state is that of O^{2-}; this observation, along with the important contribution of Fe^{3+} in the Fe $2p_{3/2}$ spectrum, confirms that the passive film developed on the carbon steel surface in the presence of 0.1 M Na-glutarate is very protective. The passive film itself might be a bilayer, with an outer region of FeOOH and and inner oxide region (Fe_3O_4 or possibly Fe_2O_3). Furthermore, as discussed above, this passive film is covered by an organic layer (chemisorbed carboxylate). This structure is consistent with the schematic representations in Figure 9. Deconvolution of the Fe $2p_{3/2}$ and O 1s spectra obtained at a takeoff angle of 50° (Figures

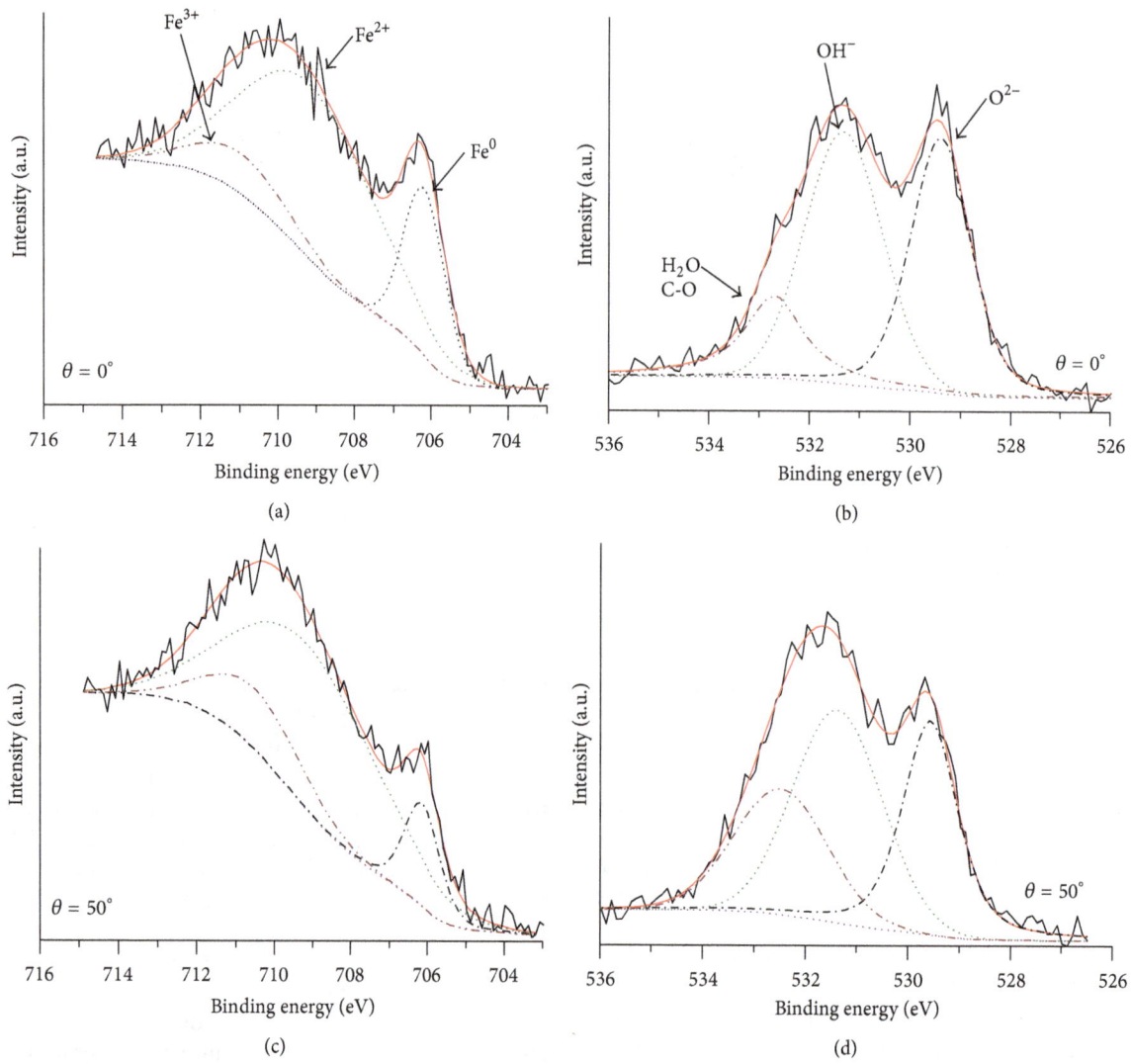

FIGURE 10: High resolution Fe $2p_{3/2}$ and O 1s spectra for native oxide film, obtained at a takeoff angle of $0°$ (a, b) and of $50°$ (c, d).

11(c) and 11(d)) shows that optimum fitting leads to the same subpeaks observed at a takeoff angle of $0°$. However, the following differences in the relative contribution of subpeaks are observed: the Fe^0 subpeak in the Fe $2p_{3/2}$ spectrum decreases and the OH^- subpeak in the O 1s spectrum increases. These observations are consistent with the fact that at a takeoff angle of $50°$ the outermost region of the film is being probed. This region is more hydrated and the response is mainly from the oxyhydroxide/oxide, with little contribution from the substrate.

4. Conclusions

Sodium glutarate showed a poor inhibitive action for corrosion of carbon steel in a 0.02 M NaCl solution when used in concentrations of 1 mM and 5 mM. However, open circuit potential and polarization curve measurements give evidence that full chemical passivation is accomplished for concentrations of 32 mM or higher; a significant improvement in protectivity is achieved. Investigation of the effect of temperature showed that increasing the temperature from 22°C to 55°C decreases the inhibition efficiency from 96% to 89%, indicating good stability of the protective film in this temperature range. Analysis of the impedance spectra for the high concentrations of glutarate was consistent with the inhibition mechanism which assumes that the carboxylates support the passivation of carbon steel in aerated solutions by plugging the defect sites and that the passivation process is enhanced by adsorption of the carboxylates on the oxide-covered surface. Such mechanism was confirmed by the XPS analysis.

Competing Interests

The authors declare that there is no conflict of interests regarding the publication of this paper.

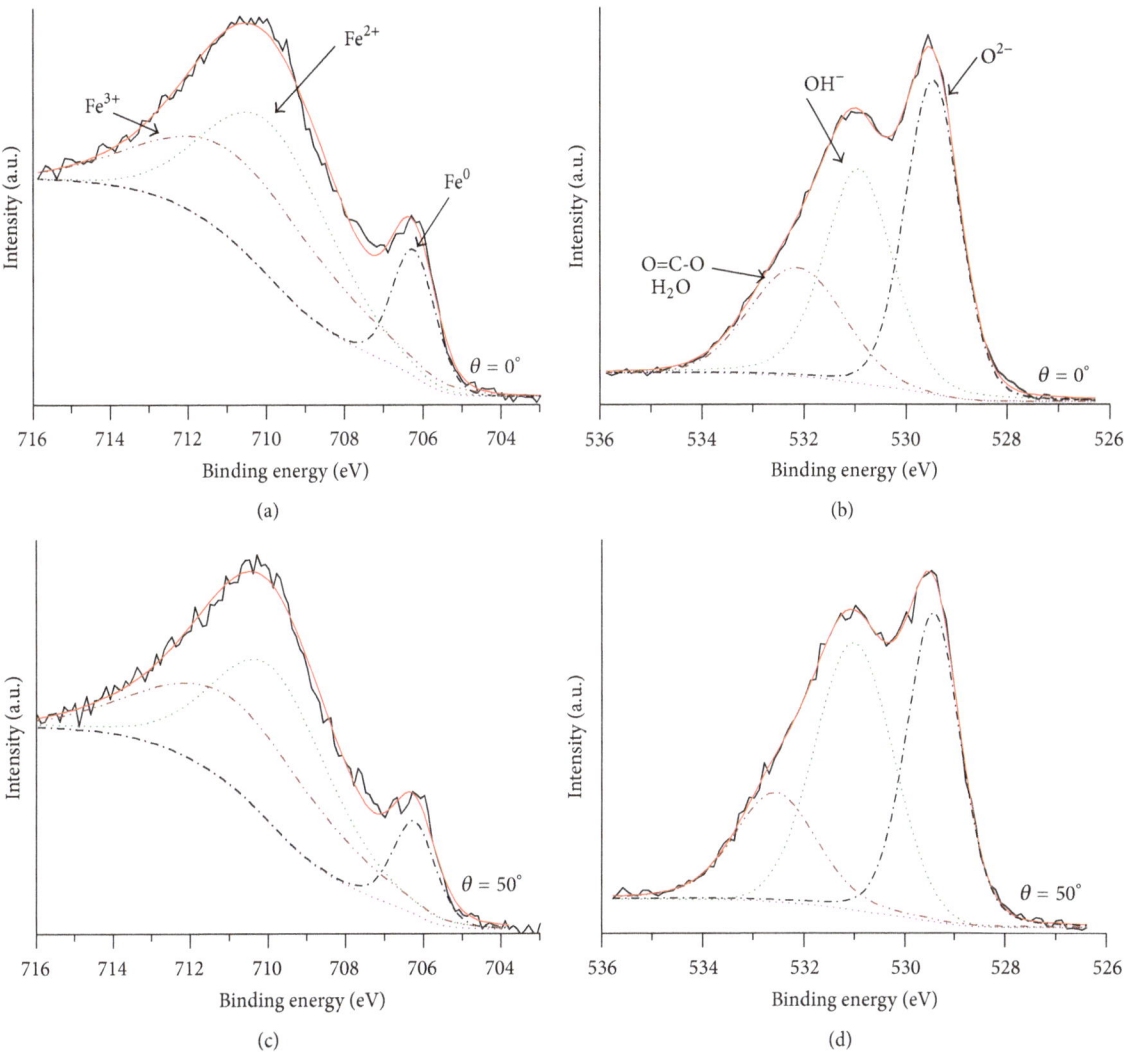

FIGURE 11: High resolution Fe $2p_{3/2}$ and O 1s spectra for carbon steel after 2 h immersion in 0.02 M NaCl containing 0.1 M Na-glutarate; spectra obtained at a takeoff angle of 0° (a, b) and of 50° (c, d).

Acknowledgments

The authors are grateful to Ms. Marbella Echeverría for technical assistance in the corrosion/electrochemistry laboratory and to Mr. W. Cauich for technical assistance with the XPS measurements. Finally, the first author is thankful to Conacyt-México for providing a scholarship for his doctoral studies.

References

[1] R. W. Revie and H. H. Uhlig, *Corrosion and Corrosion Control, An Introduction to Corrosion Science and Engineeering*, John Wiley Sons, New Jersey, USA, 4th edition, 2008.

[2] S. Turgoose, "Mechanism of corrosion inhibition in neutral environments," in *Chemical Inhibitors for Corrosion Control*, B. G. Clubley, Ed., Royal Society of Chemistry, Cambridge, UK, 1990.

[3] A. D. Mercer, "Some views on the corrosion mechanisms of inhibitors in neutral solutions," *Materials Performance*, vol. 29, no. 6, pp. 45–53, 1990.

[4] Y. I. Kuznetsov, "Physicochemical aspects of metal corrosion inhibition in aqueous solutions," *Russian Chemical Reviews*, vol. 73, no. 1, pp. 75–87, 2004.

[5] B. Sanyal, "Organic compounds as corrosion inhibitors in different environments—a review," *Progress in Organic Coatings*, vol. 9, no. 2, pp. 165–236, 1981.

[6] J.-W. Song, E.-Y. Jeon, D.-H. Song et al., "Multistep enzymatic synthesis of long-chain α,ω-dicarboxylic and ω-hydroxycarboxylic acids from renewable fatty acids and plant oils," *Angewandte Chemie International Edition*, vol. 52, no. 9, pp. 2534–2537, 2013.

[7] M. Alam, D. Akram, E. Sharmin, F. Zafar, and S. Ahmad, "Vegetable oil based eco-friendly coating materials: a review article," *Arabian Journal of Chemistry*, vol. 7, no. 4, pp. 469–479, 2014.

[8] D. M. Brasher and A. D. Mercer, "Comparative study of factors influencing the action of corrosion inhibitors for mild steel in neutral solution: I. Sodium benzoate," *British Corrosion Journal*, vol. 3, no. 3, pp. 120–129, 1968.

[9] C. L. Pace and J. E. O. Mayne, "The anomalous effect of concentrationon inhibition of the corrosion of iron by solutions of sodium benzoate," *Corrosion Science*, vol. 12, no. 8, pp. 679–681, 1972.

[10] U. Rammelt, S. Koehler, and G. Reinhard, "Synergistic effect of benzoate and benzotriazole on passivation of mild steel," *Corrosion Science*, vol. 50, no. 6, pp. 1659–1663, 2008.

[11] G. Mrowczynski and Z. Szklarska-Smialowska, "Electrochemical and eclipsometric study of iron corrosion inhibition in sodium sulphate solutions containing aliphatic acids," *Journal of Applied Electrochemistry*, vol. 9, no. 2, pp. 201–207, 1979.

[12] G. Reinhard, M. Radtke, and U. Rammelt, "On the role of the salts of weak acids in the chemical passivation of iron and steel in aqueous solutions," *Corrosion Science*, vol. 33, no. 2, pp. 307–313, 1992.

[13] U. Rammelt, S. Köhler, and G. Reinhard, "EIS characterization of the inhibition of mild steel corrosion with carboxylates in neutral aqueous solution," *Electrochimica Acta*, vol. 53, no. 23, pp. 6968–6972, 2008.

[14] U. Rammelt, S. Koehler, and G. Reinhard, "Electrochemical characterisation of the ability of dicarboxylic acid salts to the corrosion inhibition of mild steel in aqueous solutions," *Corrosion Science*, vol. 53, no. 11, pp. 3515–3520, 2011.

[15] G. T. Hefter, N. A. North, and S. H. Tan, "Organic corrosion inhibitors in neutral solutions; part 1—inhibition of steel, copper, and aluminum by straight chain carboxylates," *Corrosion*, vol. 53, no. 8, pp. 657–667, 1997.

[16] D. Lahem, M. Poelman, F. Atmani, and M.-G. Olivier, "Synergistic improvement of inhibitive activity of dicarboxylates in preventing mild steel corrosion in neutral aqueous solution," *Corrosion Engineering Science and Technology*, vol. 47, no. 6, pp. 463–471, 2012.

[17] G. R. H. Florence, A. N. Anthony, J. W. Sahayaraj, A. J. Amalraj, and S. Rajendran, "Corrosion inhibition of carbon steel by adipic acid-Zn^{2+} system," *Indian Journal of Chemical Technology*, vol. 12, no. 4, pp. 472–476, 2005.

[18] M. Manovannan and S. Rajendran, "Corrosion inhibition of carbon steel by succinic acid-Zn^{2+} system," *Research Journal of Chemical Sciences*, vol. 1, no. 8, pp. 42–48, 2011.

[19] D. Gilroy and J. E. O. Mayne, "The breakdown of the air-formed oxide film on iron upon immersion in solutions of pH 6–13," *British Corrosion Journal*, vol. 1, no. 3, pp. 102–106, 1965.

[20] H. Konno, M. Kawai, and M. Nagayama, "The mechanism of spontaneous dissolution of the air-formed oxide film on iron in a deaerated neutral phosphate solution," *Surface Technology*, vol. 24, no. 3, pp. 259–271, 1985.

[21] G. H. Cartledge, "The comparative roles of oxygen and inhibitors in the passivation of iron. I. Non-oxidizing inhibitors," *Journal of Physical Chemistry*, vol. 64, no. 12, pp. 1877–1882, 1960.

[22] D. M. Brasher, "Stability of the oxide film on metals in relation to inhibition of corrosion. II. Dual role of the anion in the inhibition of the corrosion of mild steel," *British Corrosion Journal*, vol. 4, no. 3, pp. 122–128, 2013.

[23] E. McCafferty, *Introduction to Corrosion Science*, Springer Science+Business Media, New York, NY, USA, 2010.

[24] F. Mansfeld, M. W. Kendig, and W. J. Lorenz, "Corrosion inhibition in neutral, aerated media," *Journal of the Electrochemical Society*, vol. 132, no. 2, pp. 290–296, 1985.

[25] K. Jüuttner, W. J. Lorenz, W. Paatsch, M. Kendig, and F. Mansfeld, "Bedeutung der dynamischen systemanalyse für korrosionsuntersuchungen in forschung und praxis," *Materials and Corrosion*, vol. 36, no. 3, pp. 120–130, 1985.

[26] O. Dagdag, M. El Gouri, M. Galai, M. Ebn Touhami, A. Essamri, and A. Elharfi, "Effect of temperature on the corrosion inhibition of carbon steel in 3% NaCl by hexa propylene glycol cyclotriphosphazene," *Der Pharma Chemica*, vol. 7, no. 4, pp. 284–293, 2015.

[27] G. Bhargava, I. Gouzman, C. M. Chun, T. A. Ramanarayanan, and S. L. Bernasek, "Characterization of the "native" surface thin film on pure polycrystalline iron: a high resolution XPS and TEM study," *Applied Surface Science*, vol. 253, no. 9, pp. 4322–4329, 2007.

[28] J. Wielant, T. Hauffman, O. Blajiev, R. Hausbrand, and H. Terryn, "Influence of the iron oxide acid-base properties on the chemisorption of model epoxy compounds studied by XPS," *Journal of Physical Chemistry C*, vol. 111, no. 35, pp. 13177–13184, 2007.

[29] E. Johansson and L. Nyborg, "XPS study of carboxylic acid layers on oxidized metals with reference to particulate materials," *Surface and Interface Analysis*, vol. 35, no. 4, pp. 375–381, 2003.

[30] M. De Keersmaecker, O. Van den Berg, K. Verbeken, D. Depla, and A. Adriaens, "Hydrogenated dimer acid as a corrosion inhibitor for lead metal substrates in acetic acid," *Journal of the Electrochemical Society*, vol. 162, no. 4, pp. C167–C179, 2015.

[31] P. Taheri, J. Wielant, T. Hauffman et al., "A comparison of the interfacial bonding properties of carboxylic acid functional groups on zinc and iron substrates," *Electrochimica Acta*, vol. 56, no. 4, pp. 1904–1911, 2011.

[32] A. D. Buckland, C. H. Rochester, and S. A. Topham, "Infrared study of the adsorption of carboxylic acids on haematite and goethite immersed in carbon tetrachloride," *Journal of the Chemical Society, Faraday Transactions 1: Physical Chemistry in Condensed Phases*, vol. 76, pp. 302–313, 1980.

[33] O. W. Duckworth and S. T. Martin, "Surface complexation and dissolution of hematite by C_1-C_6 dicarboxylic acids at pH = 5.0," *Geochimica et Cosmochimica Acta*, vol. 65, no. 23, pp. 4289–4301, 2001.

[34] S. J. Hug and D. Bahnemann, "Infrared spectra of oxalate, malonate and succinate adsorbed on the aqueous surface of rutile, anatase and lepidocrocite measured with in situ ATR-FTIR," *Journal of Electron Spectroscopy and Related Phenomena*, vol. 150, no. 2-3, pp. 208–219, 2006.

[35] K. D. Dobson and A. J. McQuillan, "In situ infrared spectroscopic analysis of the adsorption of aliphatic carboxylic acids to TiO_2, ZrO_2, Al_2O_3, and Ta_2O_5 from aqueous solutions," *Spectrochimica Acta A*, vol. 55, no. 7-8, pp. 1395–1405, 1999.

[36] Y. S. Hwang and J. J. Lenhart, "Adsorption of C4-dicarboxylic acids at the hematite/water interface," *Langmuir*, vol. 24, no. 24, pp. 13934–13943, 2008.

[37] S. Rajendran, V. Shribharathy, A. Krishnaveni et al., "Corrosion inhibitive property of self-assembled Nano Films formed by Adipic Acid molecules on carbon steel surface," *Elixir Thin Film Technology*, vol. 50, pp. 10509–10513, 2012.

[38] W. Temesghen and P. M. A. Sherwood, "Analytical utility of valence band X-ray photoelectron spectroscopy of iron and its oxides, with spectral interpretation by cluster and band

structure calculations," *Analytical and Bioanalytical Chemistry*, vol. 373, no. 7, pp. 601–608, 2002.

[39] A. P. Grosvenor, B. A. Kobe, M. C. Biesinger, and N. S. McIntyre, "Investigation of multiplet splitting of Fe 2p XPS spectra and bonding in iron compounds," *Surface and Interface Analysis*, vol. 36, no. 12, pp. 1564–1574, 2004.

[40] G. M. Atenas, E. Mielczarski, and J. A. Mielczarski, "Composition and structure of iron oxidation surface layers produced in weak acidic solutions," *Journal of Colloid and Interface Science*, vol. 289, no. 1, pp. 157–170, 2005.

[41] G. W. Simmons and B. C. Beard, "Characterization of acid-base properties of the hydrated oxides on iron and titanium metal surfaces," *The Journal of Physical Chemistry*, vol. 91, no. 5, pp. 1143–1148, 1987.

[42] E. McCafferty and J. P. Wightman, "Determination of the concentration of surface hydroxyl groups on metal oxide films by a quantitative XPS method," *Surface and Interface Analysis*, vol. 26, no. 8, pp. 549–564, 1998.

18

Adaptive Corrosion Protection System Using Continuous Corrosion Measurement, Parameter Extraction, and Corrective Loop

Jasbir N. Patel, Andre Chang, Haleh Shahbazbegian, and Bozena Kaminska

School of Engineering Science, Simon Fraser University, 8888 University Drive, Burnaby, BC, Canada V5A 1S6

Correspondence should be addressed to Jasbir N. Patel; jpatel@sfu.ca

Academic Editor: Michael J. Schütze

A simple current-sourced adaptive corrosion protection system (ACPS) along with a technology to extract the protection current from the Tafel plot is presented. For reliable protection of the target metal, first, the Tafel plot of the target metal is obtained. Subsequently, a novel technique proposed in this paper is used to extract the protection current from the Tafel plot. This extracted protection current is fed to the target metal to protect the metal in the existing corrosive environment. This three-part system is adaptively used to update the required protection current to effectively protect the target metal continuously. All these functionalities are integrated in a stand-alone ACPS that effectively diagnoses the corrosion status and updates the protection parameters without any manual interaction or physical modification of the set-up to offer modularity, reliability, and cost saving. To validate the technique, a laboratory scale system is realized and tested using various metal samples and various corrosive mediums. Using the experimental system, A36 metal coupons are effectively protected with protection (inhibition) efficiency of 40–100% in different corrosive mediums that can extend the life expectancy of the target metal from ~2 times to more than 100 times for the tested corrosive mediums.

1. Introduction

Despite being a well-known problem, corrosion of structures continues to be a large expense in different industries such as pipelining, aeronautics, and basic infrastructures. The corrosion of structures such as concrete bridges [1–3], concrete girders [4], and concrete structures in a marine environment [5] compromises safety and functionality of many infrastructures, leading to a costly repair [6]. Amongst other things, environmental factors such as environmental resistivity, humidity, exposure to electrolytes, and pH all play a key role in the rate at which a material such as ship structures [7], stainless steel [8], galvanized steel [9], and carbon steel [10] will corrode.

Corrosion is defined by the National Association of Corrosion Engineers (NACE) [11] as the naturally occurring deterioration of a material (usually a metal) which results from a chemical or electrochemical reaction [12] with its environment. A simple corrosion theory [13] and a physical model of the corrosion [14] are also published previously. Despite the thermodynamic tendency to undergo the oxidation reaction, there are time-proven methods that prevent and control corrosion which can reduce or eliminate its impact on public safety, economy, and environment [15, 16]. Many techniques have been developed over the years in order to protect a targeted metal. Some popular corrosion prevention techniques include use of an organic metal [17], a mechanical technique [18], a cathodic protection [19], and a photoelectrochemical approach [20]. Besides, uses of different coatings [21, 22] and electrochemical protections have proven to extend lifetime of the target metal in corrosion prone environments.

Popular methods include the use of cathodic protection systems [19, 23], one of which uses target metal as a cathode that is protected by the use of a sacrificial metal that acts as an anode (Galvanic). The impressed current cathodic protection

(ICCP) system is well explained and studied over decades [23, 24]. The effect of the alternating current (AC) for the ICCP system [25] and use of the ICCP system to protect a costal bridge structure [26] has also been described in previous studies. Though an initial installation of an ICCP system [23] is more costly, it has proven to be a cost-effective method of protection since it does not require the systematic replacement of a degraded anode.

In a typical ICCP system, appropriate control of the protection current is very important [27]. The overprotection (too much cathodic protection current) of the metal creates undesired potential throughout the structure and leads to an acceleration of the corrosion process [28]. Additionally, inappropriate ICCP system can also degrade bond strength in concrete structures [29] or the protective coating [30]. Hence, corrosion engineers must constantly evaluate and monitor existing corrosion status.

For example, in order to protect a steel structure, it has been standardized that -0.85 V (versus $Cu/CuSO_4$ reference electrode) be applied [31]; however, this potential is compensated by the instant-off potential measured in the field environment [32, 33]. However, such instant-off potential measurements are done manually in a field environment, which requires manual involvement even in remote areas. Hence, effectiveness of the corrosion protection is highly dependent on the human interaction, which is neither efficient nor cost-effective [34, 35]. Besides, capacitive spikes that appear during the current interruption can mask the instant-off potential [36, 37]. Hence, the instant-off potential measurement and the existing potential-based corrosion protection system have many limitations that should be addressed to achieve low-cost, reliable, and automated corrosion protection system. Because of the target metal properties and other environmental factors, some localized corrosion may occur in a metal structure. Effects of such nonuniform corrosion on the cracking and service life of the concrete structures have been investigated previously [38]. Even mathematical model to evaluate uniform and nonuniform corrosion induced damage in reinforced concrete is also developed [39]. Therefore, the nonuniform corrosion distribution or protection in a metal structure is a challenging problem to address.

To address these limitations, we propose a simple adaptive corrosion protection system (ACPS) to reliably protect the target metal in a wide variety of environments. The ACPS consists of measuring the electrochemical response of the target metal to obtain Tafel plot, extracting the exact protection current from the Tafel plot and adjusting the required current by using the feedback loop. This results in a diagnostic as well as protection system that can accurately monitor and protect the target metal. The feedback loop automatically measures the electrochemical behaviour (Tafel plot) analytically at the user-defined intervals and protects the target metal based on the existing corrosive conditions. The proposed closed-loop corrosion protection system offers higher reliability and precision without any adverse effects mentioned earlier. Additionally, the manual interactions are completely eliminated in the proposed system. The proposed system also optimizes the energy requirement and reduces the power waste which results in energy cost savings.

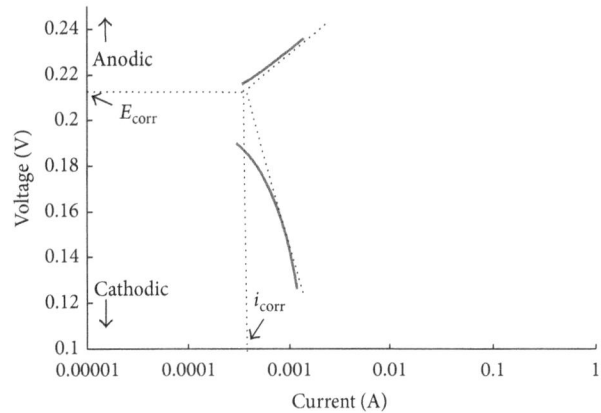

FIGURE 1: Polarization curves for iron in a quiescent solution of 0.6 M NaCl open to the air. The anodic and cathodic branches are pointed in the plot. In addition, the slopes of the anodic and cathodic branches along with interpolation of the two branches are also extracted using the dotted line. Based on this extraction, the E_{corr} and i_{corr} values are also determined.

2. Theory of the ACPS

In a typical corrosion measurement system, electrochemical response of the target metal in a corrosive environment is obtained using an electrochemical cell and the Tafel plot (see Figure 1) is obtained from the electrochemical response. The values of open-circuit potential (E_{corr}) and the corrosion current (i_{corr}) (see Figure 1) are typically extracted from the Tafel plot to calculate corrosion rate of the target metal in the given corrosive medium [40]. Additionally, the Tafel plot is accurately used to measure corrosion rate of different metals [41] and the reinforced steel [42].

It is important to note that the potential for this i_{corr} is usually the same as or very close to the open-circuit potential (E_{corr}) and i_{corr} value obtained from the Tafel plots is also the natural current flow created in a given electrochemical cell when both the oxidation and the reduction reactions are in equilibrium [43]. Hence, at this current (i_{corr}) and potential (E_{corr}), metal should be neither underprotected (not sufficient protection current or potential) nor overprotected. Thus, it is the minimum amount of current required to protect the metal from freely corroding in the present corrosion set-up. Using this minimum amount of current (i_{corr}), the optimum protection parameters can be determined by the remaining amount of corrosion that is acceptable. The proposed method opens the opportunity to control the protection current as desired by the corrosion engineer. In combination with the open-circuit voltage (E_{corr}) it is possible to predict the minimum amount of power required to protect a targeted metal. However, as we can see from the Tafel plot, the actual potential at current i_{corr} is slightly shifted towards the cathodic region from the E_{corr} potential, which ensures effectiveness of the corrosion protection at the i_{corr} current.

In summary, it is possible to precisely extract the optimum protection parameters for the present environmental condition of the target metal using the technique described

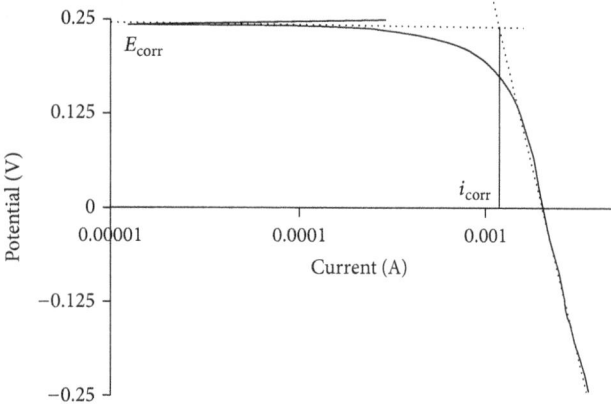

FIGURE 2: The simplified system of i_{corr} extraction. The linear part of the cathodic region is extended until it reaches the E_{corr} potential to determine i_{corr} value.

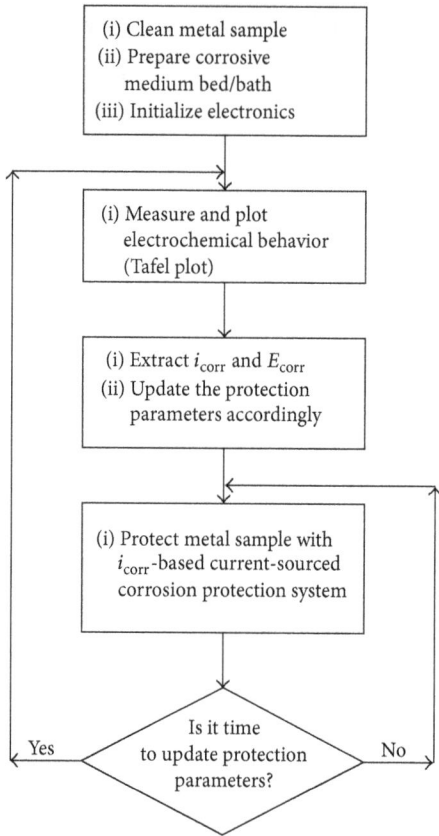

FIGURE 3: Flow chart of the proposed adaptive corrosion protection system (ACPS).

above. Also, the values of these optimum protection parameters reflect the changing environmental conditions as well as the metal corrosion status if frequent measurements can be done without changing the protection set-up. By adapting to this updated protection parameters, it is possible to protect the target metal more scientifically without the risk of underprotection or overprotection.

In a further analysis, during the electrochemical measurement (Tafel plot) of the anodic branch (Tafel plot), the target metal oxidizes slightly and hence corrodes during the anodic sweep. Even though this corrosion amount is minimal, the optimum protection parameters deviate slightly after completion of the test. This results in a slight underprotection for the optimally protected target metal. To eliminate this error, the proposed system is further simplified to minimize this anodic sweep from the electrochemical measurement to increase accuracy, reliability, and performance over an extended period (see Figure 2).

As shown in Figure 2, the electrochemical sweep is started from the anodic branch which is slightly higher than the open-circuit potential in the improved method. In this method, the value of i_{corr} is determined by extrapolating linear region of the cathodic branch and the open-circuit potential (E_{corr}) (see Figure 2). By using this method, the E_{corr} and i_{corr} values are accurately determined similar to the earlier method and reduce the small amount of metal corrosion because of the anodic sweep. This method also reduces the overall time required to measure the full behaviour and parameter extraction.

By adapting (feedback loop) the proposed current-sourced corrosion protection system and the extracted corrosion protection parameters (i_{corr} and E_{corr}), the target metal can be accurately protected with the precise amount of the corrosion protection. Unlike a classical potential-based ICCP system, the proposed method avoids overprotection or underprotection of the target metal and hence the adverse effects associated with the same. In our proposed technique, the electrochemical behaviour of the target metal is measured frequently and the optimum protection parameters

are extracted. As soon as the new optimum protection parameters are extracted, they are adapted in the protection system until the new measurement and extractions are done. We also propose to use a current-sourced protection system to effectively protect the target metal with the optimum power requirement.

3. The Proposed Adaptive Corrosion Protection System (ACPS)

The ACPS offers a novel approach over the classical potential-sourced corrosion protection system by proposing the full feedback loop based adaptive corrosion protection system.

As described in the previous section, the ACPS consists of a simple i_{corr}-based current-sourced corrosion protection system, which monitors the corrosion status at user-defined intervals and protects the target metal by adapting to the change in the corrosion status of the protected metal. Hence, the ACPS control module works as an active feedback loop system to update the protection current that is extracted from the electrochemical sweep of the protected system.

The flow chart of the ACPS is shown in Figure 3 and the exact execution steps are described as follows:

(1) The implementation of the ACPS (Figure 3) starts with the preparation of the metal to be protected and initialization of the control unit. After the initial

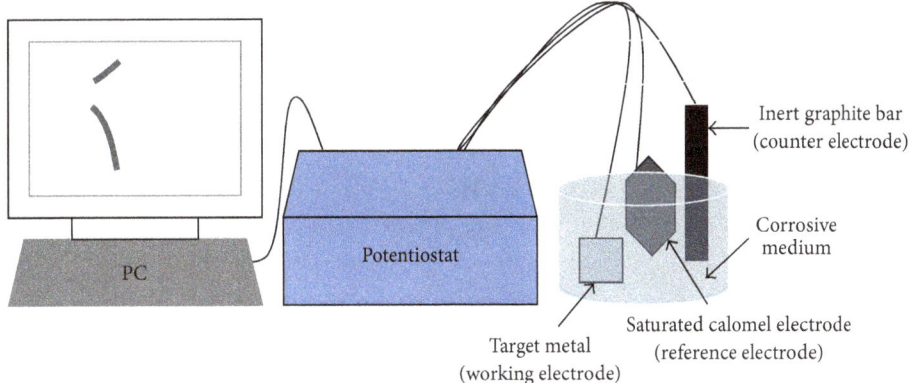

FIGURE 4: The experimental set-up to test the electrochemical behaviour.

preparation is done, the target metal and the control unit are connected together. The interval to update the protection parameters is set at this time. However, this interval can be easily updated during execution.

(2) As soon as the test parameters are determined, the electrochemical measurement is performed and the Tafel plot of the target metal is obtained.

(3) The value of i_{corr} is then extracted from the Tafel plot and the protection current is updated to adapt to changes in the corrosive state of the target metal.

(4) The metal is continuously protected with the protection current (i_{corr}) extracted in step (3) until the time interval to update the protection parameters is reached.

(5) When the time interval to update the protection parameters is reached, the complete cycle is started again from step (2).

We can see from the flow chart (Figure 3) and the steps above that the proposed control unit can be realized using a stand-alone embedded module, a LabVIEW based virtual instrument, or other low-cost systems. Additionally, a low-cost graphite bar can also be used as an electrode. Hence, the overall implementation of the proposed ACPS can be very cost-effective and it is simple to implement in a laboratory or a field environment. The ACPS system can also be used as a diagnostic tool to determine exact corrosion state of the target metal.

4. Materials and Methods

To prove the functionality of the ACPS and ability to protect a target metal effectively, systematic experiments are carried out in a laboratory environment. Each individual experimental set-up is described below with specific details and the appropriate block diagram.

4.1. Experimental Set-Up for Electrode Configuration. To determine usability of the electrochemical set-up for the ACPS, a set of experiments is established using a personal

computer (PC) controlled potentiostat (ParStat 4000 from Princeton Applied Research) (see Figure 4).

The typical electrochemical set-up (Figure 4) is implemented by connecting the target metal to the working electrode, an inert graphite bar to the counter electrode, and the saturated calomel electrode to the reference electrode of the potentiostat. The experimental set-up is prepared using a 0.6 M NaCl solution as the corrosive medium and an A36 hot-rolled steel sample (2.5 cm × 2.5 cm × 0.5 cm) as the target metal. Hence, the surface area exposed to the corrosive medium during experiments is 3125 mm^2.

The Tafel plots are obtained by sweeping potential on the working electrode with respect to the reference electrode for the given three-electrode configuration. For the experiments, the potential is swept from −0.25 V to 0.25 V with the scan rate of 2 mV/sec. The standard calomel electrode (SCE) is used as the reference electrode for all the measurements. The A36 steel sample is cleaned before each experiment to achieve repeatability in the measurements. Once complete, the i_{corr} value is extracted for each experiment and analyzed.

4.2. Experimental Set-Up for Corrosion Protection. After understanding behaviour of the electrochemical measurement and the resultant Tafel plot, experimental validation of the current-sourced protection system is performed. The extracted i_{corr} value is used as the protection current for these experiments.

To validate the principle, two different metal samples, a common galvanized hardware washer (surface area: 278.5 mm^2) and an A36 steel sample (same as previous experiment), are used. Before starting the experiment, the metal samples are polished and rinsed using isopropyl alcohol (IPA) and deionized (DI) water and dried under the N_2 stream. Two separate samples of the washer and the A36 steel are prepared. One washer and an A36 steel sample are protected using the proposed system and the second set is left in the corrosive medium to corrode freely. The 0.6 M NaCl solution is used as the corrosive medium. To extract the i_{corr} current, Tafel plot is obtained using the set-up described in Figure 4. The i_{corr} value is extracted using the potentiostat software and directly used to feed the current as a current-sourced corrosion protection system. The first set with both

TABLE 1: Different corrosive mediums used for the validation of the proposed ACPS. The concentrations of the corrosive mediums if they are important and the importance of each corrosive medium are also described.

Name of the corrosive medium	Concentration of the corrosive medium, if applicable	Comment
Sink water	—	Typical tap water in North America, which is quite corrosive
Humic acid	50 mg/L	It is a known substance to cause corrosion in the agricultural lands
Wet soil bed	—	This is a highly corrosive environment
Dry sand bed	—	This is a very weak corrosive environment

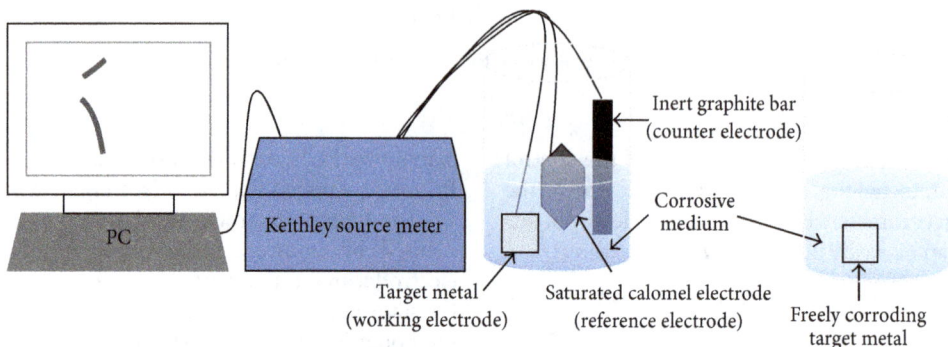

FIGURE 5: A simple experimental set-up to validate the proposed adaptive loop. A simple and cost-effective DC source meter controlled with the LabVIEW® VI is used for the set-up.

metal samples is protected for 24 hours for the proof-of-concept test. The other set is corroded freely for the same duration before all the samples are cleaned in an IPA bath and weighted again to determine weight loss of the samples.

To further validate the proposed corrosion protection system, the same experiment is extended for more than two weeks with weight measurements, electrochemical measurements, i_{corr} extraction, and current-source modification every 24 hours. The weight loss for all the samples is then analyzed. Protection (inhibition) efficiency of the system is then calculated using the following equation [44]:

$$\eta\,(\%) = \frac{W_{\text{freely corroded}} - W_{\text{protected}}}{W_{\text{freely corroded}}} \times 100, \qquad (1)$$

where $\eta(\%)$, $W_{\text{freely corroded}}$, and $W_{\text{protected}}$ represent protection (inhibition) efficiency, weight loss of the freely corroded metal sample, and weight loss of the protected metal sample, respectively.

4.3. The Adaptive Loop Set-Up. After validating the functionality of the i_{corr}-based current-sourced corrosion protection system, an automated and simple system is implemented using a simple DC source meter (Keithley Model 2400) controlled by a personal computer-based stand-alone ACPS that is implemented using the National Instrument (NI) LabVIEW based VI (see Figure 5).

In this low-cost set-up, the A36 steel sample (2.5 cm × 2.5 cm × 0.5 cm) is used for all the experiments. First, the 0.6 M NaCl solution is used as the corrosive medium. Additionally, this experiment is performed using one A36 sample

protected with the automated adaptive loop and the other A36 sample is corroded freely. The weight measurements and data collection method are the same as those in the previous subsection.

The LabVIEW VI is configured to sweep the Tafel plot of the target metal at the user-defined interval. In our experimental set-up, the Tafel plot measurement and protection parameter extraction are done every 24 hours. The value of i_{corr} is extracted from the LabVIEW VI and used by the DC source meter to protect the target metal until the i_{corr} value is extracted again. This simple and portable adaptive loop based system is also configured to obtain performance indicators of the corrosion protection continuously.

4.4. Set-Up for the ACPS Performance Study in Different Corrosive Mediums. The preliminary tests to prove the concept are performed using the 0.6 M NaCl solution as the corrosive medium. However, to further validate the effectiveness of the proposed ACPS to protect the target metal, the protection should be verified in different corrosive mediums that are typically used in the field of corrosion measurement and protection (see Table 1).

The corrosive mediums for this study are selected based on the difference of their corrosive properties and ease of availability. First, to validate our proposed principle and system, readily available sink water from a kitchen tap has been used. Secondly, the 50 mg/L humic acid solution is examined. Typically, the humic acid solution is used in research to replicate the corrosive effect of the agricultural lands. After achieving promising results from the above experiments, the ACPS tests are further expanded to use

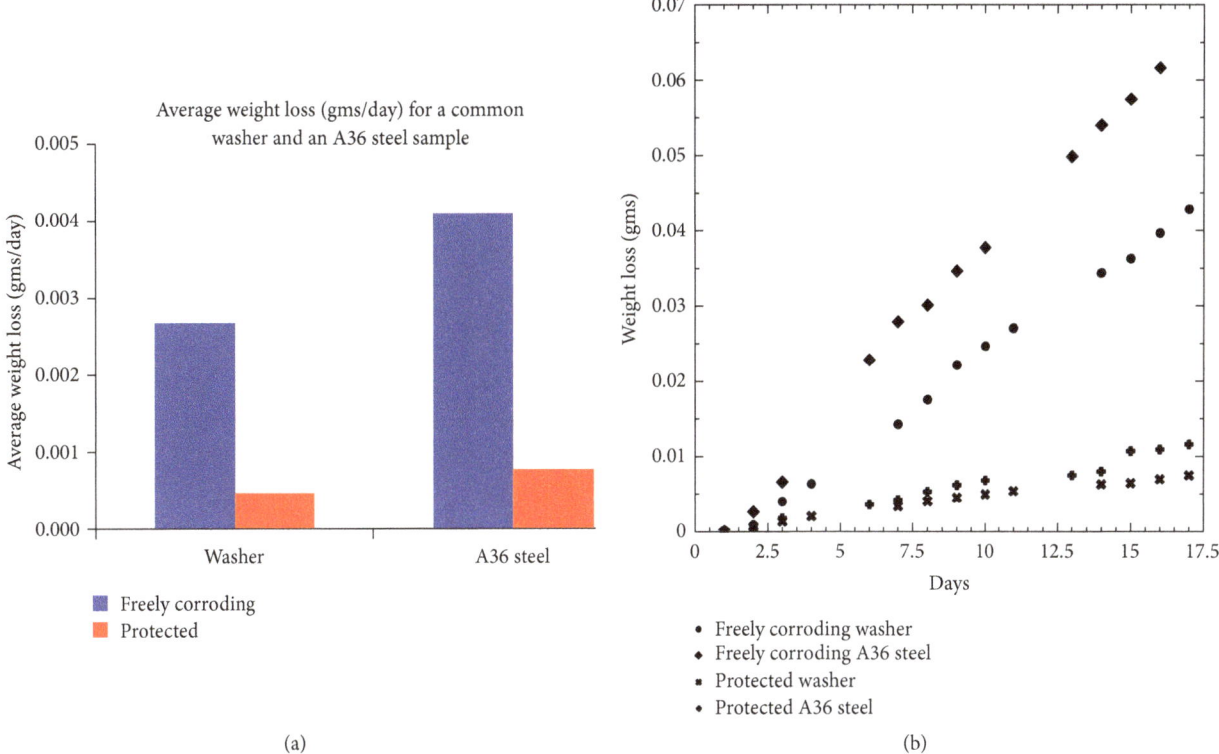

(a)

(b)

FIGURE 6: Weight loss of the freely corroding and the protected washer and A36 steel sample. Two different metal samples, a 1″ steel washer and an A36 steel metal, are immersed in the 0.6 M NaCl solution for (a) 24 hours and (b) more than two weeks. (a) The weight loss for 24 hours and (b) the weight loss for more than two weeks.

directly available garden soil and coarse sand samples. The wet soil is highly conductive medium because of the large amount of minerals in soil. Because of this, the wet soil is considered as a highly corrosive medium. On the other hand, the dry sand is poorly conductive and hence it is typically considered as a poorly corrosive medium. The different corrosive mediums selected here present different electrochemical conditions which results in different ICCP requirements to protect the target metal.

The ACPS set-up (shown in Figure 5), analysis method, data collection, and performance analysis method for these experiments remain the same as those in the previous section.

5. Results and Discussion

In this section, the main results and outcome during the laboratory experiments with the ACPS are presented. As described in the previous section, the validity of the extracted i_{corr} to protect different metal samples is examined first. The proof of accurate functioning of the adaptive loop is subsequently described followed by the ACPS performance assessment using the different corrosive mediums.

5.1. Validation of the Extracted i_{corr} as the Protection Current. After understanding behaviour of the electrochemical measurement and the resultant Tafel plot, experimental validation

TABLE 2: The weight loss per day for the freely corroding and the protected washer and A36 steel sample in the 0.6 M NaCl solution.

	Weight loss (gm/day)		
	Freely corroding	Protected	Protection (inhibition) efficiency (η)
Washer	0.00252	0.00044	83%
A36 steel	0.00385	0.00068	82%

of the extracted i_{corr} for the current-sourced protection system is done. As described in the experimental set-up in the previous section, the metal samples are weighted before and after they are protected using the current-sourced (i_{corr}-based) system. First, the samples are tested for 24 hours (see Figure 6(a)), which is followed by a long-term test (see Figure 6(b)).

During the first experiment (Figure 6(a)), the i_{corr}-based current-sourced protection system shows very good protection behaviour. Both metal samples show more than 82% efficiency of the proposed protection after 24 hours (see Table 2). The freely corroding washer and the A36 metal sample lose 0.00252 gm/day and 0.00385 gm/day weight, respectively. On the other hand, the protected washer and the A36 steel sample only lose 0.00044 gm/day and 0.00068 gm/day weight, respectively. This results in 83% and 82% reduction in weight loss in a day for the washer and the A36 steel samples, respectively. These results clearly validate the corrosion protection principle proposed in this paper. The percentages of weight loss for the A36 sample and the washer are closely related. The reason behind such a similarity is related to the type of corrosive medium used for both cases in our experiments.

For the long-term test (see Figure 6(b)), the protected metal samples showed significant reduction in weight loss in comparison to the freely corroding metal. The total weight loss after 16 days for the freely corroding washer and the A36 sample is 0.04032 gm and 0.0616 gm, respectively. On the other hand, the total weight loss after 16 days for the protected washer and the A36 sample is 0.00704 gm and 0.01088 gm, respectively. Hence, the reduction in the weight loss for the protected samples is again more than 82% for both metal samples. In the field environment, 82% protection efficiency may not be adequate to protect the target metal structure for a very long period of time. However, the proposed method will be optimized in the future to provide improved performance and higher protection efficiency. Additionally, the weight loss for the freely corroding samples as well as for the protected samples is linear for the full length of the test period. It is also important to note that all the samples show repeatable behaviour even during the weekends when they are not weighted or corrected during the test. It is also important to note that these tests are performed in a highly corrosive medium (0.6 M NaCl solution) and slight deviation in the extraction of the protection current may result in failure of the whole protection system. However, the proposed technique quickly adjusts to the changes and protects the samples effectively.

5.2. Validation of the Functional Adaptive Loop.
After validating the principle of i_{corr}-based current-sourced corrosion protection system, the subsequent approach is to implement the automated system including the feedback loop to correct the corrosion protection current (i_{corr}). It should be noted that the corrosion status measurement and the protection parameter adjustment are done at every 24-hour interval. In the future, this adjustment interval can be optimized for more consistent and repeatable performance. To validate the adaptive (feedback) loop for the proposed ACPS, the A36 steel samples in 0.6 M NaCl solution are tested for 15 days.

The weight loss for the freely corroding and the protected sample is measured every 24 hours and plotted (see Figure 7).

Similar to the previous experiment with the metal washer and the A36 steel sample (Figures 6(a) and 6(b)), the adaptive loop based corrosion protection (Figure 7) works successfully and protects the A36 metal sample using the automated measurement and extraction system described in the previous experimental section. The weight loss of the protected sample and the freely corroding sample after 15 days is 0.01 gm and 0.0574 gm, respectively. Hence, the protection (inhibition) efficiency for the adaptively protected sample is 87%, which is better than the results obtained in the manually experimented system in the previous section. These results clearly indicate that the adaptive feedback loop of the proposed ACPS performs effective corrosion protection.

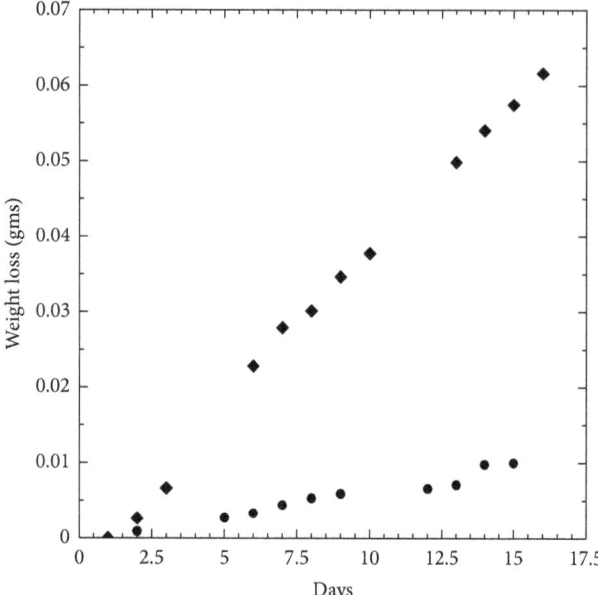

• Protected
♦ Freely corroding

FIGURE 7: The weight loss of the freely corroding and the ACPS protected A36 steel samples in the 0.6 M NaCl solution for 15 days. Two A36 samples, one corrosion protected using the proposed system and the other freely corroding, are shown. The freely corroding sample lost 0.0616 gm weight and the protected sample lost only 0.01 gm weight.

5.3. ACPS Performance with Different Corrosive Mediums.
In the previous subsection, the functionality of the adaptive feedback loop to actively protect the metal from corrosion is discussed. To further validate the proposed principle and the ACPS, the A36 steel samples are tested with different corrosive mediums as mentioned earlier. The individual results for the different corrosive mediums and a comparative chart of the corrosive mediums tested in our laboratory are shown in Figure 8.

From the charts of the different corrosive mediums for roughly two weeks (see Figure 8), it is clearly seen that the weight of the freely corroding metal is linearly decreasing which clearly shows degradation of the A36 steel sample in all corrosive mediums. However, the weight loss and hence the degradation of the metal are not present or minimal for the metal samples that are protected using the proposed ACPS. The weight loss per day for the freely corroding and the ACPS protected metal samples with the different corrosive mediums is also tabulated (see Table 3) and plotted in a clustered column chart (see Figure 8(e)).

The average weight loss per day for the freely corroding metal samples in the NaCl solution, the sink water solution, the humic acid solution, the wet soil bed, and the dry sand bed is 0.00560 gm, 0.00784 gm, 0.00752 gm, 0.00602 gm, and 0.00003 gm, respectively. On the other hand, the average weight loss per day for the ACPS protected samples in the NaCl solution, the sink water solution,

TABLE 3: The average weight loss per day for the freely corroding and the ACPS protected A36 steel sample in different corrosive mediums.

Corrosive medium	Average weight loss per day		Protection (inhibition) efficiency (η)	Increased life expectancy
	Freely corroding	ACPS protected		
0.6 M NaCl solution	0.00560	0.00071	87%	7.89
Sink water solution	0.00784	0.00051	93%	15.37
50 mg/L humic acid solution	0.00752	0.00004	99%	188.00
Wet soil bed	0.00602	−0.00063	≥100%	—
Dry sand bed	0.00003	0.00002	40%	2.50

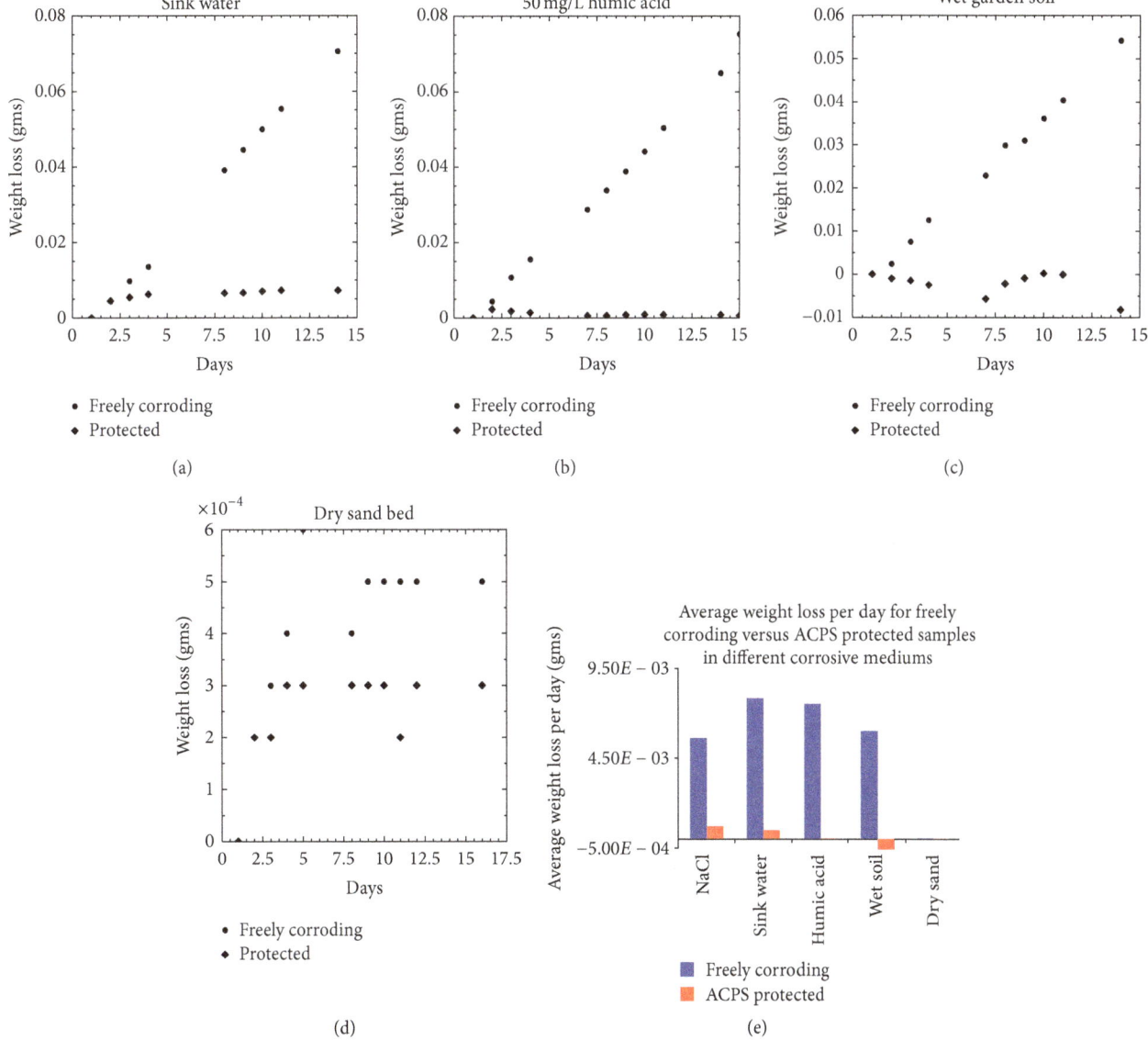

FIGURE 8: The weight loss of the freely corroding and the protected metal samples for more than two weeks in (a) the sink water solution, (b) the 50 mg/L humic acid solution, (c) the wet soil bed, and (d) the dry sand bed. Two A36 steel samples, one freely corroding and the other protected, are tested in each corrosive medium. (a) The weight loss of the freely corroding and the protected samples in the sink water solution for 13 days. (b) The weight loss of the freely corroding and the protected samples in the 50 mg/L humic acid solution for 14 days. (c) The weight loss of the freely corroding and the protected samples in the wet soil bed for 14 days. (d) The weight loss of the freely corroding and the protected samples in the dry sand bed for 12 days. (e) The calculated average weight loss per day for the freely corroding and the protected samples in all corrosive mediums.

the humic acid solution, the wet soil bed, and the dry sand bed is 0.00071 gm, 0.00051 gm, 0.00004 gm, −0.00063 gm, and 0.00002 gm, respectively. The resulting protection (inhibition) efficiency and extension in the life expectance of the target metal are tabulated in Table 3.

As we can see from Table 3 and Figure 8, the ACPS protected sample in the wet soil bed gains slight weight during the experiment. This may be because of the presence of particles or mineral deposits in the garden soil which are attached to the metal sample during the test. However, the freely corroding sample in the same environment (wet soil) linearly loses weight during the experimental period. From the experimental data, the dry sand bed showed minimal weight loss even with the freely corroding sample compared to the other corrosive mediums, which is expected because the dry sand is a poorly corrosive environment. The total weight loss of the freely corroding sample in the sand bed is about two orders of magnitude lower than the other corrosive mediums we test. Additionally, the weight loss for these samples is well within the resolution of the weighting scale and hence the weight loss data is very susceptible to error and incorrect protection (inhibition) efficiency. However, the proposed ACPS reduces metal degradation by 40%. The metal degradation behaviour in the other known lab environments (the NaCl solution, the humic acid solution, and the sink water solution) shows significant improvement with the ACPS protected system which reduces metal degradation from 87% to 99% which represents an accurate performance of the proposed system. It is also important to note that the protection (inhibition) efficiency can be improved even further using frequent corrections compared to 24 hrs used in our study.

In summary, the proposed ACPS works efficiently to reduce the metal degradation in the various corrosive environments. The proposed system is also tested using different metals (Figure 6) to validate the functionality and repeatability of the i_{corr}-based current-sourced ACPS. In the future, even better precision and performance improvement can be obtained by vigorous engineering implementation and software control.

6. Conclusions

A method of impressed current cathodic protection (ICCP) using a current-based approach as opposed to the traditional potential-based approach which is the use of a current-controlled source over a voltage-controlled source is presented.

To determine and calculate the optimum power and hence the effective protection current (i_{corr}), the Tafel plot of the target metal (such as an A36 metal sample) is obtained. The protection current from the Tafel plot is extracted and used to protect a common hardware washer and the A36 steel sample. Using the protection system, the washer and the A36 steel sample show 83% and 82% protection (inhibition) efficiency for the target metals.

To further optimize the protection parameter extraction, the experimental automated system is implemented using the LabVIEW VI on a personal computer (PC). The automated

adaptive loop is tested using the A36 steel samples in 0.6 M NaCl solution and the protection (inhibition) efficiency is 87%.

The ACPS is also validated in various corrosive mediums such as the sink water solution, the 50 mg/L humic acid solution, the wet soil bed, and the dry sand bed. The ACPS protection (inhibition) efficiency for the target metal is 93%, 99%, 110%, and 40% in the sink water, the humic acid, the wet soil bed, and the dry sand bed, respectively. Hence, the life expectancy of the target metal is increased by 7.89 times, 15.37 times, 188 times, and 2.5 times for the sink water, the humic acid, the wet soil bed, and the dry sand bed, respectively.

In summary, traditional cathodic protection applies a fixed potential to the target sample. If the sample's environmental or protection conditions change, the sample will be underprotected or overprotected, which can be detrimental. The new ACPS based approach adjusts the applied cathodic potential regularly by adjusting a constant current (i_{corr}), which can achieve the best protection for the sample.

Competing Interests

The authors declare that they have no competing interests.

Acknowledgments

The authors are thankful to Mitacs, Canada Research Chairs Program, and Natural Sciences and Engineering Research Council for their financial support to carry out this experimental work at the Simon Fraser University in Burnaby, British Columbia.

References

[1] P. Atkins and J. de Paula, *Physical Chemistry for the Life Sciences*, W. H. Freeman and Company, New York, NY, USA, 2005.

[2] M. G. Stewart and D. V. Rosowsky, "Structural safety and serviceability of concrete bridges subject to corrosion," *Journal of Infrastructure Systems*, vol. 4, no. 4, pp. 146–155, 1998.

[3] J. S. Kong and D. M. Frangopol, "Life-cycle reliability-based maintenance cost optimization of deteriorating structures with emphasis on bridges," *Journal of Structural Engineering*, vol. 129, no. 6, pp. 818–828, 2003.

[4] D. M. Frangopol, K.-Y. Lin, and A. C. Estes, "Reliability of reinforced concrete girders under corrosion attack," *Journal of Structural Engineering*, vol. 123, no. 3, pp. 286–297, 1997.

[5] A. A. Torres-Acosta and M. Martínez-Madrid, "Residual life of corroding reinforced concrete structures in marine environment," *Journal of Materials in Civil Engineering*, vol. 15, no. 4, pp. 344–353, 2003.

[6] D. M. Frangopol, K.-Y. Lin, and A. C. Estes, "Life-cycle cost design of deteriorating structures," *Journal of Structural Engineering*, vol. 123, no. 10, pp. 1390–1401, 1997.

[7] C. G. Soares, Y. Garbatov, A. Zayed, and G. Wang, "Influence of environmental factors on corrosion of ship structures in marine atmosphere," *Corrosion Science*, vol. 51, no. 9, pp. 2014–2026, 2009.

[8] H. P. Leckie and H. H. Uhlig, "Environmental factors affecting the critical potential for pitting in 18–8 stainless steel," *Journal of the Electrochemical Society*, vol. 113, no. 12, pp. 1262–1267, 1966.

[9] K. Suzumura and S.-I. Nakamura, "Environmental factors affecting corrosion of galvanized steel wires," *Journal of Materials in Civil Engineering*, vol. 16, no. 1, pp. 1–7, 2004.

[10] J. J. S. Rodríguez, F. J. S. Hernández, and J. E. G. González, "The effect of environmental and meteorological variables on atmospheric corrosion of carbon steel, copper, zinc and aluminium in a limited geographic zone with different types of environment," *Corrosion Science*, vol. 45, no. 4, pp. 799–815, 2003.

[11] L. V. Delinder, *An Introduction to Corrosion, Corrosion Basics*, NACE, Houston, Tex, USA, 1984.

[12] C. Chunan, "Corrosion electrochemistry," *Journal of Chemistry Industry*, 1994.

[13] D. A. Vermilyea and C. S. Tedmon,, "A simple crevice corrosion theory," *Journal of the Electrochemical Society*, vol. 117, no. 4, pp. 437–440, 1970.

[14] Z. P. Bazant, "Physical model for steel corrosion in concrete sea structures—theory," *Journal of the Structural Division*, vol. 105, no. 6, pp. 1137–1153, 1979.

[15] R. G. Kelly Jr., *Electrochemical Techniques in Corrosion Science and Engineering*, Mercel Dekker, Basel, Switzerland, 2003.

[16] G. A. Jacobson, *Corrosion 101*, NACE, Houston, Tex, USA, 2014, http://www.nace.org/.

[17] B. Wessling and J. Posdorfer, "Corrosion prevention with an organic metal (polyaniline): corrosion test results," *Electrochimica Acta*, vol. 44, no. 12, pp. 2139–2147, 1999.

[18] E. Heitz, "Mechanistically based prevention strategies of flow-induced corrosion," *Electrochimica Acta*, vol. 41, no. 4, pp. 503–509, 1996.

[19] C. Christodoulou, G. Glass, J. Webb, S. Austin, and C. Goodier, "Assessing the long term benefits of Impressed Current Cathodic Protection," *Corrosion Science*, vol. 52, no. 8, pp. 2671–2679, 2010.

[20] H. Park, K. Y. Kim, and W. Choi, "Photoelectrochemical approach for metal corrosion prevention using a semiconductor photoanode," *Journal of Physical Chemistry B*, vol. 106, no. 18, pp. 4775–4781, 2002.

[21] X. G. Zhang and E. M. Valeriote, "Galvanic protection of steel and galvanic corrosion of zinc under thin layer electrolytes," *Corrosion Science*, vol. 34, no. 12, pp. 1957–1972, 1993.

[22] M. Morcillo, R. Barajas, S. Feliu, and J. M. Bastidas, "A SEM study on the galvanic protection of zinc-rich paints," *Journal of Materials Science*, vol. 25, no. 5, pp. 2441–2446, 1990.

[23] J. B. Bushman, *Impressed Current Cathodic Protection System Design*, Bushman & Associates, 2010.

[24] S. Szabó and I. Bakos, "Impressed current cathodic protection," *Corrosion Reviews*, vol. 24, no. 1-2, pp. 39–62, 2006.

[25] L. Y. Xu, X. Su, and Y. F. Cheng, "Effect of alternating current on cathodic protection on pipelines," *Corrosion Science*, vol. 66, pp. 263–268, 2013.

[26] S. D. Cramer, B. S. Covino Jr., S. J. Bullard et al., "Corrosion prevention and remediation strategies for reinforced concrete coastal bridges," *Cement and Concrete Composites*, vol. 24, no. 1, pp. 101–117, 2002.

[27] R. W. Revie, *Corrosion and Corrosion Control*, John Wiley & Sons, Hoboken, NJ, USA, 2008.

[28] A. Oni, "Effects of cathodic overprotection on some mechanical properties of a dual-phase low-alloy steel in sea water," *Construction and Building Materials*, vol. 10, no. 6, pp. 481–484, 1996.

[29] J. J. Chang, "A study of the bond degradation of rebar due to cathodic protection current," *Cement and Concrete Research*, vol. 32, no. 4, pp. 657–663, 2002.

[30] S. Amami, C. Lemaitre, A. Laksimi, and S. Benmedakhene, "Characterization by acoustic emission and electrochemical impedance spectroscopy of the cathodic disbonding of Zn coating," *Corrosion Science*, vol. 52, no. 5, pp. 1705–1710, 2010.

[31] P. Pedeferri, "Cathodic protection and cathodic prevention," *Construction and Building Materials*, vol. 10, no. 5, pp. 391–402, 1996.

[32] Underground Storage Tank Branch and Mississippi Department of Environmental Quality, "Guidelines for the evaluation of underground storage tank cathodic protection systems," July 2002, http://www.neiwpcc.org/neiwpcc_docs/MS%20DEQ%20Guidelines%20for%20the%20Evaluation%20of%20UST%20Cathodic%20Protection%20Systems.pdf.

[33] T. Kodama, K. Kimura, Y. Shinoda, and N. Mochizuki, "Determination of instant-off potential of cathodically protected reinforcing steel in concrete bridges located in coastal atmosphere," in *Proceedings of the CORROSION*, NACE International, San Diego, Calif, USA, March 2006.

[34] L. V. Nielsen, K. V. Nielsen, B. Baumgarten, H. Breuning-Madsen, and H. Rosenberg, "AC induced corrosion in pipelines: detection, characterization and mitigation," in *Proceedings of the CORROSION*, NACE International, New Orleans, La, USA, March-April 2004.

[35] F. Kajiyama and Y. Nakamura, "Effect of induced alternating current voltage on cathodically protected pipelines paralleling electric power transmission lines," *Corrosion*, vol. 55, no. 2, pp. 200–205, 1999.

[36] J. Ansuini Frank and J. R. Dimond, "Field tests on an advanced cathodic protection coupon," CORROSION/2005 paper 039, 2005.

[37] N. G. Thompson and J. A. Beavers, "Measurements of IR-drop free pipe-to-soil potentials on buried pipelines," in *The Measurement and Correction of Electrolyte Resistance in Electrochemical Tests*, vol. 1056, p. 168, ASTM International, 1990.

[38] B. S. Jang and B. H. Oh, "Effects of non-uniform corrosion on the cracking and service life of reinforced concrete structures," *Cement and Concrete Research*, vol. 40, no. 9, pp. 1441–1450, 2010.

[39] Y. Zhao, A. R. Karimi, H. S. Wong, B. Hu, N. R. Buenfeld, and W. Jin, "Comparison of uniform and non-uniform corrosion induced damage in reinforced concrete based on a Gaussian description of the corrosion layer," *Corrosion Science*, vol. 53, no. 9, pp. 2803–2814, 2011.

[40] F. Mansfeld, "Tafel slopes and corrosion rates from polarization resistance measurements," *Corrosion*, vol. 19, no. 10, pp. 397–402, 1973.

[41] Z. Shi, M. Liu, and A. Atrens, "Measurement of the corrosion rate of magnesium alloys using Tafel extrapolation," *Corrosion Science*, vol. 52, no. 2, pp. 579–588, 2010.

[42] A.-H. J. Al-Tayyib, Mohammed, and S. Khan, "Corrosion rate measurements of reinforcing steel in concrete by electrochemical techniques," *ACI Materials Journal*, vol. 85, no. 3, pp. 172–177, 1988.

[43] E. McCafferty, "Validation of corrosion rates measured by the Tafel extrapolation method," *Corrosion Science*, vol. 47, no. 12, pp. 3202–3215, 2005.

[44] F. Bentiss, M. Traisnel, L. Gengembre, and M. Lagrenée, "New triazole derivative as inhibitor of the acid corrosion of mild steel: electrochemical studies, weight loss determination, SEM and XPS," *Applied Surface Science*, vol. 152, no. 3, pp. 237–249, 1999.

19

Experimental and Runge–Kutta Method Simulation to Investigate Corrosion Kinetics of Mild Steel in Sulfuric Acid Solutions

Ismaeel M. Alwaan ⓘ

Department of Materials Engineering, College of Engineering, University of Kufa, Najaf, Iraq

Correspondence should be addressed to Ismaeel M. Alwaan; ism10alw@yahoo.com

Academic Editor: Ramana M. Pidaparti

The mild steel is extensively used in different industrial applications and the biggest problem in the application of mild steel is corrosion. In this work, the reaction kinetics of mild steel with sulfuric acid at different concentrations and at different temperatures were studied in combination with the experimental data and theoretical approach using the Runge–Kutta method. The results revealed that the rate of reaction constant for temperatures in the range of 30–50°C was changed from 2618 to 2793 L^3/mol^3.h, respectively. The order of reaction of mild steel was 4th order in all temperature ranges. The enthalpy, entropy, and Gibbs free energy of mild steel reaction at a temperature of 298 K were estimated. The activation energy (E/R) of the reaction was 4.829 K. It was concluded that the sulfuric acid reaction with mild steel occurred easily and the inhibitors should be used in these systems.

1. Introduction

Mild steel alloy has been extensively utilized in manufacture as a substance for reaction containers, pipes, etc. [1]. Poor corrosion resistance of mild steel in corrosive electrolytes has largely prevented its implementation. Acid solutions are commonly used in different processes such as acid picking, cleaning, and descaling, which may cause the corrosion of metals [2, 3]. The alloy of mild steel destroyed by corrosive materials has led to significant economic losses and has created huge problems in industrial instrument security [4].

The investigation of mild steel corrosion and iron is a major theoretical issue and has attracted significant attention. Many scientists are conducting research on mild steel corrosion. The corrosion inhibitors of two imidazoline derivatives have been investigated for mild steel and the chloride-substituted was found better as compared with the fluoride-substituted [5]. The effect of 4,6-diamino-2-pyrimidinethiol (4D2P) inhibitor on mild steel oxidation in hydrochloride acid media was studied [6]. The rind, seed, and peel extract of watermelon were studied as corrosion inhibitor for mild steel in hydrochloride acid media [7].

Computational methods were extensively used for the purpose of designing new inhibitors with excellent inhibition characteristics. The simulations investigation was adopted to investigate corrosion-resisting aluminum and stainless steel pipes using 3D finite element model [8]. The sodium phosphate, sodium nitrite, and benzotriazole inhibitors were used to simulate steel metal in concrete pore solutions [9]. Thixoforging and simulation of complex parts of aluminum alloy AlSi7Mg were investigated [10]. A mathematical model was carried out for the sulfuric acid and ferric ion diffusion and the copper sulfide mineral leaching process [11]. Carbon fiber, carbon/carbon, and some modified carbon/carbon blends were exposed to a simulated atomic oxygen ambience to study their attitude in low earth orbit [12]. The reactions between silicon and nitrogen were studied using the shrinking core model [13]. Monte Carlo simulation technique was adopted to study the adsorption behavior of furan derivatives on mild steel face in hydrochloric acid [14]. Corrosion inhibition mechanism of two-mercaptoquinoline Schiff based on mild steel surface is investigated by quantum chemical calculation and molecular dynamics simulation [15].

TABLE 1: The chemical structure of mild steel specimen (weight percentage).

Elements	C	Si	Mn	P	S	Cr	Mo	Cu	Fe
Weight percentage (wt %)	0.19	0.26	0.64	0.06	0.05	0.08	0.02	0.27	Bal.

Runge–Kutta method was extensively used for solving the different model of corrosion. Runge–Kutta, Euler–Maruyama, and Milstein methods were utilized to investigate the relationship between the time and pit corrosion depth in nuclear power plant piping systems [16]. The model is suggested to predict the precipitation allocation of corrosion output by using five-order Runge–Kutta format [17]. The Runge–Kutta method was used to solve the two-phase homogeneous model numerically and the major attitude of activated corrosion outputs [18]. A new partial differential model for monitoring and detecting copper corrosion products by sulfur dioxide (SO_2) pollution is proposed using Runge–Kutta method [19].

The present paper explores corrosion kinetics of mild steel in sulfuric acid solutions using weight loss techniques. The effect of temperature (30–50°C) on corrosion is thoroughly assessed and discussed. Thermodynamic parameters were also calculated and discussed. The Runge–Kutta method was furthermore applied in an endeavor to obtain insights into the mechanism of corrosion of mild steel face at the molecular level.

2. Experimental

Mild steel sheet was mechanically press scissor into pieces of measuring $3 \times 2 \times 0.1$ cm. These pieces were utilized without polishing. However, surface curing of the pieces included cleaning, degreasing in absolute ethanol, and drying in acetone. Solutions of (0.1-0.5 M) H_2SO_4 were provided by dilution of 97% sulfuric acid (weight percentage) utilizing bidistilled water. The chemical structure of this alloy specimen is illustrated in Table 1.

Weight loss tests were conducted using beakers (100 ml) of test solutions maintained at "30°C" for different concentrations (0.1, 0.2, 0.3, 0.4, and 0.5 M) under total immersion conditions. All tests were made in aerated solutions. Weight loss of the specimens was determined by keeping them in test solutions for a time period range of 1–5 days. After completing a duration of treatment time specimens were scrubbed with a bristle brush under running water in order to remove the corrosion product. Specimens were then dried and reweighed. The weight loss was taken as the difference between the weight at a given time and the initial weight and is determined by using LP 120 digital balance with sensitivity of ±1 mg. The tests were performed in triplicate to guarantee the reliability of the results and the mean value of the weight loss is reported. Weight loss allowed calculation of the mean corrosion rate in mg/cm² h.

3. Theoretical

3.1. The Runge–Kutta Method [20, 21]. The analytical details of the Runge–Kutta method can be outlined with reference to (1) below and an initial condition ($y = y_0$ at $x = x_0$). It is desired to find the value of (y) when ($x = x_0 + h$) where (h) is some given constant:

$$\frac{dy}{dx} = f(x, y) \tag{1}$$

According to the Runge–Kutta method, it can be shown analytically that the ordinate at $x = x_0 + h$ to the curve through (x_0, y_0) is given by

$$y = y_0 + \frac{1}{6}(K1 + 4K2 + K3) \tag{2}$$

where K_1, K_2, and K_3 are given by the equations

$$K1 = h.f(x_0, y_0) \tag{3}$$

$$K2 = h.f\left(x_0 + \frac{1}{2}h, y_0 + \frac{1}{2}K1\right) \tag{4}$$

$$K3 = h.f(x_0 + h, y_0 + 2K2 - K1) \tag{5}$$

Formula (2) is known as the third-order Runge–Kutta formula because the term corresponding to the third derivative term in the Taylor series for y expanded about (x_0, y_0) is correct [20, 21].

A mathematical model for first-order ordinary differential equation can be found to be used in Rung–Kutta method. The mathematical model represents the relationship between the temperature and the rate of reaction as explained below:

The transition state equation [22, 23] is

$$r = \frac{RT}{Nh} * \text{EXP}\left(\frac{\Delta S_a^o}{R}\right) * \text{EXP}\left(\frac{-\Delta H_a^o}{RT}\right) \tag{6}$$

where r is the rate of reaction, ΔH^o_a is the enthalpy of activation at standard condition, ΔS^o_a is the entropy of activation at standard condition, h is Planck's constant, and N is the Avogadro number.

Equation (6) is rearranged to obtain

$$\frac{r}{T} = M * \text{EXP}\left(\frac{-\Delta H_a^o}{RT}\right) \tag{7}$$

where $M = R/Nh * \text{EXP}(\Delta S^o_a/R)$

Take the (ln) function for both sides of (7) to get

$$\ln\left(\frac{r}{T}\right) = \frac{-\Delta H_a^o}{R * T} + \ln(M) \tag{8}$$

$$\ln(r) = \ln(T) - \frac{\Delta H_a^o}{R * T} + \ln(M) \tag{9}$$

Derivate (9) to find

$$\frac{1}{r}\frac{dr}{dT} = \frac{1}{T} + \frac{\Delta H_a^o}{R.T^2} \tag{10}$$

TABLE 2: Rate of reaction (mg/cm^2.h) of mild steel in sulfuric acid at 0.1 M concentration and at different temperature (30-50°C).

Temperature(°C)	1/T (K^{-1})	Rate(r) (mg/cm^2.h) AT C=0.1 M H$_2$SO$_4$	r/T(mg/cm^2.h.K)
30	0.003299	0.175	0.000577
35	0.003245	0.20275	0.000658
40	0.003193	0.2305	0.000736
45	0.003143	0.25825	0.000812
50	0.003095	0.286	0.000885

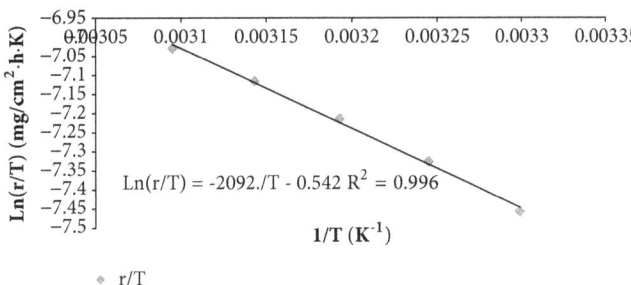

◆ r/T

FIGURE 1: Log(r/T) versus (1/T) of mild steel in sulfuric acid at (0.1 M) concentration and at different temperature (30–50°C).

Multiply (10) by (r) and the result is

$$\frac{dr}{dT} = \frac{r}{T} + \frac{r.\Delta H_a^o}{R.T^2} \qquad (11)$$

Equation (11) is the first-order ordinary differential equation which represents the relationship between the temperature (T) and the rate of reaction (r). This equation can be used in Rung–Kutta method to calculate the rate of reaction at different temperature (30–50°C) for different concentration.

4. Results and Discussion

The corrosion of mild steel was studied in combination between the theoretical and experimental data to analyse the kinetics of reaction with sulfuric acid (H$_2$SO$_4$) at various concentrations (0.1–0.5 M). The theoretical data were obtained by numerical methods, especially by the Runge- Kutta method [20, 21]. The corrosion rate of mild steel experimentally was determined using the relation:

$$R = \frac{\Delta w}{A.t} \qquad (12)$$

where Δw is the mass loss, A is the area, and t is the immersion of period time.

The enthalpy ΔH_a^o can be found using (11). The enthalpy ΔH_a^o was found by plotting ln(r/T) versus 1/T for the experimental data shown in Table 2 according to (8) to get a straight line. The slope of straight line represents ΔH_a^o/R and the intercept is ln(M), where M is equal to (R/Nh * EXP(ΔS_a^o/R) as shown in Table 2 and Figure 1.

Figure 1 shows that the equation of the straight line is

$$\ln\left(\frac{r}{T}\right) = -\frac{2092}{T} - 0.542 \qquad (13)$$

Therefore (ΔH^o/R) is equal to 2092 K and the ln(M) is equal to 0.542 and ΔS= 196.68 J/mol. The value of ΔH^o/R can be used in (11) to calculate the rate of reaction at different temperature (30, 35, 40, 45, and 50°C) and at different concentration (0.1, 0.2, 0.3, 0.4, and 0.5 M) by using Rung–Kutta method as shown in Table 3.

Gibbs free energy at temperature 298 K was calculated by the following Gibbs free energy equation [24]:

$$\Delta G = \Delta H - T.\Delta S$$

$$\Delta G^o = 17392.89\,\text{J/mol} - 298\,\text{K} \times 196.679\,\text{J/mol.K} \qquad (14)$$

$$= -41217.32\,\text{J/mol}$$

The minus sign in Gibbs free energy indicates that the reaction of mild steel with sulfuric acid was a spontaneous reaction [24].

To calculate the order of reaction (n) for mild steel in sulfuric acid ln(r) versus ln(C) was plotted according to the following equation [24]:

$$r = \frac{dc}{dt} = KC^n \qquad (15)$$

where (r) is the rate of reaction, (K) is the rate of reaction constant, and (n) is the order of reaction. Take the ln function for both sides of (15) to get

$$\ln(r) = n\ln(C) + \ln(K) \qquad (16)$$

If ln(r) versus ln(C) is plotted according to (16), a straight line is obtained. The slope of straight line represents the order of reaction (n) and the intercept is the rate of reaction constant ln(K) as shown in Table 3 and Figure 2.

TABLE 3: The rate of reaction (mg/cm^2.h) of mild steel in sulfuric acid at different concentration (0.1-0.5 M) and at different temperature (30-50°C).

Concentration H$_2$SO$_4$ (M)	Rate of reaction (mg/cm^2.h) at				
	30°C*	35°C	40°C	45°C	50°C
0.1	0.175	0.198961	0.225336	0.25427462	0.285929
0.2	22.48125	25.55936	28.94761	32.6652073	36.73163
0.3	44.7875	50.91976	57.66988	65.0761399	73.17734
0.4	67.09375	76.28016	86.39215	97.4870725	109.623
0.5	89.94	102.2545	115.8097	130.682624	146.951

* denotes experimental data and other data are theoretical data that were found by Runge–Kutta method.

TABLE 4: The rate of reaction constant of mild steel in sulfuric acid at different temperature (30-50°C).

Temperature (°C)	1/T (K)	Rate of reaction constant K
30	0.0032987	2617.56559
35	0.00324517	2659.78348
40	0.00319336	2705.38633
45	0.00314317	2749.02065
50	0.00309454	2793.35874

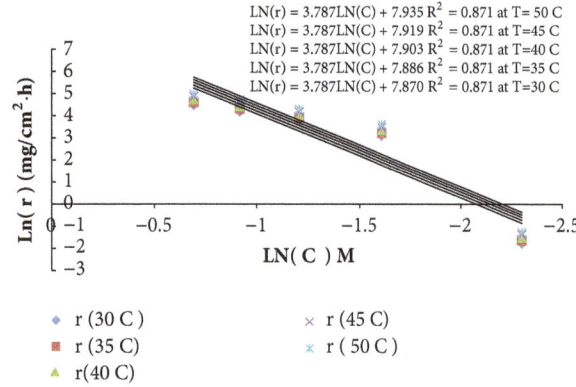

FIGURE 2: ln(r) versus ln(C) of mild steel in sulfuric acid at different concentration (0.1–0.5 M) and at different temperature (30-50°C).

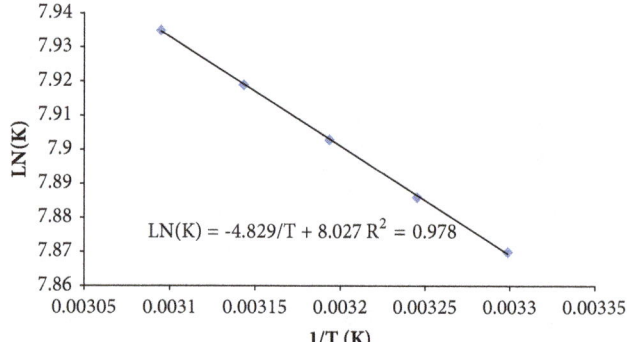

FIGURE 3: ln(K) versus (1/T) of mild steel in sulfuric acid at different temperature (30-50°C).

The rate of reaction equations at different temperatures (30–50°C) was found as shown in Figure 2:

$$\ln (r) = 3.787 \ln (C) + 7.935 \quad at\ T = 50\,C \quad (17)$$

$$\ln (r) = 3.787 \ln (C) + 7.919 \quad at\ T = 45\,C \quad (18)$$

$$\ln (r) = 3.787 \ln (C) + 7.903 \quad at\ T = 40\,C \quad (19)$$

$$\ln (r) = 3.787 \ln (C) + 7.886 \quad at\ T = 35\,C \quad (20)$$

$$\ln (r) = 3.787 \ln (C) + 7.870 \quad at\ T = 30\,C \quad (21)$$

The order of reaction of sulfuric acid with mild steel is a 4th-order reaction as shown in (17), (18), (19), (20), and (21) at the range of temperature (30–50°C). The rate of reaction constant increases with an increase in temperature because the kinetic energy of molecules increases with increase in

temperature [24, 25] as shown in Table 4 and Figure 3. To find the activation energy (E), ln(K) versus (1/T) was plotted according to the following equations [24] below and the results are shown in Figure 3 and Table 3:

$$K = Ae^{-E/RT} \quad (22)$$

$$\ln (K) = -\frac{E}{RT} + \ln (A) \quad (23)$$

The plot of ln(K) versus (1/T) according to (23) obtains a straight line, the slope of which represents the activation energy of a reaction (E/R) and the intercept is equal to the preexponential factor (A).

The activation energy (E/R) for the reaction of mild steel with sulfuric acid was estimated from the slope as shown in Table 3 and Figure 3 and was found equal to 4.829 K for the temperature range of 30-50°C and the preexponential factor (A) was 3062.54 as shown in the following equation:

$$\ln(K) = -\frac{4.829}{T} + 8.027 \qquad (24)$$

The value of activation energy was very low, indicating that mean of the reaction takes place easily and spontaneously, which is in agreement with the result of minus value of Gibbs free energy.

5. Conclusion

The corrosion kinetics of mild steel with sulfuric acid in combination with the experimental data and theoretical approach using the Runge–Kutta method was investigated. The rates of sulfuric acid reaction with mild steel were increased with increased temperatures. Moreover, the rates of reaction constant (K) at temperature range of 30–50°C were 2618–2793 L^3/mol^3.h, respectively. The reaction order of mild steel was 4th order in all ranges of temperature (30–50°C). The enthalpy and entropy of reaction were 17.393 KJ/mol and 196.68 J/mol, respectively. The value of Gibbs free energy was minus value (-41.217 KJ/mol), and therefore it was concluded that the reaction of mild steel with sulfuric acid was spontaneous. The activation energy of the reaction of mild steel with sulfuric acid was calculated and it was very low (E/R = 4.829 K) at different temperatures and at different concentration of sulfuric acid, which leads to concluding that the reaction of a mild steel with sulfuric acid readily occurred. Runge–Kutta simulation technique can be used to simulate the corrosion of mild steel surface in different concentrations of H_2SO_4.

Conflicts of Interest

The author declares that he has no conflicts of interest.

References

[1] M. A. Migahed, A. M. Abdul-Raheim, A. M. Atta, and W. Brostow, "Synthesis and evaluation of a new water soluble corrosion inhibitor from recycled poly(ethylene terphethalate)," *Materials Chemistry and Physics*, vol. 121, no. 1-2, pp. 208–214, 2010.

[2] D. D. N. Singh, T. B. Singh, and B. Gaur, "The role of metal cations in improving the inhibitive performance of hexamine on the corrosion of steel in hydrochloric acid solution," *Corrosion Science*, vol. 37, no. 6, pp. 1005–1019, 1995.

[3] G. E. Badr, "The role of some thiosemicarbazide derivatives as corrosion inhibitors for C-steel in acidic media," *Corrosion Science*, vol. 51, no. 11, pp. 2529–2536, 2009.

[4] D. Asefi, M. Arami, and N. M. Mahmoodi, "Electrochemical effect of cationic gemini surfactant and halide salts on corrosion inhibition of low carbon steel in acid medium," *Corrosion Science*, vol. 52, no. 3, pp. 794–800, 2010.

[5] K. Zhang, B. Xu, W. Yang, X. Yin, Y. Liu, and Y. Chen, "Halogen-substituted imidazoline derivatives as corrosion inhibitors for mild steel in hydrochloric acid solution," *Corrosion Science*, vol. 90, pp. 284–295, 2015.

[6] R. Yıldız, "An electrochemical and theoretical evaluation of 4,6-diamino-2-pyrimidinethiol as a corrosion inhibitor for mild steel in HCl solutions," *Corrosion Science*, vol. 90, pp. 544–553, 2015.

[7] N. A. Odewunmi, S. A. Umoren, and Z. M. Gasem, "Watermelon waste products as green corrosion inhibitors for mild steel in HCl solution," *Journal of Environmental Chemical Engineering*, vol. 3, no. 1, pp. 286–296, 2015.

[8] G.-F. Sui, J.-S. Li, H.-W. Li, F. Sun, T.-B. Zhang, and H.-Z. Fu, "Investigation on the explosive welding mechanism of corrosion-resisting aluminum and stainless steel tubes through finite element simulation and experiments," *International Journal of Minerals, Metallurgy, and Materials*, vol. 19, no. 2, pp. 151–158, 2012.

[9] J.-J. Shi and W. Sun, "Electrochemical and analytical characterization of three corrosion inhibitors of steel in simulated concrete pore solutions," *International Journal of Minerals, Metallurgy, and Materials*, vol. 19, no. 1, pp. 38–47, 2012.

[10] K.-K. Wang, "Thixo-forging and simulation of complex parts of aluminum alloy AlSi7Mg," *International Journal of Minerals, Metallurgy, and Materials*, vol. 17, no. 1, pp. 53–57, 2010.

[11] S.-H. Yin, A.-X. Wu, S.-Y. Wang, and H.-J. Wang, "Simulation of solute transportation within porous particles during the bioleaching process," *International Journal of Minerals, Metallurgy, and Materials*, vol. 17, no. 4, pp. 389–396, 2010.

[12] X.-C. Liu, L.-F. Cheng, L.-T. Zhang, X.-G. Luan, and H. Mei, "Behavior of pure and modified carbon/carbon composites in atomic oxygen environment," *International Journal of Minerals, Metallurgy, and Materials*, vol. 21, no. 2, pp. 190–195, 2014.

[13] S.-W. Yin, L. Wang, L.-G. Tong, F.-M. Yang, and Y.-H. Li, "Kinetic study on the direct nitridation of silicon powders diluted with α-Si_3N_4 at normal pressure," *International Journal of Minerals, Metallurgy, and Materials*, vol. 20, no. 5, pp. 493–498, 2013.

[14] K. F. Khaled and A. El-Maghraby, "Experimental, Monte Carlo and molecular dynamics simulations to investigate corrosion inhibition of mild steel in hydrochloric acid solutions," *Arabian Journal of Chemistry*, vol. 7, no. 3, pp. 319–326, 2014.

[15] S. K. Saha, P. Ghosh, A. Hens, N. C. Murmu, and P. Banerjee, "Density functional theory and molecular dynamics simulation study on corrosion inhibition performance of mild steel by mercapto-quinoline Schiff base corrosion inhibitor," *Physica E: Low-dimensional Systems and Nanostructures*, vol. 66, pp. 332–341, 2015.

[16] G. S. Fontes, P. F. F. e Melo, and A. S. D. M. Alves, *Nuclear Engineering and Design*, vol. 293, p. 485, 2015.

[17] D. Lu, Y. Liu, and X. Zeng, "Experimental and numerical study of dynamic response of elevated water tank of AP1000 PCCWST considering FSI effect," *Annals of Nuclear Energy*, vol. 81, pp. 73–83, 2015.

[18] L. Li, J. Zhang, W. Song, Y. Fu, X. Xu, and Y. Chen, "CATE: a code for activated corrosion products evaluation of water-cooled fusion reactor," *Fusion Engineering and Design*, vol. 100, pp. 340–344, 2015.

[19] F. Clarelli, B. De Filippo, and R. Natalini, "Mathematical model of copper corrosion," *Applied Mathematical Modelling:*

Simulation and Computation for Engineering and Environmental Systems, vol. 38, no. 19-20, pp. 4804–4816, 2014.

[20] S. T. Karris, *Numerical Analysis Using MATLAB and Spreadsheets*, Orchard Publications, 2nd edition, 2004.

[21] V. G. Jenson and G. V. Jeffreys, *Mathematical Methods in Chemical Engineering*, Academic press, London, UK, 2nd edition, 1981.

[22] A. Ostovari, S. M. Hoseinieh, M. Peikari, S. R. Shadizadeh, S. J. Hashemi, and Sci. Corros, "Corrosion inhibition of mild steel in 1 M HCl solution by henna extract: a comparative study of the inhibition by henna and its constituents (Lawsone, Gallic acid, α-d-Glucose and Tannic acid)," *Corrosion Science*, vol. 51, no. 9, pp. 1935–1949, 2009.

[23] A. K. Singh and M. A. Quraishi, "Effect of Cefazolin on the corrosion of mild steel in HCl solution," *Corrosion Science*, vol. 52, no. 1, pp. 152–160, 2010.

[24] J. M. Simth, *Chemical Engineering Kinetics*, Singapore, McGraw-Hill, 3rd edition, 1981.

[25] A. A. Taker, *Physical Chemistry For Polymer*, Mosal University Editions, Mosul, Iraq, 1984.

20

Pitting Corrosion of the Resistance Welding Joints of Stainless Steel Ventilation Grille Operated in Swimming Pool Environment

Mirosław Szala (ID) **and Daniel Łukasik**

Department of Materials Engineering, Faculty of Mechanical Engineering, Lublin University of Technology, Nadbystrzycka 36D, 20-618 Lublin, Poland

Correspondence should be addressed to Mirosław Szala; m.szala@pollub.pl

Academic Editor: Kai Wang

This work focuses on the pitting corrosion of ventilation grilles operated in swimming pool environments. The ventilation grille was made by resistance welding of stainless steel rods. Based on the macroscopic and microscopic examinations, the mechanism of the pitting corrosion was confirmed. Chemical composition microanalysis of sediments as well as base metal using scanning electron microscopy and energy-dispersive spectroscopy (SEM-EDS) method was carried out. The weldments did not meet the operating conditions of the swimming pool environment. The wear due to the pitting corrosion was identified in heat affected zones of stainless steel weldment and was more severe than the corrosion of base metal. The low quality finish of the joints and influence of the welding process on the weld metal microstructure lead to accelerated deposition of corrosion effecting elements such as chlorine.

1. Introduction

The aesthetic appearance and the ease of keeping the surface clean, good strength properties, ductility, weldability, and corrosion resistance make the use of stainless steels encouraging for elements of the construction and equipment of swimming pools like ventilation grilles. Although polymer materials such as ABS (acrylonitrile butadiene styrene) and PVC (polyvinyl chloride) are relatively cheaper than stainless steel and can also be operated in pool conditions, none of them matches all the mentioned beneficial properties of stainless steel. Stainless steel elements are used to make swimming pool equipment, such as stairs, ladders, ventilation systems (e.g., ventilation grilles), barriers, drainage grills, or other decorative elements. These components can be made from stainless steel grades such as 1.4301, 1.4404, 1.4539, and 1.4547 [1–4], and their chemical composition is presented in Table 1.

According to the environmental corrosivity categories described by the standard [5], the interior of the swimming pool is classified as C4, on a 5-degree scale of corrosiveness (C1: very low, C5: a very large corrosivity category). That is why one of the basic criteria for the metal materials selection used to build structures and elements of the swimming pool is their resistance to corrosion phenomena. Hence, the operating conditions of metal elements in the pool environment are unstable and difficult to describe, which can prevent grades dedicated to specific applications from meeting the requirements. The high price of more corrosion resistant stainless steel grades containing more expensive alloy components often prompts designers to choose cheaper, more available materials with inferior corrosion resistance. For example, it is a common mistake to use cheaper steel grade like 1.4301 instead of, for example, 1.4401 or 1.4404. The first of these materials is widely used in the construction of food processing equipment and industrial installations, but it corrodes easily in the swimming pool environment [6]. Grade 1.4404, due to the presence of molybdenum and nitrogen, is described in the literature, as dedicated to the production of equipment and design of swimming pools [1–4]. However, in specific swimming pool applications, elements manufactured

TABLE 1: Chemical composition of selected stainless steels according to EN 10088-1 standard [8].

Name	Number	C	N	Cr	Mo	Ni	Other
X8CrNi18-8	1.4301[a]	≤0,070	≤0,11	17,5~19,5	-	8,0~10,5	-
X2CrNi18-9	1.4307[a]	≤0,030	≤0,11	17,5~19,5	-	8,0~10,5	-
X5CrNiMo17-12-2	1.4401[a]	≤0,070	≤0,11	16,5~18,5	2,0~2,5	10,0~13,0	-
X2CrNiMo17-12-2	1.4404[a]	≤0,030	≤0,11	16,5~18,5	2,0~2,5	10,0~13,0	-
X1CrNiMoCu25-20-5	1.4539[b]	≤0,020	≤0,15	19,0~21,0	4,0~5,0	24.0~26.0	Cu = 1,2~2,0
X1CrNiMoCuN20-18-7	1.4547[c]	≤0,020	0,18~0,25	19,5~20,5	6,0~7,0	17,5~18,5	Cu = 0,5~1,0

[a]Si ≤ 1,0; Mn ≤ 2,0; P_{max} = 0, 045; S ≤ 0,015; [b]Si ≤ 0,7; Mn ≤ 2,0; P_{max} = 0,030; S ≤ 0,010; [c]Si ≤ 0,7; Mn ≤ 1,0; P_{max} = 0,030; S ≤ 0,010.

from stainless steel grade 1.4404 undergo pitting corrosion, as described in [7].

Works [1–4, 6, 7, 9] describe examples of negative consequences of corrosion phenomena, observed for stainless steels used for swimming pool equipment. The effects of the general or pitting corrosion are usually losses related to the necessity of exchanging corroded elements, such as the case of corrosion of the ventilation duct [6] and drain [7] operated in the pool. In the case of load-bearing elements, pitting corrosion can cause the development of stress corrosion [9, 10]. Pitting is an extremely localized attack that is manifested by holes, or pits, in the metal surface as well as especially in heat affected zones of weld joints [11–13]. Pitting is a particularly insidious form of corrosion since it is difficult to detect until the structure has been severely attacked [4, 6, 7, 12, 14].

In preventing the occurrence of corrosion phenomena of steel elements used in swimming pools, the technological nature of their construction and assembly (e.g., welding) plays an important role. Designing elements should be carried out based on principles such as maintaining high surface smoothness, making structures without sharp corners and undercuts, designing self-cleaning structures, protecting gaps against accumulation of water and sediments in them, using appropriate joining techniques with high care, and the quality of the welding joints finish, so that the joints would have corrosion resistance similar to the base material [1, 12].

Stainless steel corrosion in elements operated in swimming pool environment is considered as a real problem. The literature survey indicates that there is little information about pitting corrosion of stainless steel resistance welding joints. Authors focus mainly on investigating the corrosion of fusion welding processes (e.g., [11, 15–17], [18, p. 304]), none of them influenced by pool environment. The conducted literature review indicates only one paper related to pitting corrosion of resistance welding weldment used in pools [6]. Hence, it can be acclaimed that the state of the art in the field of pitting corrosion resistance of stainless steel weldments operated at swimming pool environments is not entirely investigated.

The aim of the article is to investigate the corrosion of a ventilation grille operated in a swimming pool environment. The grille was made by resistance welding of stainless steel shaped rods.

2. Material and Methodology of Research

The ventilation grille was made of welded stainless steel rods. The tested element was part of the swimming pool ventilation system. The grille was mounted indoors and located at the inlet of the ventilation duct on the wall and had no direct contact with the pool water. The swimming pool atmosphere was characterized by significant humidity resulting from evaporation of pool water. The grille was operated for just a few months until it detected visible corrosion changes and its disassembly.

The temperature of the ventilation system elements was lower than the air temperature in the swimming pool. Chlorinated pool water had a pH close to 7.0. This led to the condensation of water vapor on the ventilation grille, creating conditions for corrosion phenomena to occur on its surface.

Figure 1 shows the examined element and was taken after dismantling the element, dictated by the change of the colour of the grille surface to "rusty." The presence of corrosion deposit products and pits was observed mainly in areas of joint heat affected zones (HAZ).

In order to recognize the corrosion mechanism, macroscopic and microscopic observations of the worn areas of ventilation grille focused on heat affected zones of resistance welded grille were made. Macroscopic examinations based on the observation of the surface and joint areas were made using a Nikon SMZ 1500 stereomicroscope. To reveal the topography of pitting surfaces, the samples were cleaned using ultrasonic washer in order to remove corrosion products and the steel grille was cross-sectioned on a metallographic cutter (Brillant 200) and then investigated using a Nikon MA200 metallographic microscope. Phenom World ProX scanning electron microscope with chemical composition of the steel and corrosive sediments analysis using SEM-BSD (backscatter electron detector) and SEM-topo modes were carried out.

3. Results

The object of our investigation was the ventilation grille presented in Figure 1. The weldment areas were mostly severely affected by corrosion, which is visible in optical photographs given in Figure 2 as well as in metallographic weldment cross section in Figure 3. Additionally, in photographs acquired

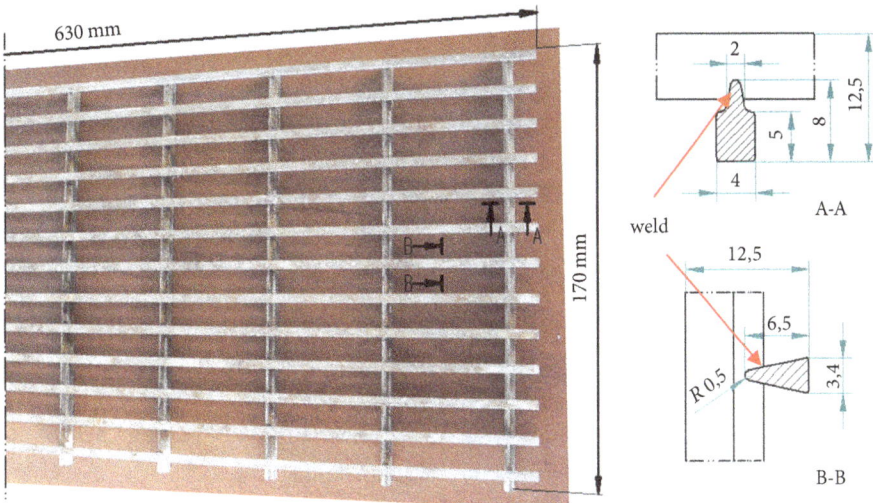

FIGURE 1: Ventilation grille made by resistance welding of stainless steel rods, dimensions in mm.

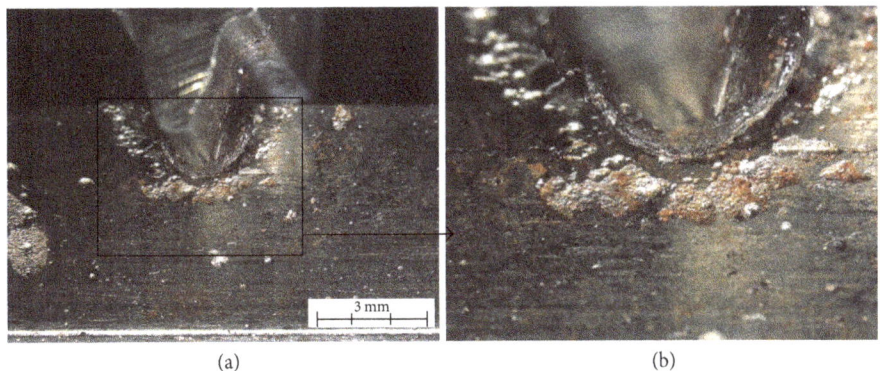

FIGURE 2: Resistance welding weldment corrosion (a) and magnified HAZ (b), stereoscopic microscope.

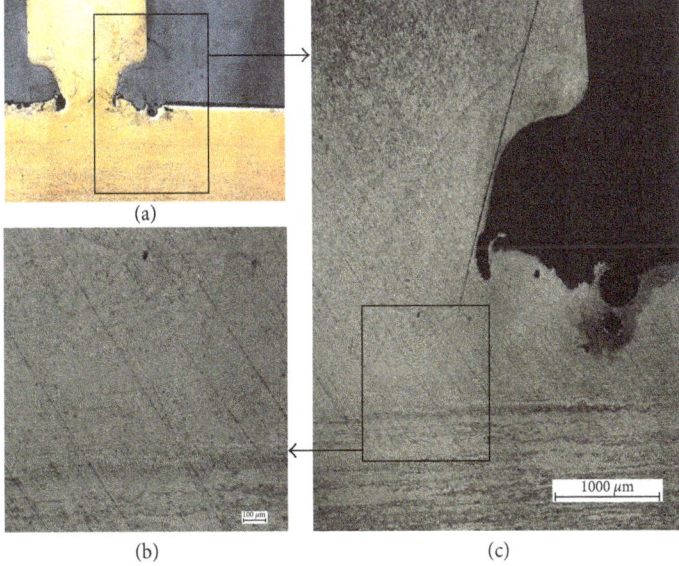

FIGURE 3: Corrosion pits observed in the cross section of joint (a) in the HAZ; (b, c) magnified selected area of microstructure of base metals and heat affected zone presented, metallographic microscope.

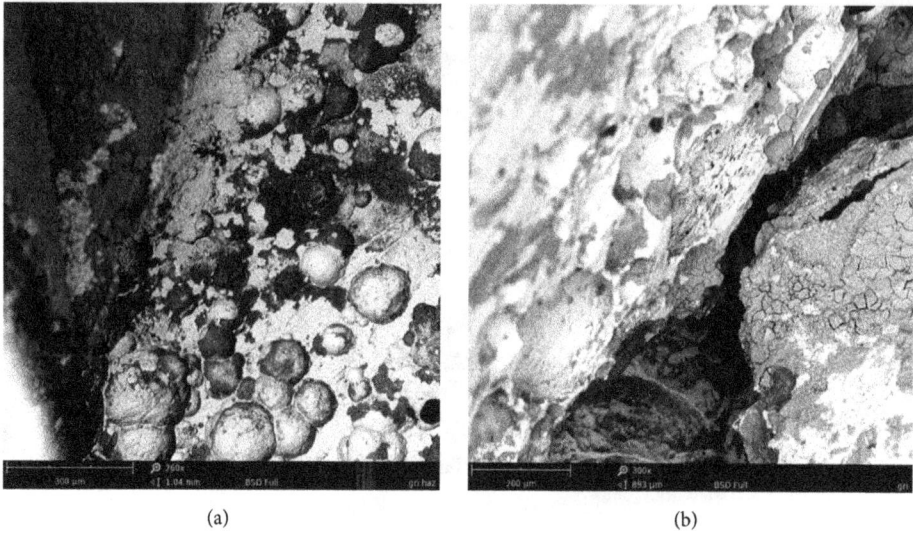

(a)   (b)

FIGURE 4: The effect of pitting observed in the HAZ (a) and crack between grille rods (b), SEM-BSD.

by the scanning electron microscope, corrosion pits and corrosive sediments are identified, Figure 4. Furthermore, Figure 5, Figure 6, and Table 2 contain the results of chemical composition of base metal and corrosive sediments.

4. Discussion

Stainless steels achieve corrosion resistance thanks to the passive layer of oxides spontaneously formed on the surface of the products. The passive film can be damaged and its restoration depends on the specific characteristics of the material and the environment. In the absence of a passive layer or its damage, the process of corrosion damage is intensified. The investigated grille was manufactured by electric resistance welding, as shown in Figures 1 and 2.

In addition, microstructure of weldment, which differs from the base metal, mostly affects corrosion resistance of stainless steel as presented in Figure 3(a). Results of metallographic investigations of the weldment acknowledged the anisotropy of the structure of welded rods, which derives from metal forming processes, so the area of the weldment that is plastically deformed and influenced by resistance welding process hence contains different microstructure from base metal (Figures 3(b) and 3(c)). In [11, 12], phenomena associated with welded joints, such as dendritic segregation, precipitation of secondary phases, formation of unmixed zones, recrystallization, and grain growth, are listed as influencing the corrosion resistance of the heat affected zone (HAZ). The operating conditions of the ventilation system resulted in deposition of chlorine condensate on the ventilation grille. The cyclic nature of the process led to an increase in the concentration of chlorine in selected places on the surface of the grille. This intensified the occurrence of local corrosion phenomena, including pitting corrosion. It is common knowledge that, at a higher temperature, the oxide film ceases to form due to the lower solubility of oxygen in the solution and the penetration potential decreases with

increasing temperature. Garcia et al. [11] gave information that fusion welding processes can cause local changes in the composition of the welded material, which can alter the stability of the passive layer and its corrosion behaviour [15]. In addition, from the microstructural point of view, the formation of δ-ferrite in stainless steel is another parameter taken into account, since it can be prejudicial because of its susceptibility to attack in a corrosion medium [11, 16], such as a swimming pool, a chlorine-rich environment. According to the literature, the pitting corrosion resistance in pool environment of stainless steel element can be increased using steels with a molybdenum content above 2.5% and chromium above 17% by weight, containing nitrogen.

Leda claims [19] that near nonmetal inclusions, grain boundaries, dislocations, surface nonuniformities, and the probability of corrosion attack increase. The metallurgical purity and homogeneity of steel elements have a great influence on corrosion resistance [13]. Also, the lack of nonmetallic inclusions, especially those located on the surface, in stainless steel such as sulphur are factors that improve corrosion resistance [20]. On the contrary, in the investigated grille, areas of poor weldment finish and surface nonuniformities were identified, visible in Figure 2. They were produced during the resistance welding process due to heat imputed, force used in resistance welding process, and atmosphere. These areas must have been finishing, machined, and chemically etched after welding. Poor metal finish of weldment accelerates concentration of corrosive deposits (e.g., chlorine-rich) that can influence corrosion of weldment. In order to prevent the corrosion of the ventilation grille to remove chlorine-containing deposits from its surface, it is possible to periodically wash its surface. We also proposed that grilles produced by resistance welding process could not met requirements of harsh operating pool environment due to difficulties connecting to finishing treatment of weldment.

The formation of corrosion pits on stainless steel elements in the atmosphere of the swimming pool is caused mostly

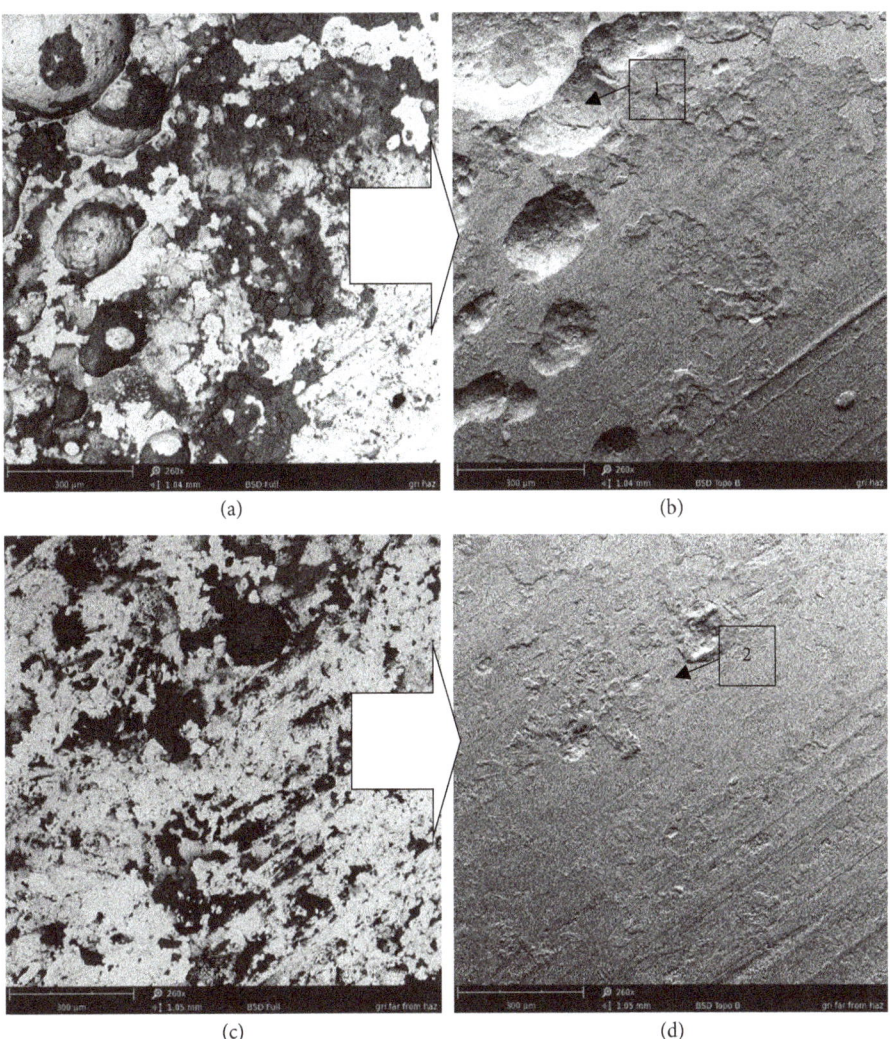

FIGURE 5: Comparison of grille surfaces observed in the heat affected zone HAZ (a, b) and in distance of HAZ (c, d). SEM-BSD and SEM-BSD-topo modes. Points 1 and 2 are locations of chemical analysis; results presented in Figure 6 and Table 2.

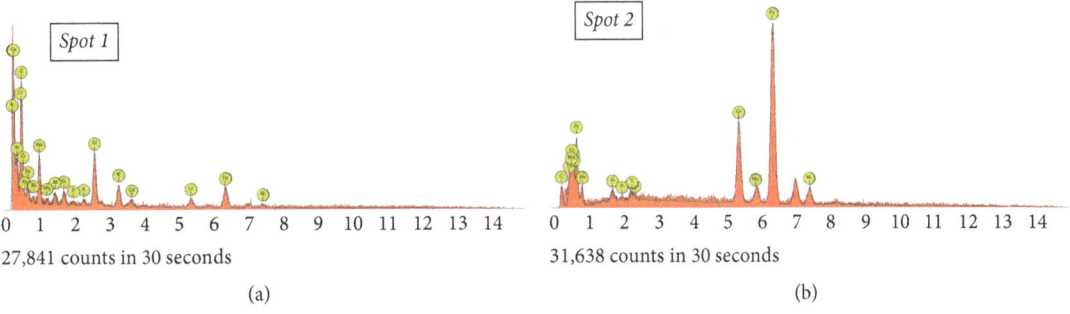

FIGURE 6: Spectrogram (SEM-EDS) obtained due to chemical analysis conducted in spots marked in Figure 5: (a) a corrosive sediment and (b) surface of base metal.

by increased temperature and high concentration of chlorine ions in the air. As a result of pitting corrosion attack, the tested ventilation grille corrosion products and corrosion pits were noted, as shown in Figures 2, 3, 4, and 5, especially in the heat affected zone (HAZ). The dimensions of pits on the surface of the element reached 1 mm. The depths of pits observed on the cross-sectional surface of the test piece were approx. 0.3 mm (Figure 2). On the contrary, base metal was less corroded (Figure 3). On the whole, the morphology of identified pits indicates pitting corrosion attack [6, 7, 10].

Examination of HAZ acknowledged visible severe pitting action, Figures 3 and 4(a), and generation of deep cracks

TABLE 2: The results of chemical composition analysis at spot 1 (sediment) and spot 2 (metal surface) in Figure 5.

Element	Spot 1: sediment		Spot 2: base metal	
	Concentration [wt.%]	Error [%]	Concentration [wt.%]	Error [%]
Aluminium	1.3	7.8	-	-
Calcium	0.8	10.5	-	-
Carbon	8.3	2.0	1.0	6.9
Chlorine	5.3	3.0	-	-
Chromium	2.1	7.4	16.8	2.3
Iron	8.0	4.2	67.0	1.5
Magnesium	0.9	13.5	-	-
Manganese	-	-	1.9	10.5
Molybdenum	-	-	1.8	13.1
Nickel	2.0	12.6	10.3	5.2
Nitrogen	11.0	3.6	-	-
Oxygen	44.4	2.2	-	-
Phosphorus	0.3	18.0	0.2	51.6
Potassium	2.6	4.8	-	-
Silicon	1.0	7.4	0.8	14.2
Sodium	11.6	3.4	-	-
Sulfur	0.5	12.9	0.2	16.4

in between welded rods, Figures 3 and 4(b). Inside crack corrosion pits and corrosive deposits were identified. While comparing areas presented in Figure 5, it is clearly visible that the HAZ was affected by corrosion compared to the base metal. In the vicinity of the HAZ, the occurrence of pits and corrosion deposits is evident.

Results of chemical analysis of the corrosion deposits and welded steel are presented qualitatively in Figure 6(a) and quantitatively in Table 2. Spectroscopic analysis of base metal (Figure 6(b), Table 2) indicated that the investigated element could have been manufactured from stainless steel grade 1.4401 or 1.4404 (Table 1); however, it is known that the content of carbon could not be properly identified properly by SEM-EDS method. Deposits contain chemical elements which came from swimming pool environment and are used for swimming pool water treatment (e.g., Cl, Ca, Table 2). In particular, chlorine compounds play a major role in the pitting corrosion processes [6, 7] and content of chlorine in deposits equals 5.3%. In addition, the relatively high temperature of water and air in the swimming pool causes evaporation of chlorinated pool water. Ventilation systems in swimming pools often work in a closed circuit, which makes the metal elements used at the pool exposed to the effects of moist air characterized by increased concentration of chlorine ions or chlorine compounds (chloramines) [4, 7]. The mentioned factors definitely favour the occurrence of corrosion phenomena, mainly general, stress, and pitting corrosion. However, to characterize a corrosive sediment precisely, for further understanding the nature of the sediments, the XRD (X-ray diffraction) investigations should be employed. The intensification of corrosion phenomena was also influenced by poor weldment finish that accelerates deposition of corrosive elements (e.g., chlorine). In order to prevent ventilation grille from corrosion, the systematic

removal of chlorine-containing sediments from its surface must be taken into account.

5. Conclusions

The case study on the subject of the pitting corrosion of resistance welding joints found in the stainless steel ventilation grille utilised in a swimming pool environment was investigated in the work. The following conclusions can be drawn on the basis of the conducted analysis.

The results confirmed that pitting corrosion was the main corrosion mechanism of the stainless steel ventilation grille. Heat affected zone (HAZ) of resistance welded joint of grille operated in swimming pool environment was lower than resistance of stainless steel based material.

The investigated ventilation grille was manufactured from stainless steel type Cr-Ni-Mo, probably grade 1.4401 or 1.4404. Pitting corrosion affects mostly the weldment area due to its microstructure which differs from the base metal. The intensification of corrosion phenomena was also influenced by poor weldment finish which accelerates deposition of corrosive elements (e.g., chlorine).

The environment as well as the operating conditions of an element affects the corrosive wear of stainless steel grille. The increase in concentration of chlorine deposited on the corroded grille areas has been identified. Chlorine plays a major role in the pitting corrosion processes of stainless steel.

In order to prevent ventilation grille from corrosion, the removal of chlorine-containing sediments from its surface must be taken into account.

Conflicts of Interest

The authors declare that there are no conflicts of interest regarding the publication of this paper.

References

[1] N. Baddoo and P. Cutler, "Stainless steel in indoor swimming pool buildings," *Technical Note: Swimming Pools*, 2007.

[2] "Stress corrosion cracking failure," http://www.corrosion-doctors.org/Forms-SCC/swimming.htm.

[3] C. Houska and J. Fritz, *Successful Stainless Swimming Pool Design*, Nickel Institute, 2017.

[4] "Stainless steel in swimming pool buildings - selecting and using stainless steel to cope with changes in swimming pool design," Nickel Development Institute, 2010, https://www.nickelinstitute.org/~/Media/Files/TechnicalLiterature/StainlessSteelinSwimmingPoolBuidlings_12010_.pdf.

[5] ISO 12944-1:2001, "Paints and varnishes — Corrosion protection of steel structures by protective paint systems — Part 1: General introduction".

[6] P. Sedek, J. Brózda, and J. Gazdowicz, "Pitting corrosion of the stainless steel ventilation duct in a roofed swimming pool," *Engineering Failure Analysis*, vol. 15, no. 4, pp. 281–286, 2008.

[7] M. Szala, K. Beer-Lech, and M. Walczak, "A study on the corrosion of stainless steel floor drains in an indoor swimming pool," *Engineering Failure Analysis*, vol. 77, pp. 31–38, 2017.

[8] "PN-EN 10088-1:2014 - Stainless steels. Part 1: List of stainless steels".

[9] J. Woodtli and R. Kieselbach, "Damage due to hydrogen embrittlement and stress corrosion cracking," *Engineering Failure Analysis*, vol. 7, no. 6, pp. 427–450, 2000.

[10] D. A. Horner, B. J. Connolly, S. Zhou, L. Crocker, and A. Turnbull, "Novel images of the evolution of stress corrosion cracks from corrosion pits," *Corrosion Science*, vol. 53, no. 11, pp. 3466–3485, 2011.

[11] C. Garcia, F. Martin, P. de Tiedra, Y. Blanco, and M. Lopez, "Pitting corrosion of welded joints of austenitic stainless steels studied by using an electrochemical minicell," *Corrosion Science*, vol. 50, no. 4, pp. 1184–1194, 2008.

[12] J. N. DuPont, J. C. Lippold, and S. D. Kiser, *Welding Metallurgy and Weldability of Nickel-Base Alloys*, John Wiley & Sons, Inc, 2009.

[13] S. J. Kim, S. G. Hong, and M.-S. Oh, "Effect of metallurgical factors on the pitting corrosion behavior of super austenitic stainless steel weld in an acidic chloride environment," *Journal of Materials Research*, vol. 32, no. 7, pp. 1343–1350, 2017.

[14] I. Lenart and M. Szala, "Korozja wżerowa eksploatowanej na pływalni kratki wentylacyjnej wykonanej ze stali nierdzewnej," in *PraceNaukowe Młodych Badaczy: TYGIEL*, M. Szala, Ed., pp. 208–223, Lublin: Politechnika Lubelska, 2013.

[15] E. Zumelzu, J. Sepúlveda, and M. Ibarra, "Influence of microstructure on the mechanical behaviour of welded 316 L SS joints," *Journal of Materials Processing Technology*, vol. 94, no. 1, pp. 36–40, 1999.

[16] R. Chen, P. Jiang, X. Shao, G. Mi, and C. Wang, "Effect of magnetic field applied during laser-arc hybrid welding in improving the pitting resistance of the welded zone in austenitic stainless steel," *Corrosion Science*, vol. 126, pp. 385–391, 2017.

[17] J. Kim, B. Lee, W. Hwang, and S. Kang, "The Effect of Welding Residual Stress for Making Artificial Stress Corrosion Crack in the STS 304 Pipe," *Advances in Materials Science and Engineering*, vol. 2015, pp. 1–7, 2015.

[18] X. Li, B. Gong, C. Deng, and Y. Li, "Failure mechanism transition of hydrogen embrittlement in AISI 304 K-TIG weld metal under tensile loading," *Corrosion Science*, vol. 130, pp. 241–251, 2018.

[19] H. Leda, "Materialy inzynierskie w zastosowaniach biomedycznych," *Wydawnictwo Politechniki Poznanskiej*, 2012.

[20] E. A. Abd El Meguid, N. A. Mahmoud, and S. S. Abd El Rehim, "Effect of some sulphur compounds on the pitting corrosion of type 304 stainless steel," *Materials Chemistry and Physics*, vol. 63, no. 1, pp. 67–74, 2000.

21

Electrolyte Composition for Distinguishing Corrosion Mechanisms in Steel Alloy Screening

Ingmar Bösing, Jorg Thöming, and Michael Baune

Center for Environmental Research and Sustainable Technology (UFT), University of Bremen, Leobener Straße 6, Bremen, Germany

Correspondence should be addressed to Ingmar Bösing; ingmar.boesing@uni-bremen.de

Academic Editor: Francisco Javier Perez Trujillo

The formation and breakdown of passive layers due to pitting corrosion are a major cause of failure of metal structures. The investigation of passivation and pitting corrosion requires two different electrochemical measurements and is therefore a time consuming process. To reduce time in material characterization and to study the interactions of both mechanisms, here, a combined experiment addressing both phenomena is introduced. In the presented electrolyte the different corrosion mechanisms are distinguished and investigated by cyclic voltammograms and polarization scans. The measurements show a passive area, metastable pit growth, and pitting corrosion as well as repassivation. The pitting corrosion is separated from additional dissolution processes and the standard deviation of the corrosion potential is smaller than in other electrolytes. Both passivation and pitting corrosion can be observed in one measurement without additional corrosion attacks. The deviation between different measurements of the same steel is small; this is helpful for the screening of similar materials.

1. Introduction

The broad range of existing corrosion phenomena can be investigated through numerous electrochemical methods. A great deal of different set-ups is known to study various material and corrosion parameters [1], but the complex nature and the interactions between experimental factors, such as electrolyte composition or temperature and the different corrosion phenomena, like passivation, pitting corrosion or inhibitions, impede studying the single mechanisms separately.

Exploring large sets of samples requires high throughput methods that allow scanning a wide range of mechanical, physical, or chemical parameters. In corrosion science, high throughput methods are used to investigate wide fields of corrosion on different materials [2], coatings [3], corrosion phenomena such as corrosion inhibitors [4], pretreatments [5], and corrosion mechanism [6] like pitting corrosion [7]. Typically, each corrosion phenomenon measurement requires its own set-up. Polarization scans in aggressive media (containing halides), for instance, are performed to investigate pitting corrosion. In contrast to this, to study passivation processes, cyclic voltammograms are recorded in passivation promoting electrolytes.

Pitting corrosion is a localized attack on different materials like iron, chromium, nickel, cobalt, and stainless steel that results in deep pits in the material. These pits can lead to component failure or act as an initiation for cracking [8]. Particularly the passivated and technically essential austenitic steels can be affected by pitting corrosion. In an environment that favors pitting corrosion, halide ions or other components are present that lead to a breakdown of passivity on impurities and imperfections of the passive layers [9]. In the present manuscript we focus on chloride ions that are present in media such as seawater and pharmaceutical solutions, to force a breakdown of passivity.

The passivity of iron-based materials is caused by a formation of an oxide layer on the surface of the metal. A higher chromium content in the alloy leads to a more protective passive film through the formation of an interlinked chromium oxide network [10, 11]. The passive film is formed either by a spontaneous reaction or by (electrochemical) surface treatments, such as cyclic potentiodynamic polarization [12]. The protectiveness and thermodynamic stability of the passive layer depend on different aspects like temperature and pH-value [11, 13]. The passivation of iron is faster and better in environments with a high pH-value. In alkaline solutions (pH > 11.5), the passivation of iron usually happens spontaneously [14]. The passive layer is commonly not thicker than a few nanometers and protects the metal surface against reactions with the environment. While the formation of passive layers is a crucial mechanism that prevents many construction materials from general corrosion, it makes the same alloys susceptible against pitting corrosion [8].

The passive film formed on stainless steel consists of two parts, an inner chromium oxide layer and an outer iron-oxide layer [15, 16]. If the chromium content is high enough, an interlinked chromium oxide network forms within the passive films and makes it insoluble [14, 17, 18].

The breakdown of passivity can result from different mechanisms, from which the following three are mainly discussed in the literature [19]: in the penetration mechanism the aggressive ions (halide) penetrate through the passive layer at imperfections and react with the metal [20, 21]. According to the film breakdown mechanism, there are breaks within the film that give the halide anions direct access to the metal surface [22, 23]. The adsorption of aggressive anions on the oxide surface can lead to surface tension that results in additional breaks in the passive layer and allows anions to get in contact with the bare metal [24, 25]. The third mechanism assumes that the adsorption of the halides results in the transfer of metal cations from the oxide to the electrolyte. In this manner, the passive layer becomes thinner and is finally completely removed at certain points [26, 27].

After pit initiation/breakdown of passivity pit growth occurs. Two types of pits are distinguished. At the beginning of pitting, metastable pits occur; these pits can be repassivated. Under certain circumstances (e.g., critical age [28], pit characteristics [29], and high external potentials) and when the potential difference between the active and passive region is high enough, stable pit growth starts.

To study different corrosion phenomena and mechanisms, sample preparation, sample design, choice of the electrolyte, and experimental setting play an important role [30]. So far, the investigation of passivation requires different electrolytes than the investigation of crevice corrosion or pitting corrosion.

A large amount of techniques exists to form and analyze the passive layers on stainless steel. As an electrochemical technique, cyclic voltammetry in different electrolytes (e.g., sulphuric acid [31], borate buffer [11], sodium chloride [32], or phosphate buffer [12]) is commonly used. The formation of a passive layer on iron requires an aqueous environment that favors the formation of an insoluble iron-oxide complex (e.g., by a high pH-value). In addition to electrolytes that show a clear passivation behavior, one can also use electrolytes that represent the application areas of the tested materials, for example, phosphate-buffered saline for medical use or concrete [33]. The peak current at the active-passive region, the passive current, and the beginning of the transpassive region are some values that are used to interpret the formation and stability of the passive layers. For information about the pitting corrosion resistance, potentiodynamic polarization curves in sodium chloride solution are usually performed. The corrosion potential E_{corr}, the pitting potential E_{pit}, noticeable by a rapid increase of the current density, and the repassivation potential E_r describe the corrosion behavior in aggressive media [25].

Especially when the investigated materials only show slight differences, the statistical nature of corrosion phenomena makes clear statements about the studied materials difficult. In this paper, an electrolyte composition is described consisting of phosphate buffer and sodium chloride which makes it possible to investigate both passivation and pitting corrosion of steel alloys in one single measurement. In order to illustrate that differences between the analyzed alloys become much clearer in the proposed electrolyte compared to commonly applied electrolytes, we first investigated the mechanism separately: we performed cyclic voltammograms in pure phosphate buffer to depict the passivation and polarization scans in sodium chloride solution to describe the pitting corrosion processes. In a next step we compared this with CVs and polarization scans in phosphate buffer containing a specific amount of sodium chloride and underlined our findings with microscope recordings. This allows us to demonstrate the advantage of the proposed electrolyte using two stainless steels (AISI 304 and AISI 420) and a bearing steel with a very low chromium content (AISI 5210).

2. Experimental

The cyclic voltammetry measurements were done in 0.1 M phosphate buffer (pH = 7.5) and phosphate buffer containing different amounts of NaCl. The polarization scans were recorded in 3.5 wt.% NaCl solution and phosphate buffer containing 3.5 wt.% NaCl. All measurements were carried out at room temperature. A three-electrode cell was used for the experiments. A platinum electrode was used as counter electrode and a Ag/AgCl electrode as reference electrode. All potentials in this paper are referred to the Ag/AgCl electrode.

As working electrodes, we used three different standard alloys, with a 2-dimensional circular surface (10 mm diameter), all of them embedded in a Teflon holder. The chemical compositions of the working electrode materials were expected to be in the standard range (Table 1).

Before each measurement, the working electrode was wet polished with SiC emery paper up to 2000 grit. Afterwards, the electrodes were sonicated in deionized water and degreased in ethanol.

The cyclic voltammetry measurements were carried out with a scan velocity of 100 mVs^{-1} from −800 mV to 1100 mV for different numbers of cycles. The potentiodynamic polarization scans were carried out with a scan velocity of 1 mVs^{-1} and started at −700 mV while ending at 700 mV. For all experiments, the PGU 2A-OEM potentiostat (IPS) was used.

3. Results and Discussion

3.1. Cyclic Voltammetry in Phosphate Buffer. To understand the passivation behavior of the different alloys, we carried out cyclic voltammograms in 0.1 molar phosphate buffer. From the anodic and cathodic peaks as well as from the differences between the cycles and the different alloys, conclusions about the passivation process can be drawn. The peak location is related to the reacting species and the peak height to the

TABLE 1: Standard range of chemical composition of the working electrodes (wt.%) [34].

Alloy	C	Si	Mn	P	S	Cr	Ni	N
AISI 5210	0.93–1.05	0.15–0.35	0.25–0.45	≤0.025	≤0.015	1.35–1.6		
AISI 420	0.43–0.5	≤1.0	≤1.0	≤0.4	≤0.03	12.5–14.5		
AISI 304	≤0.7	≤1.0	≤2.0	≤0.045	≤0.03	17.5–19.5	8.0–10.5	≤0.1

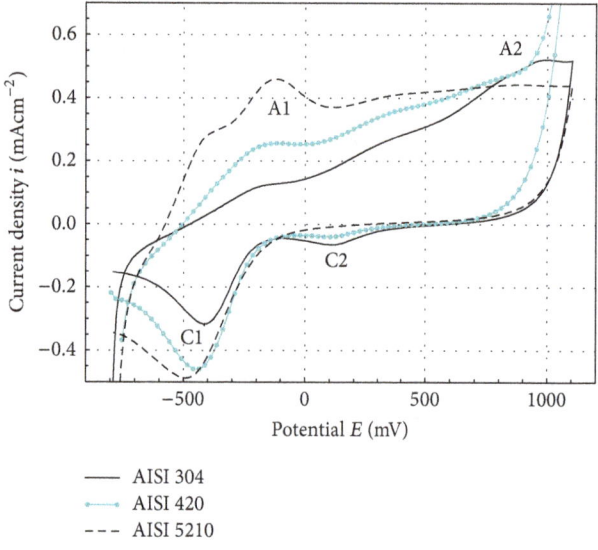

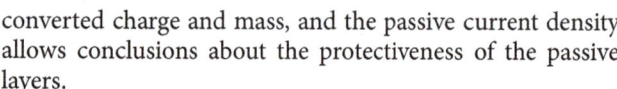

FIGURE 1: 1st cycle of cyclic voltammograms for AISI 5210, AISI 420, and AISI 304 recorded in phosphate buffer. The scans are recorded from −800 mV to 1100 mV with a scan velocity of 100 mVs⁻¹. All potentials are measured against the Ag/AgCl electrode.

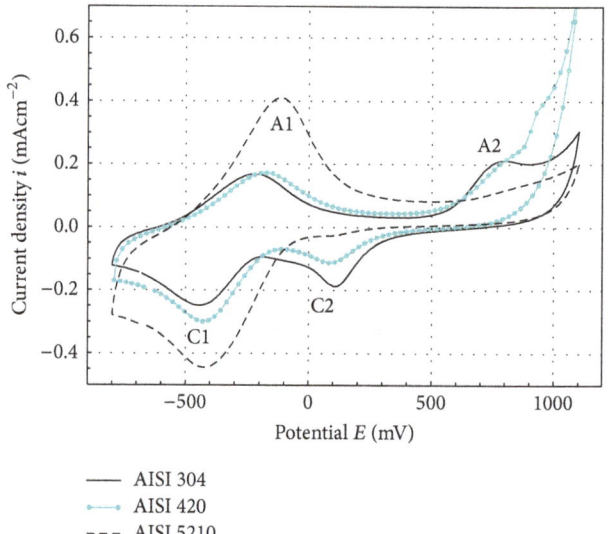

FIGURE 2: 10th cycle of cyclic voltammograms for AISI 5210, AISI 420, and AISI 304 recorded in phosphate buffer. The scans are recorded from −800 mV to 1100 mV with a scan velocity of 100 mVs⁻¹. For each electrode the 10th cycle is displayed. All potentials are measured against the Ag/AgCl electrode.

converted charge and mass, and the passive current density allows conclusions about the protectiveness of the passive layers.

Figure 1 shows the first cycle of the cyclic voltammograms of the three alloys. The CVs display oxidation peaks for the iron and chromium species during anodic polarization and reduction peaks of these species during cathodic polarization. The peak heights are related to the chromium contents and thus also to the protectiveness of the passive layers.

The anodic polarization scan shows different oxidation peaks and shoulders and no passive area. Two peaks (A1 and A2) are of particular interest. The peaks A1 are associated with the oxidation of Fe(II) species to a Fe_3O_4/FeO hydrated layer that forms on a preexisting Cr(III) oxide network. The peaks A2 belong to the oxidation of Cr(III) to Cr(VI)-species. When the polarization direction changes, a small peak can be seen for the reduction of the Cr(IV)-species to Cr(III) oxide (C2) and at more negative potentials a peak (C1) is associated with the reduction of the Fe_3O_4/FeO layer to Fe(III) [11, 12, 35, 36].

With increasing chromium content in the different alloys the oxidation peaks A1 decrease, while the oxidation peaks A2 increase. The area under the peaks is related to the transferred charge during the oxidation processes. The larger the area is the more the charge is transferred during the oxidation of the specific species. The same applies to the reduction peaks.

There is nearly no passive region, but the current between peak A1 and peak A2 increases with decreasing chromium content. AISI 420 already shows rapid increase in current density which is related to dissolution and oxygen evolution beginning at 900 mV. The absence of a clear passive region reveals the lack of the protective passive layer formed in the first cycle.

At higher cycles, the oxidation/reduction peaks are better visible and the growth of the passive layer leads to a passive region. The height of the active/passive region and of the passive current density is connected to the ability to passivate (Figure 2).

Again, with increasing chromium content peak A1 decreases while peak A2 increases. The current density in the passive region is the lowest for AISI 304 (highest chromium content) and increases with decreasing chromium content, but the differences are small.

In the first cycle, the peaks are not clear and there is nearly no passive region. This indicates the absence of protection through the Fe_3O_4/FeO hydrated layer formed in the first cycle. At the first cycle the Cr(IV)-species is not interlinked in the Fe-oxide layer and soluble. The high oxidation peak A2 and the very small reduction Cr(IV)-reduction peak underline this. After the first cycles, the Cr(IV)-oxide gets "arrested" [12] (and insoluble) in the Fe-oxide and the protective passive layer is formed [14].

The small active/passive region, presented in Figure 2, of AISI 304 and AISI 420 depicts a faster passivation than the much wider active-passive region of AISI 5210. However, the smaller anodic charge below A1 also means less reaction to form the passive layer. On the other hand, the larger Cr(III)-oxidation peak A2 indicates higher chromium content while the passivation improves. Interestingly, AISI 420 shows a transpassive region dedicated to the evolution of oxygen and the dissolution of iron which neither AISI 304 nor AISI 5210 show. An explanation of this behavior requires further information.

The cyclic voltammograms point out the general ability of the different alloys to passivate and form a protective passive layer on the surface of the materials. All alloys show passivation peaks and a passive region, but the differences between the passive current of AISI 304 and AISI 420 are very small.

3.2. Potentiodynamic Polarization Scans in NaCl Solution.

As stated above, in order to gather more information on the corrosion resistance, the stability and protectiveness of the passive layer, and pitting behavior additional experiments are necessary.

Polarization scans in sodium chloride solution were recorded to determine the characteristic corrosion potential E_{corr} and the pitting potential E_{pit} as well as the region of metastable pit growth. In combination with the CVs we were able to get more information about the interaction between passivation and pitting.

Figure 3 shows the potentiodynamic polarization scans of AISI 304, AISI 420 and AISI 5210 in 3.5 wt.% NaCl solution. The corrosion potential E_{corr} increases with increasing chromium content. After passing the corrosion potential, AISI 5210 shows direct dissolution, noticeable by the rapid increase of current density. There is neither a passive region nor a region of metastable pit growth as for stainless steels. In contrast, AISI 304 and AISI 420 show metastable pit growth overlapped by an increase in current density related to constant metal dissolution. The region of metastable pit growth lies between the corrosion potential (AISI 304: −275.19 mV) and the beginning of stable pit growth (AISI 304: 37.95 mV). The characteristic potentials of the potentiodynamic polarization scans in NaCl are listed in Table 2.

The polarization scans in sodium chloride disclose clear disadvantages: there is not a passive area to draw conclusions about passivation, nor is the pit growth separated by additional dissolution processes. Further, both stainless steels show similar corrosion potentials, although their corrosion resistance is quite different.

3.3. Cyclic Voltammetry in Phosphate Buffer with NaCl.

To overcome the mentioned limitations and in order to provide a better understanding of passivation processes, the breakdown of passivity, and corrosion characteristics, we suggest an electrolyte composition of phosphate buffer and sodium chloride. While polarizing in phosphate buffer, a more protective passive layer can form on stainless steel. By adding sodium chloride, the Cl⁻-ions in the electrolyte can force dissolution, breakdown of the passive film, and pitting corrosion.

TABLE 2: Characteristic values of polarization scans in 3.5 wt.% NaCl solution.

Alloy	E_{corr}	E_{pit}	Metastable pit growth
AISI 304	−275.19 mV	37.95 mV	−275.19 mV–37.95 mV
AISI 420	−319.6 mV	−126.18 mV	−258.22 mV–−80.99 mV
AISI 5210	−405.59 mV		

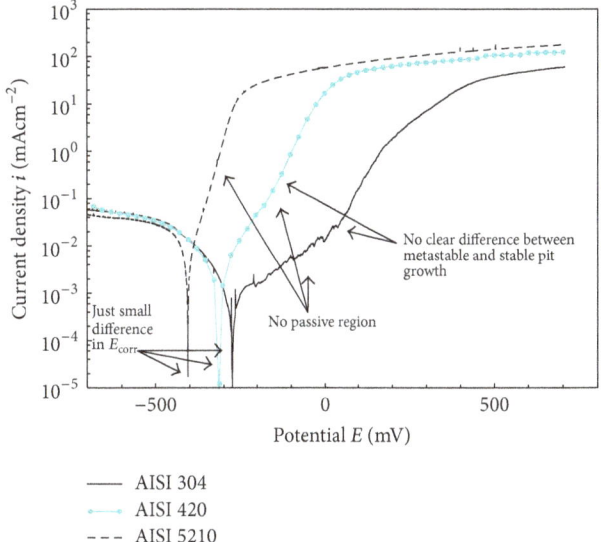

— AISI 304
⋯•⋯ AISI 420
- - - AISI 5210

FIGURE 3: Potentiodynamic polarization scan in 3.5 wt.% NaCl solution. The measurements are performed from −700 mV to 700 mV with a scan velocity of 1 mVs⁻¹. All potentials are recorded against the Ag/AgCl electrode.

By adding different amounts of sodium chloride, stable pit growth and passivation processes are visible. With increasing amounts of sodium chloride, the dissolution due to pitting corrosion shifts to more negative potentials and the cathodic polarization curves change their appearance.

Additionally small amount of sodium chloride (0.2 wt.% and 0.3 wt.%) results in a slightly higher current density during the anodic polarization in the first cycles. The cathodic polarization curve also shows different behavior. By adding sodium chloride, anodic current is measurable during the beginning of the cathodic polarization. With an increasing amount of sodium chloride, the anodic current grows to a hysteresis (Figure 4). The increase of anodic current matches the occurrence of stable pits during the anodic scan at high potentials. These pits can still grow during the cathodic polarization in regions of metastable pit growth [35, 37].

At low sodium chloride concentrations, the anodic peaks are still visible for the oxidation of Fe-species to Fe_3O_4/FeO layers and Cr(III) to Cr(IV)-species and the equivalent reduction reactions. At higher NaCl concentrations (Figure 4: 1 wt.% NaCl), the dissolution processes overlay the other reactions.

For a small amount sodium chloride, the shape of cycle 10 and cycle 20 is similar to the cycles without sodium chloride (compare Figure 5 cycle 10 and cycle 20 with Figure 2). This exemplifies that the alloy can be repassivated and form a

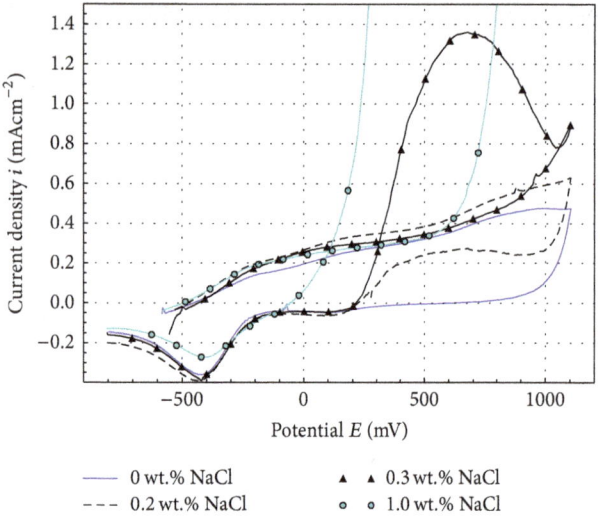

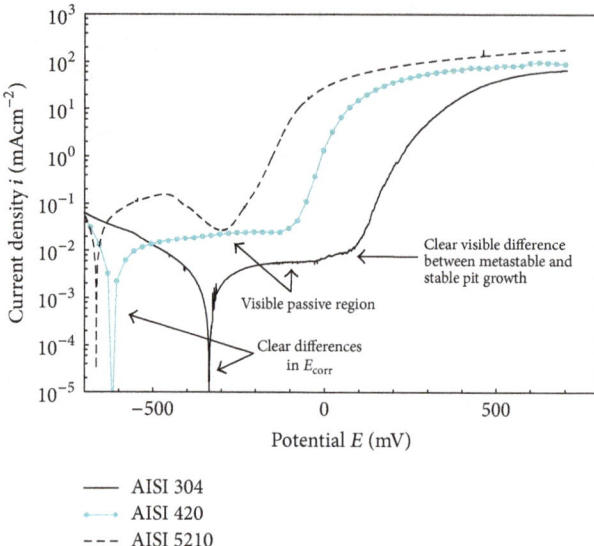

FIGURE 4: First cycle of cyclic voltammograms of AISI 304 in phosphate buffer with different contents of NaCl. The CVs are recorded from −800 mV to 1100 mV with a scan velocity of 100 mVs⁻¹. All potentials are measured against a Ag/AgCl electrode.

FIGURE 6: Potentiodynamic polarization scans of AISI 304, AISI 420, and AISI 5210 in phosphate buffer with 3.5 wt.% NaCl.

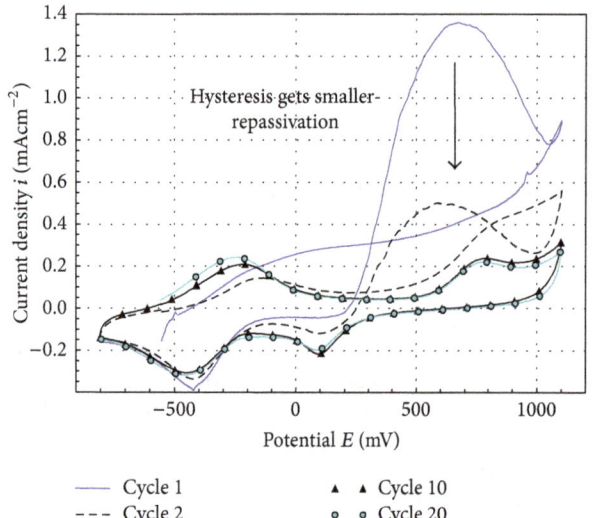

FIGURE 5: Cyclic voltammogram of AISI 304 in phosphate buffer with 0.3 wt.% NaCl. The scans are recorded from −800 mV to 1100 mV with a scan velocity of 100 mVs⁻¹. All potentials are measured against a Ag/AgCl electrode.

protective passive layer against pitting corrosion, even in an aggressive environment.

The combination of phosphate buffer, an electrolyte that enhances passivation processes, and sodium chloride, which leads to pitting corrosion, shows both the passivation of the stainless steel and the growth of the passive layer as well as the breakdown of passivity and the growth of pits on the material surface. The good passivation behavior is also visible through the repassivation of the surface at higher cycles.

At higher amounts of sodium chloride it is more difficult to observe the passivation, because the passive area shrinks and the dissolution overlays characteristics of the

cyclic voltammograms. For investigations of high amounts of sodium chloride in phosphate buffer on the passivation of different alloys, we used potentiodynamic polarization scans.

3.4. Potentiodynamic Polarization Scans in Phosphate Buffer with NaCl. The advantage of showing both passivation and pitting corrosion is even more distinct for polarization scans in phosphate buffer containing 3.5 wt.% sodium chloride. The CVs already clarify that passivation and pitting corrosion can be separated. During the polarization scans, a clear passive region is visible. This region is followed by metastable pit growths, without further dissolution. At higher potentials the pitting potential E_{pit} is clearly separated from the other processes (Figure 6).

Similar to measurements in pure sodium chloride solution (Figure 3) the corrosion potential E_{corr} increases with increasing chromium content. By adding phosphate buffer, the differences between the alloys become more obvious. After passing E_{corr} all three alloys display different behavior than in pure NaCl solution. Both stainless steels, AISI 304 and AISI 420, show a clear passive region. The passive current of AISI 304 is significantly lower than the passive current of AISI 420. After the passive region, a region of metastable pit growth is visible, followed by stable pit growth and metal dissolution, noticeable by a rapidly increasing current density.

All characteristic values for the potentiodynamic polarization scan in phosphate buffer with sodium chloride are listed in Table 3. From the measurements it becomes clear that, in contrast to pure NaCl solution, there is passivation in phosphate buffer with 3.5 wt.% NaCl. The cyclic voltammograms already show that there are different passivation peaks in phosphate buffer. But even without multiple cycles through all potentials for the formation of the passive layers (multiple formations of $Fe_3O_4/FeOH$ and an interlinked Cr(IV)-network with in the layer) a clear passive region is visible.

TABLE 3: Characteristic values for potentiodynamic polarization scans in phosphate buffer with 3.5 wt.% NaCl.

Alloy	E_{corr}	E_{pit}	Metastable pit growth
AISI 304	−336.3 mV	78.4 mV	−12.7 mV–78.4 mV
AISI 420	−617.2 mV	−85.3 mV	−172.7 mV–−85.3 mV
AISI 5210	−663.7 mV		

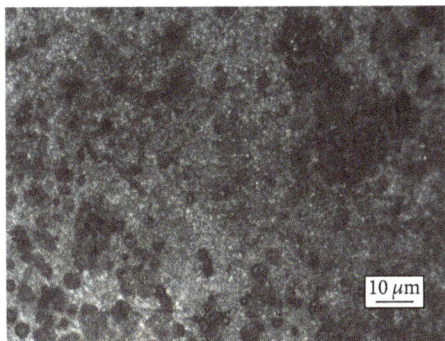

FIGURE 7: Microscope recordings of AISI 420 after polarization scan in 3.5 wt.% NaCl solution.

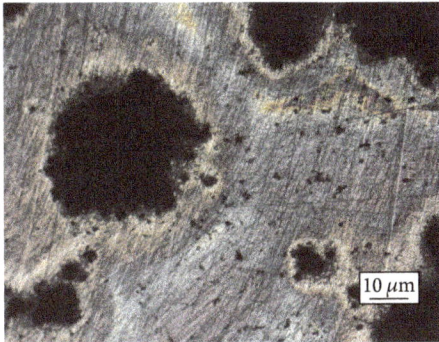

FIGURE 8: Microscope recording of AISI 420 after polarization scan in phosphate buffer containing 3.5 wt.%.

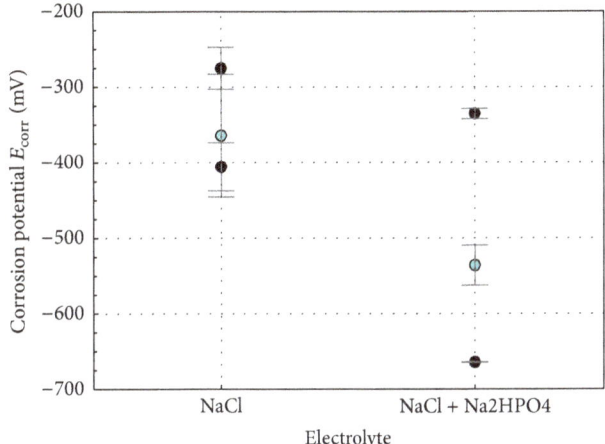

FIGURE 9: Corrosion potentials and its standard deviations of AISI 304, 420, and 5210 measured in NaCl solution and phosphate buffer with NaCl. $\sigma = \sqrt{(1/(n-1))\sum(x_i - \overline{x})^2}$, $n = 4$.

During the anodic polarization of the working electrode the negative phosphate ions (PO_4^{-3}) adsorb on the electrode surface. Due to the competitive adsorption of the phosphate and chloride ions, less of the aggressive Cl^- ions can reach the surface and the pitting potential E_{pit} moves towards positive potentials [38].

While E_{pit} is higher in phosphate buffer containing sodium chloride the corrosion potential E_{corr} is more negative. This illustrates the thermodynamic equilibrium shifts towards the more negative potentials.

At polarization scans in phosphate buffer with additional sodium chloride the region of metastable pit growth is not overlaid by additional dissolution processes. In contrast to that, the region of metastable pit growth, when doing polarization scans in pure sodium chloride, is overlaid by additional dissolution processes. To verify these findings, we took microscope recordings of the surfaces after polarization scans and compared the corrosion attacks.

The polarization scan in pure sodium chloride solution leads to small pits and wide-area attacks (compare Figure 7). This is in good agreement with the observations from Figure 3 where the region of pit growth is overlapped by general metal dissolution. The polarization scans in phosphate buffer with sodium chloride result in big pits surrounded by intact metallic surface and very small pits (Figure 8). The microscope recordings underline the potential of the electrolyte combination to separate the pitting corrosion from additional corrosion processes.

3.5. Determination of the Reliability of the Measurements. In order to investigate different materials that differ only slightly from one another, it is essential that the deviations of measurements on the same materials stay small and values of different materials do not overlap. To ensure the workability of the electrolyte combination, we calculated the standard deviation of the corrosion potentials in pure NaCl solution as well as in phosphate buffer containing NaCl.

Figure 9 shows the corrosion potentials and its standard deviation of AISI 304, AISI 420, and AISI 5210 for the two different potentiodynamic polarization scans. The corrosion potential for the polarization scans in pure NaCl solution is higher than in phosphate buffer containing NaCl. The standard deviation for the corrosion potential in phosphate buffer with NaCl is significantly smaller than in pure NaCl solution. This allows a better prediction about single corrosion behavior.

4. Conclusion

The polarization scans in phosphate buffer with sodium chloride uncover the ability of the material to be passivated and its susceptibility to pitting corrosion. In the case of pitting corrosion, the metastable pit growth is separated from additional corrosion phenomena and can be investigated by itself. The microscope recordings underline these findings.

Just small differences between the corrosion potentials in sodium chloride solution are enlarged, and a clear passive region, a region of metastable pit growth, and an exact starting point for stable pit growth are visible. The standard deviation of the characteristic electrochemical values was significantly smaller than in pure NaCl solution. This is very helpful for the investigation of similar alloys while screening new materials.

The cyclic voltammetry in the electrolyte solution shows both the passivation of the surfaces and the pitting corrosion. The presence of the hysteresis is an additional sign for stable pit growth. After the first cycles, the shape of the scan changed and the last cycles resemble the last cycles in pure phosphate buffer. Hence, repassivation occurs even in the presence of aggressive ions and a surface treatment by cyclic voltammetry in phosphate buffer can enhance the pitting corrosion resistance. By adding a small amount of sodium chloride to the phosphate buffer, the examination of passivation and the breakdown of passivity and the repassivation become possible.

The suggested electrolyte composition is an easily applicable tool for the investigation of passivity and passivation breakdown. The ability to separate metastable pit growth from additional dissolution and the greater separation of the characteristic values in combination with a smaller standard deviation make it a promising electrolyte composition for material research in high throughput steel screening.

Conflicts of Interest

The authors declare that there are no conflicts of interest regarding the publication of this paper.

Acknowledgments

Financial support of subproject D03 "Electrochemical High Throughput Characterization of Metallic Micro Samples" of the Collaborative Research Center SFB 1232 "Farbige Zustände" by the German Research Foundation (DFG) is gratefully acknowledged.

References

[1] P. Marcus and F. Mansfeld, *Analytical Methods in Corrosion Science and Engineering*, CRC Press, 2006.

[2] M. Balasubramanian, V. Jayabalan, and V. Balasubramanian, "Optimizing pulsed current parameters to minimize corrosion rate in gas tungsten arc welded titanium alloy," *The International Journal of Advanced Manufacturing Technology*, vol. 39, no. 5-6, pp. 474–481, 2008.

[3] J. He, J. Bahr, B. J. Chisholm et al., "Combinatorial materials research applied to the development of new surface coatings X: a high-throughput electrochemical impedance spectroscopy method for screening organic coatings for corrosion inhibition," *Journal of Combinatorial Chemistry*, vol. 10, no. 5, pp. 704–713, 2008.

[4] B. D. Chambers and S. R. Taylor, "Multiple electrode methods to massively parallel test corrosion inhibitors for AA2024-T3," *NACE - International Corrosion Conference Series*, pp. 066781–0667814, 2006.

[5] D. L. Schulz, R. A. Sailer, C. Braun et al., "Trimethylsilane-based pretreatments in a Mg-rich primer corrosion prevention system," *Progress in Organic Coatings*, vol. 63, no. 2, pp. 149–154, 2008.

[6] N. D. Budiansky, F. Bocher, H. Cong, M. F. Hurley, and J. R. Scully, "Use of coupled multi-electrode arrays to advance the understanding of selected corrosion phenomena," *Corrosion*, vol. 63, no. 6, pp. 537–554, 2007.

[7] I. Annergren, D. Thierry, and F. Zou, "Localized electrochemical impedance spectroscopy for studying pitting corrosion on stainless steels," *Journal of The Electrochemical Society*, vol. 144, no. 4, pp. 1208–1215, 1997.

[8] G. S. Frankel, "Pitting corrosion of metals: a review of the critical factors," *Journal of The Electrochemical Society*, vol. 145, no. 6, pp. 2186–2198, 1998.

[9] K.-H. Tostmann, "Lochkorrosion," in *Korrosion*, pp. 80–82, WILEY-VCH, Weinheim, Germany, 2001.

[10] R. F. Steigerwald, "The corrosion behavior of some Fe-Cr alloys," *Metallurgical Transactions*, vol. 5, no. 11, pp. 2265–2269, 1974.

[11] C. Pallotta, N. De Cristofano, R. C. Salvarezza, and A. J. Arvia, "The influence of temperature and the role of chromium in the passive layer in relation to pitting corrosion of 316 stainless steel in NaCl solution," *Electrochimica Acta*, vol. 31, no. 10, pp. 1265–1270, 1986.

[12] Z. Bou-Saleh, A. Shahryari, and S. Omanovic, "Enhancement of corrosion resistance of a biomedical grade 316LVM stainless steel by potentiodynamic cyclic polarization," *Thin Solid Films*, vol. 515, no. 11, pp. 4727–4737, 2007.

[13] N. Ramasubramanian, N. Preocanin, and R. D. Davidson, "Analysis of passive films on stainless steel by cyclic voltammetry and auger spectroscopy," *Journal of The Electrochemical Society*, vol. 132, no. 4, pp. 793–798, 1985.

[14] P. Schmuki, "From bacon to barriers: a review on the passivity of metals and alloys," *Journal of Solid State Electrochemistry*, vol. 6, no. 3, pp. 145–164, 2002.

[15] M. Da Cunha Belo, N. E. Hakiki, and M. G. S. Ferreira, "Semiconducting properties of passive films formed on nickel-base alloys type alloy 600: influence of the alloying elements," *Electrochimica Acta*, vol. 44, no. 14, pp. 2473–2481, 1999.

[16] N. B. Hakiki, S. Boudin, B. Rondot, and M. Da Cunha Belo, "The electronic structure of passive films formed on stainless steels," *Corrosion Science*, vol. 37, no. 11, pp. 1809–1822, 1995.

[17] K. Sieradzki and R. C. Newman, "A percolation model for passivation in stainless steels," *Journal of The Electrochemical Society*, vol. 133, no. 9, pp. 1979-1980, 1986.

[18] D. E. Williams, R. C. Newman, Q. Song, and R. G. Kelly, "Passivity breakdown and pitting corrosion of binary alloys," *Nature*, vol. 350, no. 6315, pp. 216–219, 1991.

[19] H. Böhni, "Breakdown of passivity and localized corrosion processes," *Langmuir*, vol. 3, no. 6, pp. 924–930, 1987.

[20] T. P. Hoar, D. C. Mears, and G. P. Rothwell, "The relationships between anodic passivity, brightening and pitting," *Corrosion Science*, vol. 5, no. 4, pp. 279–289, 1965.

[21] U. R. Evans, "The passivity of metals. Part I. The isolation of the protective film," *Journal of the Chemical Society (Resumed)*, pp. 1020–1040, 1927.

[22] K. J. Vetter and H.-H. Stehblow, "Entstehung und gestalt von korrosionslöchern bei lochfraß an eisen und theoretische folgerungen zur lochkorrosion," *Ber. Bunsen-Gesellschaft Phys. Chem*, pp. 1024–1035, 1970.

[23] N. Sato, "A theory for breakdown of anodic oxide films on metals," *Electrochimica Acta*, vol. 16, no. 10, pp. 1683–1692, 1971.

[24] T. P. Hoar, "The production and breakdown of the passivity of metals," *Corrosion Science*, vol. 7, no. 6, pp. 341–355, 1967.

[25] J. Soltis, "Passivity breakdown, pit initiation and propagation of pits in metallic materials–review," *Corrosion Science*, vol. 90, pp. 5–22, 2015.

[26] J. M. Kolotyrkin, "Pitting corrosion of metals," *Corrosion*, vol. 19, no. 8, pp. 261–268, 1963.

[27] T. P. Hoar and W. R. Jacob, "Breakdown of passivity of stainless steel by halide ions," *Nature*, vol. 216, no. 5122, pp. 1299–1301, 1967.

[28] T. Shibata and T. Takeyama, "Stochastic theory of pitting corrosion," *Corrosion*, vol. 33, no. 7, pp. 243–251, 1977.

[29] D. E. Williams, J. Stewart, and P. H. Balkwill, "The nucleation, growth and stability of micropits in stainless steel," *Corrosion Science*, vol. 36, no. 7, pp. 1213–1235, 1994.

[30] S. R. Taylor, "The investigation of corrosion phenomena with high throughput methods: a review," *Corrosion Reviews*, vol. 29, no. 3-4, pp. 135–151, 2011.

[31] K. Osozawa, K. Bohnenkamp, and H.-J. Engell, "Potentiostatic study on the intergranular corrosion of an austenitic chromium-nickel stainless steel," *Corrosion Science*, vol. 6, no. 9-10, pp. 421–433, 1966.

[32] Y. Yi, P. Cho, A. Al Zaabi, Y. Addad, and C. Jang, "Potentiodynamic polarization behaviour of AISI type 316 stainless steel in NaCl solution," *Corrosion Science*, vol. 74, pp. 92–97, 2013.

[33] X. Shang, Y. Zhang, N. Qu, and X. Tang, "Electrochemical analysis of passivation film formation on steel rebar in concrete," *International Journal of Electrochemical Science*, vol. 11, no. 7, pp. 5870–5876, 2016.

[34] Deutsche Edelstahlwerke, https://www.dew-stahl.com/service/technische-bibliothekbroschueren/werkstoffdatenblaetter/.

[35] J. Morales, P. Esparza, R. Salvarezza, and S. Gonzalez, "The pitting and crevice corrosion of 304 stainless steel in phosphate-borate buffer containing sodium chloride," *Corrosion Science*, vol. 33, no. 10, pp. 1645–1651, 1992.

[36] S. Omanovic and S. G. Roscoe, "Effect of linoleate on electrochemical behavior of stainless steel in phosphate buffer," *Corrosion*, vol. 56, no. 7, pp. 684–693, 2000.

[37] C. A. Acosta, R. C. Salvarezza, H. A. Videla, and A. J. Arvia, "The pitting of mild steel in phosphate-borate solutions in the presence of sodium sulphate," *Corrosion Science*, vol. 25, no. 5, pp. 291–303, 1985.

[38] S. A. M. Refaey, S. S. Abd El-Rehim, F. Taha, M. B. Saleh, and R. A. Ahmed, "Inhibition of chloride localized corrosion of mild steel by PO43-, CrO42-, MoO42-, and NO2- anions," *Applied Surface Science*, vol. 158, no. 3, pp. 190–196, 2000.

Permissions

All chapters in this book were first published in IJC, by Hindawi Publishing Corporation; hereby published with permission under the Creative Commons Attribution License or equivalent. Every chapter published in this book has been scrutinized by our experts. Their significance has been extensively debated. The topics covered herein carry significant findings which will fuel the growth of the discipline. They may even be implemented as practical applications or may be referred to as a beginning point for another development.

The contributors of this book come from diverse backgrounds, making this book a truly international effort. This book will bring forth new frontiers with its revolutionizing research information and detailed analysis of the nascent developments around the world.

We would like to thank all the contributing authors for lending their expertise to make the book truly unique. They have played a crucial role in the development of this book. Without their invaluable contributions this book wouldn't have been possible. They have made vital efforts to compile up to date information on the varied aspects of this subject to make this book a valuable addition to the collection of many professionals and students.

This book was conceptualized with the vision of imparting up-to-date information and advanced data in this field. To ensure the same, a matchless editorial board was set up. Every individual on the board went through rigorous rounds of assessment to prove their worth. After which they invested a large part of their time researching and compiling the most relevant data for our readers.

The editorial board has been involved in producing this book since its inception. They have spent rigorous hours researching and exploring the diverse topics which have resulted in the successful publishing of this book. They have passed on their knowledge of decades through this book. To expedite this challenging task, the publisher supported the team at every step. A small team of assistant editors was also appointed to further simplify the editing procedure and attain best results for the readers.

Apart from the editorial board, the designing team has also invested a significant amount of their time in understanding the subject and creating the most relevant covers. They scrutinized every image to scout for the most suitable representation of the subject and create an appropriate cover for the book.

The publishing team has been an ardent support to the editorial, designing and production team. Their endless efforts to recruit the best for this project, has resulted in the accomplishment of this book. They are a veteran in the field of academics and their pool of knowledge is as vast as their experience in printing. Their expertise and guidance has proved useful at every step. Their uncompromising quality standards have made this book an exceptional effort. Their encouragement from time to time has been an inspiration for everyone.

The publisher and the editorial board hope that this book will prove to be a valuable piece of knowledge for researchers, students, practitioners and scholars across the globe.

List of Contributors

Ladan Khaksar and Gary Whelan
Department of Mechanical Engineering, Faculty of Engineering and Applied Science, Memorial University of New foundland, St. John's, NL, Canada A1B 3X5

John Shirokoff
Department of Process Engineering, Faculty of Engineering and Applied Science, Memorial University of New foundland, St. John's, NL, Canada A1B 3X5

Syarizal Fonna, M. Ridha and Syifaul Huzni
Department of Mechanical Engineering, Syiah Kuala University, Jalan Tgk Syech Abdul Rauf 7, Banda Aceh 23111, Indonesia

Israr M. Ibrahim
Tsunami & Disaster Mitigation Research Center (TDMRC), Syiah Kuala University, Jalan Tgk Abdul Rahman, Gp. Pie, Meuraxa District, Banda Aceh 23111, Indonesia

A. K. Ariffin
Department of Mechanical and Materials Engineering, Universiti Kebangsaan Malaysia, 43600 Bangi, Selangor, Malaysia

Guowei Li and Sidi Kabba Bakarr
College of Civil and Transportation Engineering, Hohai University, Nanjing, 210098, China

Jingqiu Wang
Key Laboratory ofMinistry of Education for Geomechanics and Embankment Engineering, Hohai University, Nanjing 210098,China

Xue Liu
Guangdong Nanyue Transportation Investment Construction Co., Ltd., Guangzhou 510000, China

Chengyu Hong
Department of Civil Engineering, Shanghai University, Shanghai, 200444, China

M. Krishna Prasad
GMR Institute of Technology, Rajam 532127, India

K. Srinivasa Rao
Andhra University College of Engineering, Visakhapatnam 530003, India

Madhusudhan Reddy
Defence Metallurgical Research Laboratory, Hyderabad 500066, India

Gosipathala Sreedhar
CSIR-Central Electrochemical Research Institute, Karaikudi 630 006, India

Ahmad Zaki
Department of Civil Engineering, Universitas Abdurrab, Pekanbaru, 28291 Riau, Indonesia

Megat Azmi Megat Johari, Wan Muhd Aminuddin and Wan Hussin
School of Civil Engineering, Engineering Campus, Universiti Sains Malaysia, Nibong Tebal, 14300 Penang, Malaysia

Yessi Jusman
Department of Electrical Engineering, Faculty of Engineering, Universitas Muhammadiyah Yogyakarta Kasihan, Bantul 55183, Yogyakarta, Indonesia

Rodrigo Monzon Figueredo, Mariana Cristina de Oliveira, Leandro Jesus de Paula, Heloisa Andréa Acciari and Eduardo Norberto Codaro
School of Engineering, São Paulo State University (UNESP), Guaratinguetá, SP, Brazil

Jinjin Zhang, Jin Yang and Feilong Ye
Department of Building Engineering, Oujiang College,Wenzhou University,Wenzhou 325035, China

Hui Chen
Department of Building Engineering, Oujiang College, Wenzhou University, Wenzhou 325035, China
Department of Structural Engineering, Tongji University, Shanghai 200092, China

H. J. Zhou, Y. F. Zhou, Y. N. Xu, Z. Y. Lin, F. Xing and L. X. Li
Guangdong Provincial Key Laboratory of Durability for Marine Civil Engineering, Shenzhen University, Shenzhen 518060, China

A. López-Ortega and R. Bayón
IK4-TEKNIKER, Eibar, Spain

J. L. Arana
Department of Metallurgical and Materials Engineering, University of the Basque Country, Spain

Vinod P. Raphael and Shaju K. Shanmughan
Department of Chemistry, Government Engineering College, Thrissur, Kerala 680009, India

Joby Thomas Kakkassery
Research Division, Department of Chemistry, St. Thomas' College (Autonomous), Thrissur, Kerala 680001, India

Qingmiao Ding, Xiao Chu, Tao Shen and Xiaoxiao Yu
Airport College, Civil Aviation University of China, Tianjin, China

Rogelio Ramos, Benjamin Valdez-Salas, Roumen Zlatev and Michael Schorr Wiener
Engineering Institute, Autonomous University of Baja California, Boulevard Benito Juarez, Insurgentes Este, 21280 Mexicali, BC, Mexico

Jose María Bastidas Rull
National Center of Metallurgical Research (CENIM) Madrid,The Spanish State Council for Scientific Research (CSIC), Madrid, Spain

Herdi Susanto
Department of Mechanical Engineering, Faculty of Engineering, Teuku Umar University Meulaboh 23681 Aceh Barat, Indonesia

Syifaul Huzni and Syarizal Fonna
Department of Mechanical and Industrial Engineering, Faculty of Engineering, Syiah Kuala University, Darussalam, Banda Aceh 23111, Indonesia

Takuma Asabe, Muhammad Rifai, Motohiro Yuasa and Hiroyuki Miyamoto
Department of Mechanical Engineering, Doshisha University, Kyoto 610-0394, Japan

Hanan Farhat
College of the North Atlantic-Qatar, Doha 24449, Qatar

Kaveh Andisheh, Allan Scott and Alessandro Palermo
Department of Civil Engineering, University of Canterbury, Private Bag 4800, Christchurch 8140, New Zealand

G. Chan-Rosado and M. A. Pech-Canul
Departamento de Física Aplicada, Cinvestav-Mérida, Km. 6 Ant. Carr. a Progreso, AP73, Cordemex, 97310 Mérida, YUC, Mexico

Jasbir N. Patel, Andre Chang, Haleh Shahbazbegian and Bozena Kaminska
School of Engineering Science, Simon Fraser University, 8888 University Drive, Burnaby, BC, Canada V5A 1S6

Ismaeel M. Alwaan
Department of Materials Engineering, College of Engineering, University of Kufa, Najaf, Iraq

Mirosław Szala and Daniel Aukasik
Department of Materials Engineering, Faculty of Mechanical Engineering, Lublin University of Technology, Nadbystrzycka 36D, 20-618 Lublin, Poland

Ingmar Bösing, Jorg Thöming and Michael Baune
Center for Environmental Research and Sustainable Technology (UFT), University of Bremen, Leobener Straße 6, Bremen, Germany

Index

A

Acid Corrosion, 101, 107, 194

Acidic Chemical Bath, 1-2, 8

Adaptive Corrosion Protection System, 184, 186

Adaptive Loop, 188-190, 192

Adipic Acid, 173, 179, 182

B

Basalt-glass Fibre Reinforced Polymer, 15

Boundary Element Method, 10, 13, 95

C

Carbon Dioxide Corrosion, 1

Carbon Steels, 40-41

Chloride Stress Corrosion Cracking, 142

Chloride-induced Corrosion, 150, 152, 161, 164-165, 168

Corrosion Kinetics, 9, 87, 166, 177, 195-196, 199

Corrosion Standard Tests, 41

Corrosion-induced Deterioration Models, 150

Crack Length Ratio, 40-41, 44

Crack Sensitivity Ratio, 41

Cracking Resistance, 44

Cubic Ferrous Sulfide, 1

D

Decohesion, 41

Dicarboxylates, 172-173, 182

Discrete Wavelet Transform, 118-119

Dispersive Energy Spectrometry, 40-41

E

Electrochemical Impedance Spectroscopy, 83-84, 86-87, 93, 97-98, 109-111, 116, 173, 214

Electrolyte Composition, 208-209, 211, 214

Electronegativity, 101, 107, 110

Energy Dispersive X-ray, 1, 3

F

Fes Films, 1-3

Fibre Reinforced Polymer, 15-16, 22, 170

G

Greigite, 1-2

Ground Penetrating Radar, 14, 30, 38-39

H

Half-cell Potential Technique, 10-12, 30, 125-126

Hot Corrosion, 23-29

Hydrochloride Acid, 195

Hydrogen-induced Cracking, 40-41, 46

Hydroxylation, 16

I

Impedance Spectroscopy, 83-84, 86-87, 93, 95, 97-98, 109-111, 116, 173, 193, 214

Iron Sulfide, 1, 8-9, 43-44

L

Lanthanum Aluminates, 23

Linear Polarization Resistance, 1-2

Localized Corrosion, 8, 40, 84, 87, 108, 110, 118, 121, 123, 185, 214-215

M

Mackinawite, 1-6, 8-9, 40, 44-45

Magnesium Sacrificial Anode, 113, 115-117

Marcasite, 1-2, 7

Molecular Dynamics Simulation, 195

Monte Carlo Simulation Technique, 195

N

Neutral Solutions, 172, 181-182

O

Open Circuit Potential, 78-79, 81, 83-84, 86-88, 90-91, 93, 95, 111, 113-114, 173-174, 180

Open Circuit Potential Registration, 79, 93

Optic Fibre Bragg Grating, 17

P

Passive Materials, 77-78, 80-81, 88, 92, 95-96

Perovskite, 23-24, 28

Phenylhydrazone, 101-102

Pitting Corrosion, 57, 117-118, 123, 139, 154, 159, 166-168, 201-202, 204-209, 211-215

Polarization, 1-2, 10-12, 78-82, 85-88, 91, 93-95, 99, 101-103, 108-109, 111, 113-114, 166, 172-174, 177, 180, 185, 193, 208-215

Precipitation, 1-2, 7-9, 141, 196, 204

R

Rebar Corrosion, 10-11, 13-14, 30, 33, 36, 38, 59, 69, 71, 167-168

Reference Electrode, 78-79, 88, 102-103, 112, 118, 123, 126-127, 173, 187-188, 209

Reinforced Concrete, 10, 14, 16, 33, 36-39, 47, 57, 59, 74-76, 125-127, 150, 152, 159, 161-171, 185, 192-193

Resistance Welding Joints, 201, 206

Runge-kutta Method, 195-196, 198-199

S

Sacrificial Anode, 111-117

Saturated Camomel Electrodes, 78

Scanning Vibrating Electrode Technique, 118

Scanning Vibrating Reference Electrode, 118, 123

Severe Plastic Deformation, 134, 140-141

Sodium Chloride, 30-31, 87, 209, 211-215

Sodium Glutarate, 172-174, 177-178, 180

Stainless Steel, 8, 22, 81, 83, 89, 92, 94, 96-100, 110, 123, 135, 140, 143-144, 146, 148-149, 166, 184, 193, 195, 199, 201-204, 206-209, 211-212, 214-215

Steel Bars, 39, 47-50, 54, 56-58, 60, 64, 74-76, 151, 155, 158, 161-162, 164-165, 167-168

Stress Corrosion Cracking, 46, 117, 134, 140-142, 149, 207

Strontium Vanadates, 26

Sulfur Dioxide, 196

Sulfuric Acid Solutions, 195-196

T

Thermal Barrier Coatings, 23, 28-29

Tribocorrosion, 77-83, 86-100

V

Ventilation Grille, 201-206

W

Working Electrode, 2, 78-79, 88-89, 102, 112, 173, 187-188, 209, 213

X

X-ray Diffraction, 1, 3, 24, 40-41, 167, 206

X-ray Photoelectron Spectroscopy, 95, 173, 182

www.ingramcontent.com/pod-product-compliance
Lightning Source LLC
Chambersburg PA
CBHW082045190326
41458CB00010B/3460